Human Movement Explained

Kim Jones MSc, MCSP, Dip TP, SRP
Lecturer, Department of Physiotherapy Studies, Keele University, Staffordshire, UK

Karen Barker MSc, MCSP, SRP
Superintendent Physiotherapist, Nuffield Orthopaedic Centre NHS Trust, Oxford

BUTTERWORTH
HEINEMANN

Butterworth-Heinemann Ltd
Linacre House, Jordan Hill, Oxford OX2 8DP

℞ A member of the Reed Elsevier plc group

OXFORD LONDON BOSTON
MUNICH NEW DELHI SINGAPORE SYDNEY
TOKYO TORONTO WELLINGTON

First published 1996

British Library Cataloguing in Publication Data
Jones, Kim
   Human Movement Explained –
   (Physiotherapy Practice Explained Series)
   I. Title II. Barker, Karen III. Series 612.76

ISBN 0 7506 1747 0

Library of Congress Cataloguing in Publication Data
Jones, Kim
   Human movement explained / Kim Jones, Karen Barker. – 1st ed.
   p. cm.
   Includes bibliographical references and index.
   ISBN 0 7506 1747 0
   1. Exercise therapy. 2. Human locomotion. 3. Medical rehabilitation. 4. Physical
therapy. I. Barker, Karen. II. Title.
   [DNLM: 1. Exercise Therapy—methods. 2. Movement—physiology.
   WB 541 J77h 1995]
   RM725.J66 1995
   615.8'2—dc20
   DNLM/DLC
   for Library of Congress   95-31861
                                       CIP

Typeset by Keytec Typesetting Ltd, Bridport, Dorset
Printed and bound by Hartnolls Ltd, Bodmin, Cornwall

Contents

Preface   vii

Acknowledgements   ix

Part I   The control of human movement   1

1 Mechanical basis of movement   3

2 Skeletal basis of movement   34

3 Musculoskeletal basis of movement   50

4 Neurophysiological aspects of human movement   66

Part II   Exercise in rehabilitation   95

5 Classification of exercise   97

6 Exercise prescription   106

7 Fitness testing   120

8 Group exercise   158

9 Balance and proprioception   175

10 Strength   196

11 Isokinetic exercise   224

12 Flexibility   243

Part III   Locomotion and ergonomics   259

13 Relaxation   261

14 Posture 275

15 Gait analysis 297

16 Walking aids and orthotics 325

17 Ergonomic approach to lifting and handling 354

18 Clinical measurement 385

Index 403

Preface

The aim of this book is to provide a complete introductory text for movement studies which can be used by undergraduate students of physiotherapy and the allied professions.

Currently there is no single core text which totally fulfils the syllabus requirements and students have to rely on a number of books, each of which covers only selective sections of the syllabus.

Movement studies is a core subject for first-year physiotherapy students. It principally involves learning the concepts and techniques of rehabilitation. Walking aid provision and gait re-education, fitness testing, posture and its correction, relaxation techniques and the prescription of exercise are but a few subjects which fall under its auspices.

The book has been written by a university lecturer in physiotherapy studies and an experienced clinician, both of whom have an in-depth knowledge and interest in biomechanics, neurology, ergonomics and orthopaedics and their application to human movement and physiotherapy practice. This educationalist/clinician author combination has resulted in a book that covers both syllabus requirements and facilitates the application of knowledge to current clinical practice.

The book is written in three parts and assumes that students already have a basic knowledge of anatomy and physiology. Part I provides information on biomechanical principles and introduces concepts associated with skeletal, musculoskeletal and neurophysiological bases of the control of human movement.

Part II provides a concise but comprehensive guide to the principles and types of exercise which can be used in rehabilitation. It describes examples of exercise regimens and basic fitness tests which can be performed in the physiotherapy department. Finally, Part III incorporates important locomotor and ergonomic areas which are often taught under the 'movement umbrella'.

The book is not intended to be a complete rehabilitation manual. It is an introductory text providing information on many areas of rehabilitation in which the qualified physiotherapist may work. To this end, each chapter is comprehensively referenced with recent material to allow the interested reader to pursue a subject area in greater detail.

Throughout, endeavours have been made to use current clinical terminology, SI units of measurement and to illustrate key theoretical concepts with clinical examples where possible.

It is hoped that this book will be helpful long after student days are over and that the practising physiotherapist will continue to find it a useful reference for planning exercise programmes and treatment regimens for movement disorders in everyday clinical practice.

KJ
KB

Acknowledgements

Writing a book requires support, interest, constructive criticism and patience from family, friends and colleagues. In this regard, thanks are due to: the staff of the Department of Physiotherapy Studies at Keele University and the staff and colleagues of the Nuffield Orthopaedic Centre, Oxford. In addition we have appreciated and acted on the comments of: Lois Allen, Rosemary Bell, Janet Bonn, Marie Carter, Dr Simon Ellis, Helen Frost, Jill Guymer, Janet Hancock, Jane Holmes, Christine Hornby, Jennifer Hough, Alison Hughes, John Low, Claire Montgomery, Pat Moore, Ann Reed, Jean Richards, Debbie Simm, John Tidswell, Malcolm Warren-Forward and Karen Yeomans.

Special thanks go to Tim Rickard and John Buckley for their respective expert guidance in the preparation of the isokinetic and fitness testing chapters; to Chattanooga UK, Plysu, Duffield Medical and Sensor Medics who went to great lengths to provide photographs for us. Also to Marian Tidswell for her support and encouragement throughout and to Nigel Palastanga, the instigator of this venture!

Lastly to our families for taking second place for many months – thank you Ian, Fiona, Andrew, Ann and Sonny.

Part I
The control of human movement

1.

Mechanical basis of movement

The study of movement and its application in therapeutic exercise is based on mechanical principles. Before movement can be studied, it is essential that the basic principles of mechanics are grasped and understood. This chapter is concerned with the force, time and distance relationships that are applied to human body movement, i.e. biomechanics as well as the pure study of forces and their effects, or mechanics. Mechanics may be studied by two different approaches – kinetics and kinematics. The former is concerned with forces and their effects, whereas the latter is qualitative and concerned with geometry and temporal qualities of motion. It may also be divided into the study of statics and dynamics, i.e. the body at rest and in motion.

It is intended that at the end of this chapter the reader should have a firm grasp of the mechanical principles that movement therapy is based upon. These include:

Force	Centre of gravity
Resolution and composition of forces	Equilibrium
	Stability
Motion	Friction
Levers	Pressure
Angle of pull	Energy, work and power
Torque	Stress and strain
Pulleys	Fluid mechanics
Pendulums	Definitions of mechanical terms

FORCE

Force is a concept that cannot be seen, although its effect can be felt, e.g. by squeezing a rubber ball. It is an agent that tends to produce change in the state of rest or motion of an object (Enoka, 1988). Force can be designated as applied force (contact), inertial force (motion) and gravitational force (weight). In all cases force cannot exist on its own and there must be a balance to maintain a state of equilibrium. Brancazio (1984) categorizes force as either contact or non-contact force.

Contact forces are pushes and pulls exerted by one object in direct contact with another, e.g. friction, muscle forces and ground reaction forces. Non-contact forces refer to effects not due to direct contact, such as gravity and the attraction and repulsion of electrically charged particles.

A force has four component parts:

- magnitude
- action line
- direction
- point of application.

With the above qualities, force is a vector, having both magnitude and direction. A force vector is a line drawn to scale of a certain length to represent the magnitude, at some inclination to represent the direction and in a particular position relative to the body to identify the point of application. An arrow head will provide the direction of the force (Le Veau, 1977).

Effects of forces

Forces will act together to produce an effect. The effects of forces can be used to analyse human movement by determining the net effect of several forces acting upon the body and the functional effect of that force, e.g. rotation or stabilization. Forces may be:

Linear

When forces act along the same straight line of action they are said to be collinear in a linear system of forces. This is the simplest combination of forces. They may act in the same direction (Figure 1.1a) or opposite to each other (Figure 1.1b).

Parallel

When the action lines of the forces under consideration are parallel to each other and lie in the same plane they form a parallel force system. The forces do not have the same line of action and tend to produce a rotatory effect if they are not in equilibrium (Figure 1.1c).

Force couple

This is a specialized example of a parallel force system which occurs when two forces of equal magnitude act at a distance from each other and in opposite directions. Under these circumstances they will produce a turning effect. As long as the two forces are equal and opposite there will be no linear displacement or acceleration and the resultant of

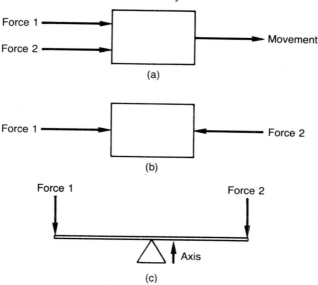

Fig. 1.1 Forces: (a) linear tension force; (b) linear compression force; (c) parallel force

the two forces will be zero, but the object still turns. In the human body, there is no true example of a force couple, but the body may act to form one, e.g. the force couple formed by the thumb and index finger in unscrewing a lid from a jar.

Concurrent force systems

When forces action lines meet at a point they are said to operate in a concurrent force system. The forces may be applied from several different angles so that projections of their lines of action will cross. This intersection may occur inside or outside the body. For example, if one views the right hand side of the body, in standing the line of gravity falls anterior to the ankle joint so that gravitational forces pull the body in a clockwise direction; forces produced by the tension of the posterior tibial muscles act in an anticlockwise direction to maintain an erect stance (Gowitzke and Milner, 1988).

Parallel force systems in equilibrium

In many cases forces will be in parallel and lie in the same plane, but also be in equilibrium. This effect can be utilized therapeutically, especially in the application of orthotics. Most bracing works upon the principle of three-point pressure. The supporting forces of the orthosis are arranged so that two forces pressing against the trunk or limb are opposed by a third acting between them. The third force must be equal in magnitude to the two forces acting in the same direction. For

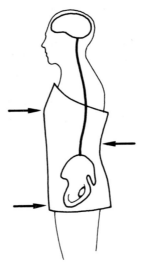

Fig. 1.2 Three-point pressure – parallel forces in equilibrium

example, in a Milwaukee brace to correct spinal malalignment, a three-point pressure system with forces acting in a parallel force system is seen (Figure 1.2).

Resultant of forces

Where many forces are acting on an object, the resultant force, or the final effect, will need to be calculated. The resultant is the single force obtained by combining all of the given forces. If all forces acting on the body produce a resultant force of zero, the body is in equilibrium.

Analysis of forces

Forces may be analysed to find the resultant effect of several forces (composition). Alternatively, the resultant may be analysed to find its component forces (resolution). The analysis may be done by graphic means or trigonometry.

Composition of forces – parallelogram method

The rule of parallelogram of forces states that if two forces act on a body at the same point from two different directions, the body will move in a direction which will be the diagonal of a parallelogram drawn from the point of application of the forces (Figure 1.3a). This can be seen in the body in the action of the anterior and posterior fibres of deltoid. A vector is drawn to represent the force produced by the anterior fibres of deltoid (Figure 1.3b). Using the same scale, a vector is drawn for the posterior fibres, both starting from the same point, representing the muscle origin. The other two sides of the parallelogram are constructed and a line is drawn to represent the deltoid insertion to the opposite side of the parallelogram. This line is the

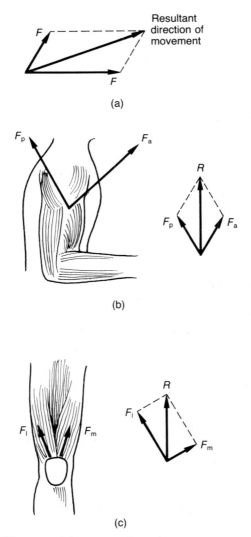

Fig. 1.3a–c Parallelogram of forces (F_a, force from anterior fibres of deltoid; F_p, force from posterior fibres of deltoid; F_l, force from vastus lateralis; F_m, force from vastus medialis)

resultant force and its magnitude and direction can be measured (Figure 1.3c).

Another example of the resultant of forces is the three abductors of the hip. Inman (1947) determined the action lines of tensor fascia lata, gluteus medius and gluteus minimus, each of which is a muscle with a different shape, mass and angle of pull, inserting on or near the greater trochanter. Inman postulated that each individual muscle would contribute to overall abduction proportionally to their individual mass. Thus their relative force values were calculated to be:

Tensor fascia lata	1
Gluteus minimus	2
Gluteus medius	4

On this basis, and knowing the lines of action of the muscles, the single resultant of these forces could be calculated (Figure 1.4).

Similar examples of muscle forces combining to produce a resultant force occur in examples such as the medial and lateral heads of gastrocnemius which produce a resultant force that pulls upwards on the Achilles tendon and the combined action of quadriceps to produce a resultant force that moves the patella proximally.

Resolution of forces

Given a resultant force, it is equally possible to resolve it into its component parts. In Figure 1.5, the action of the middle fibres of

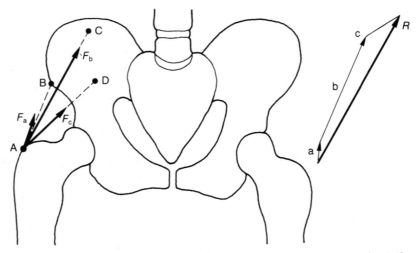

Fig. 1.4 Resolution of hip abductor forces (AB, action line of tensor fascia lata; AC, action line of gluteus medius; AD, action line of gluteus minimus; F_a, force vector of tensor fascia lata; F_b, force vector of gluteus medius; F_c, force vector of gluteus minimus)

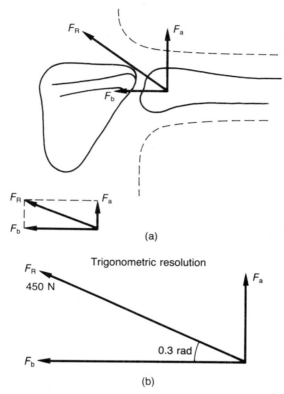

Fig. 1.5a,b Resultant forces (F_a, normal component perpendicular to long bone axis; F_b, tangential component along bone axis; F_R, resultant)

deltoid are studied. Force A represents the proportion of muscle force that acts perpendicular to the long axis of the bone and therefore produces rotation. This is sometimes called the 'normal' component. Force B represents the proportion of muscle force directed along the axis of the bone towards the joint and therefore acts to compress the joint (tangential component). By using vectors, the resultant force is calculated.

This may also be resolved by trigonometry if the angle between the vectors is known. In Figure 1.6, a man tries to push a heavy object; he will tilt his body to be inclined to the vertical and rely on friction between himself and the floor to produce a horizontal force. This horizontal force requires a vertical applied force, i.e. body weight, in order to provide horizontal friction force between the floor and the man. The resultant force vector will be inclined. If the angle between the forces is known, the use of trigonometry will determine the value of the component forces. In the example of the middle fibres of deltoid, we know that the resultant force is 450 Newtons (N). Figure 1.6 shows

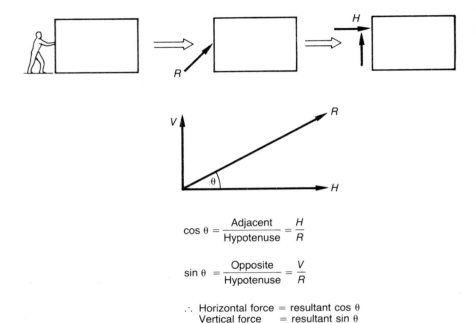

$$\cos \theta = \frac{\text{Adjacent}}{\text{Hypotenuse}} = \frac{H}{R}$$

$$\sin \theta = \frac{\text{Opposite}}{\text{Hypotenuse}} = \frac{V}{R}$$

∴ Horizontal force = resultant cos θ
Vertical force = resultant sin θ

Fig. 1.6 Resolution of forces by trigonometry

the trigonometric resolution of the component forces which act at an angle of 0.3 rad or 17.2° to each other:

$$F_a/450 = \sin 0.3 \qquad\qquad F_b/450 = \cos 0.3$$
$$F_a = 450 \times \sin 0.3 \qquad F_b = 450 \times \cos 0.3$$
$$F_a = 450 \times 0.2955 \qquad F_b = 450 \times 0.9553$$
$$F_a = 133 \text{ N} \qquad\qquad F_b = 430 \text{ N}.$$

Thus the normal component has a magnitude of 133 N and the tangential component a magnitude of 430 N.

Free-body diagrams

In order to visualize and analyse human motion, the effect of all the forces acting on the body must be considered. The use of a free-body diagram makes this less complex by drawing a simplified diagram, usually a stick man, of the system isolated from the environment (Figure 1.7). All the external forces that influence the system are indicated on the diagram by vectors and labelled. Thus the external forces on the body are readily visualized in context of the object being studied (Luttens and Wells, 1982).

For further examples the reader is referred to Chapter 3 of Le Veau (1977) or Chapter 7 of O'Connell and Gardner (1972).

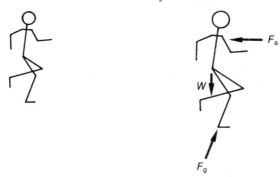

Fig. 1.7 Free-body diagram to analyse forces (F_a, force of air resistance; W, weight; F_g, ground reaction force)

MOTION

Motion and force are closely related. Motion always involves changing the position of a body. This may be:

- *Linear motion* – all parts of the body move the same distance in the same direction in the same amount of time.
- *Angular/rotatory motion* – movement of a body about a fixed point so that all parts of the body travel in arcs as they move through the same angular displacement.
- *Curvilinear motion* – movement of a body along a curved path, e.g. the trajectory of a missile.

Sir Isaac Newton (1642–1727) produced three laws governing motion, the laws of inertia, acceleration and action–reaction.

Newton's first law of inertia

This states that a body will remain at rest or in uniform motion until acted upon by an applied force. In other words, a force is required to stop, start or alter motion. In reality, because of gravity, forces act continuously on bodies and a change in motion occurs when there is a net imbalance of forces. Inertia refers to the difficulty with which an object's motion is altered. This varies with the mass of the object, i.e. the amount of matter it is composed of. If two objects are travelling at the same velocity it will be more difficult to alter the motion of the one with the greatest mass, hence it is described as having a greater inertia.

Newton's second law of acceleration

This states that the change of momentum of a body is proportional to the applied force acting on it and takes place in the direction in which the force acts; acceleration of the body is inversely proportional to the mass of that body. Thus, momentum is the quantity of motion possessed by a body and is the product of mass and velocity:

$$\text{Momentum} = \text{Mass} \times \text{Velocity}$$

As in human movement, mass is constant and the applied force is proportional to the rate of change of momentum or acceleration. So acceleration is proportional to the force divided by the mass.

$$\text{Acceleration} = \text{Force/Mass}$$

$$\text{Force} = \text{Mass} \times \text{Acceleration}$$

Units of measurement: mass is measured in kilograms; acceleration is measured in metres per second per second; force is measured in newtons. Thus, a newton must be equivalent to a kilogram metre per second squared, i.e.

$$F(N) = M(kg) \times A(m/s^2)$$

$$1\,N = 1\,kg/s^2$$

Conceptually, Newton's law implies a cause–effect relationship, whereby force may be regarded as the cause, as it represents the interaction between a system and the environment. In contrast, the mass × acceleration represents the effect, as it indicates the kinematic effects of the interaction on the system.

Newton's third law: action–reaction

This states that for every action (applied force) there is an equal and opposite reaction. This emphasizes the idea that force represents an interaction between an object and its surroundings. If a person jumps from the ground, they exert a force against the ground and, conversely, the ground responds with an equal and opposite force (ground reaction force). Newton's third law requires that these two forces are equivalent in magnitude but opposite in direction. As an example, if you press your finger against the same finger on the opposite hand, the right finger will feel a force applied by the left finger, but the left finger will also feel a force applied by the right finger. In other words, each is both applying a force and coping with a reaction force.

In its application, Newton's law creates a situation where a force has to be applied to a body in order to produce an acceleration for that body. There is thus a loop effect whereby a force is applied to a mass; the mass may accelerate and will produce a reaction in the form of inertia. The applied force causes acceleration and a reaction to maintain equilibrium or produce motion (Figure 1.8).

The study of human movement involves many instances of masses

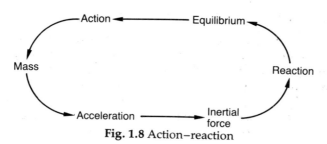

Fig. 1.8 Action–reaction

undergoing acceleration from either internally produced forces such as muscle tension or from external forces such as ground reaction forces.

LEVERS

A lever is a device for transmitting force. It is a straight bar or rigid structure of which one point, the fulcrum or axis, is fixed, another is connected with the force or weight to be resisted or acted on, and a third is connected with the force or power applied (MacDonald, 1973) (Figure 1.9).

Levers are categorized according to the relationship between the positions of the fulcrum, effort and the resistance.

Mechanical advantage

The distance from the fulcrum to the effort force will determine the effort arm. The distance of the resistance or load to the fulcrum will determine the resistance arm. The effectiveness of any force will depend on: (a) the magnitude of the force, and (b) the length of the lever arm on which it acts.

The mechanical advantage of a lever refers to the ratio of the amount of effort expended to that of the work performed. It may be determined by dividing the force arm distance by the resistance arm distance (O'Connell and Gardner, 1972).

$$MA = df/dr$$

Thus the mechanical advantage of a lever will depend upon the relative locations of the forces.

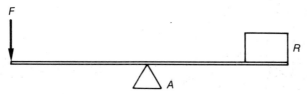

Fig. 1.9 A lever (*F*, force; *A*, axis; *R*, resistance)

Human Movement Explained

First-class lever

The first-class lever has the fulcrum located between the effort and the resistance (Figure 1.10a). It is the most versatile of the lever systems. Depending upon the relative distances of the force and resistance from the fulcrum, the amount of force required to move the load may be large or small. The direction of movement of the lever is always changed, so if the force is applied downwards the resistance will move upwards and vice versa. Examples of first-class order of levers in the body are triceps extending the elbow, the foot being plantar flexed on the ankle joint and the posterior cervical muscles extending the head across the atlanto-occipital joints (Figure 1.10b).

Second-class lever

A second-class lever has the resistance located between the fulcrum and the effort (Figure 1.11a). Consequently the effort arm will always be greater than that of the resistance. An example of such a lever system is a man pushing a wheelbarrow. This is a particularly powerful type of lever as the weight to be moved is nearer to the fulcrum than the effort, thus less force is required and mechanical advantage is gained. Movement is powerful but slow. There is much controversy about whether second-class levers exist in the human body, but an example is when one is standing on the toes while leaning against a

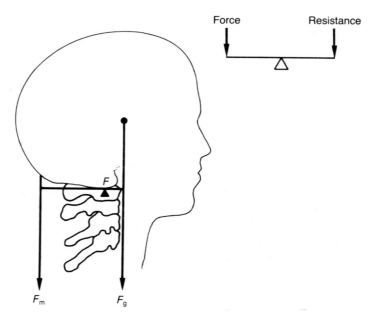

Fig. 1.10a,b First-class lever (F, fulcrum = atlanto-occipital joint; F_m, force of posterior cervical muscles; F_g, weight due to gravitational force tending to flex head)

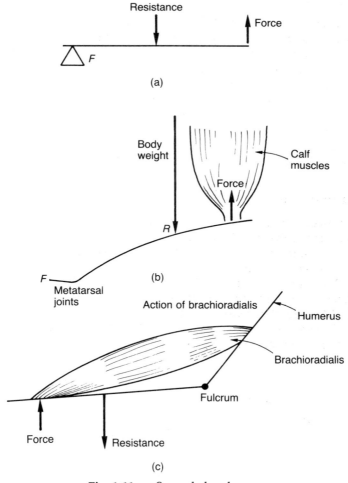

Resistance

Force

F

(a)

Body
weight

Calf
muscles

Force

R

F

Metatarsal
joints

(b)

Action of brachioradialis

Humerus

Brachioradialis

Fulcrum

Force

Resistance

(c)

Fig. 1.11a–c Second-class lever

supporting structure. The tarsal and metatarsal bones are stabilized by muscle action to form a lever where the metatarsophalangeal joints form the fulcrum (Figure 1.11b), the resistance or load is transmitted through the talus and the effort is applied by the calf muscles through the Achilles tendon; without a supporting structure the subject would lose balance and fall backwards. Another example is the action of brachioradialis in flexing the elbow joint when the forearm is in the mid position (Figure 1.11c).

Third-class lever

The third-class lever system has the effort located between the fulcrum and the resistance (Figure 1.12a). Hence the effort arm is always less than that of the resistance arm. This system of levers is therefore less powerful but gains in the speed and range of movement that is afforded. There are numerous examples of third-class levers in the body, such as biceps flexing the forearm (Figure 1.12b), iliopsoas flexing the hip and the hamstrings flexing the knee.

Angle of pull

Mechanically, a force is most effective when it is applied perpendicular to the lever. In the body, the direction of force exerted by a muscle will depend upon the relationship of the long axis of the moving bone to the insertion of the muscle. The angle of pull is the angle between the long axis of the bone and the line of pull of the muscle (Figure 1.13a). Because of the relative configuration of the muscle and axis of the bone, the angle of pull of a muscle is frequently small. If the angle of pull is very small, then little movement will be produced but the muscles will pull the bones lengthwise towards their proximal joint and act to stabilize them. For example, the angle of pull of coraco-brachialis means that it acts as a stabilizer to the shoulder joint (Figure 1.13b).

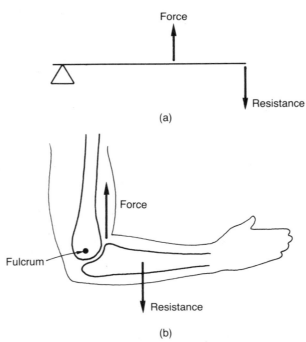

Fig. 1.12a,b Third-class lever

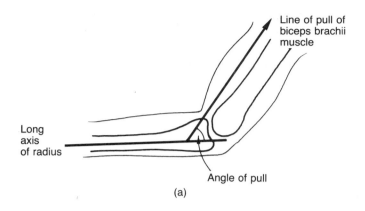

(a)

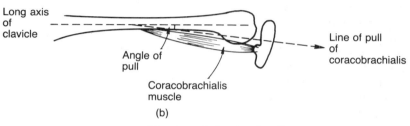

(b)

Fig. 1.13a,b Angle of pull

Advantages and disadvantages of long lever arms

In the human body, all muscular moment arms are short in relation to the levers that they move, as they are practically all first- or third-order levers. The disadvantage of this is that a force greater than the load must be produced if the load is to be moved.

The advantages of such a system are that a very small movement of the short end of the lever is magnified in proportion to the length of the lever, producing a large range of motion distally. Thus at the shoulder, a very small shortening of pectoralis major of 2 cm will produce a large arc of movement for the upper extremity of 83° (Gowitzke and Milner, 1988). As the distal end moves a much larger distance than the proximal muscle attachment in the same amount of time, the distal end also moves proportionally faster. To conserve energy or muscle force and for increased speed, we can shorten the length of the moment arm by flexing the distal joints and thus decrease the force necessary to move the limb.

TORQUE

In any lever system, forces will frequently generate a turning effect, as with a door handle. This turning effect is dependent on the value of

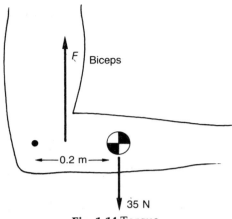

Fig. 1.14 Torque

the applied force, together with the effective leverage or distance perpendicular to the line of action of the force. The turning effect is called torque or moment of force. It is defined as a force multiplied by the perpendicular distance from the line of action of the force to the centre of rotation of the object:

$$\text{Moment} = \text{Force} \times \text{Distance}$$

An example is shown at the elbow (Figure 1.14). The weight of the forearm acts to try to extend the elbow with a force of 35 N acting on a lever arm of 0.2 m. Thus there is a tendency to try to extend the arm of $35 \times 0.2 = 7$ Nm. This moment must be counteracted by the force generated by the biceps brachii muscle. Thus the force of biceps multiplied by its lever arm must equal 7 Nm if the system is to stay balanced.

Torque curves

The magnitude of the rotary component of a muscle can be found at different joint angles and plotted to give a curve that will show the change in the magnitude of the rotary component with each change in joint angle. This is called the torque curve and will be described in more detail in Chapter 11.

PULLEYS

The pulley

A pulley consists of a grooved wheel which turns on an axle with a rope or cable running over it. Pulleys are usually mounted on a block and may be mounted in combination. They are also found anatomically in the human body. There are three types of pulley:

- single, fixed pulley
- single, movable pulley
- combination pulleys.

Single fixed pulley

This is a first-class lever which will change the direction but not the magnitude of a force acting on the pulley rope (Figure 1.15a). The force remains the same on either side of the pulley, irrespective of the angle of pull, as the effort arm equals the resistance arm and both equal the radius of the pulley. In the body, a single fixed pulley acts to fulfil two purposes: (a) to change the direction of force to increase a muscle's angle of pull, and (b) to change the direction of force to produce a different movement.

For example, the angle of pull of gracilis is increased by the pulley effect of the femoral and tibial medial condyles which the muscle passes over before inserting into the tibia (Figure 1.15b). Figure 1.15c

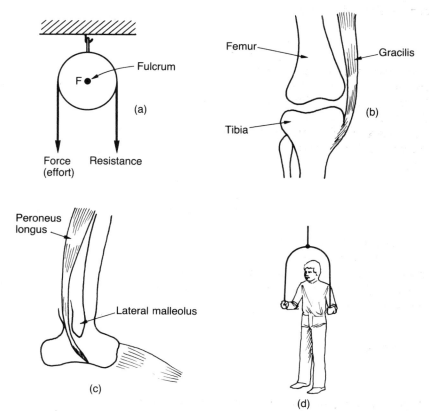

Fig. 1.15 Fixed pulley

shows peroneus longus which passes behind the lateral malleolus, which acts as a pulley before inserting to the base of the first meta-tarsal. It thus acts as a plantar flexor – without the effect of the pulley, the force would pass in front of the ankle joint and act as a dorsiflexor. A clinical example of such a pulley system would be the performance of auto-assisted shoulder exercises (Figure 1.15d).

Single movable pulley

This acts as a second-class lever (Figure 1.16). The effort force acts upon the effort arm (the diameter of the pulley). The load acts on the resistance arm which is the radius of the pulley. As the effort arm is twice that of the resistance, the mechanical advantage will be 2. There are no anatomical examples of single, movable pulleys, although they will be observed in the clinical setting in the application of traction to orthopaedic patients.

Combination pulleys

Pulleys may be combined either as fixed or movable pulleys. They are frequently used in sling suspension (Figure 1.17). A fixed pulley is attached to an overhead support frame. A rope attaches a movable pulley to the fixed pulley, the distance between these being altered by changing the length of the rope. The load (or limb) is attached to the movable pulley, giving a mechanical advantage of 2.

Thus pulleys may serve three functions:

- to change the direction of a force
- to modify the effect of a force
- to obtain a mechanical advantage

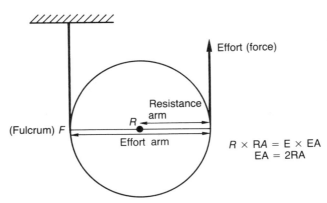

Fig. 1.16 Movable pulley

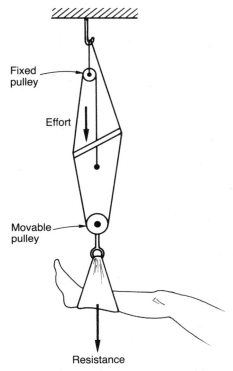

Fig. 1.17 Combination pulleys

PENDULUMS

A pendulum is a heavy particle suspended by a weightless thread and free to move to and fro about a horizontal axis (Figure 1.18). At rest the weight will lie vertically below the point of suspension. If the weight is

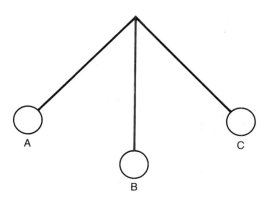

Fig. 1.18 Pendulum

displaced it will swing to and fro at a regular frequency. One complete swing in each direction is termed an oscillation (A–C) and a swing from the vertical to either side will determine the amplitude of the pendulum (AB). A force is required to displace the weight to start the pendulum in motion, and it will continue to oscillate until it is brought to rest by an opposing force or by resistance of the air. In the body the limbs act as natural pendulums. If the knee is held in extension the lower limb will swing from the hip joint at a frequency of approximately 1.9 sec and the lower leg will swing around the knee joint at a frequency of approximately 1.1 sec (MacDonald, 1973). Thus the longer the pendulum, the slower the frequency.

Pendular movements

Pendular movements may be used therapeutically, as they will enable range of motion to be obtained with little exertion of energy. For example, to increase shoulder range of motion in someone with a painful frozen shoulder, the patient may be encouraged to lean forwards and to perform pendular swinging exercises (Codman's exercises – Codman, 1934) (Figure 1.19).

The effect of the pendulum may be increased by holding a weight in the hand, as this lowers the centre of gravity and therefore effectively lengthens the arm of the pendulum. Pendular movements will also be used in axial suspension where the limb is supported vertically above the joint to be moved. By a pendular effect, a wide range of movement may be obtained.

Fig. 1.19 Codman's pendular exercises

GRAVITATIONAL FORCES AND EQUILIBRIUM

Centre of gravity

Gravitational forces act on all particles in a mass, but have the net effect of acting at only one point, the centre of that mass or the centre of gravity. If the object is of uniform shape and density, such as a cube, the centre of gravity will be at the geometric centre. However, few objects are uniform and in the body the centre of gravity will change with the body posture that is assumed. The centre of gravity may be defined as the point at which the sum of all moments is equal to zero, and is the point at which the body is balanced. In upright standing it is generally accepted that the centre of gravity lies within the pelvis anterior to the body of the second sacral vertebra.

Traditional methods of attempting to determine the exact location of the centre of gravity include the use of a balance board and knife edge, segmental photography, weighing of cadavers (Dempster, 1955) and immersion in water (Contini *et al.*, 1963).

Segmental centres of gravity

The centre of gravity of each body segment may also be calculated. It is located at approximately 4/7th of the distance above the segment's distal end. Contini and Drillis (1966) studied volunteers in New York to produce a table of segmental centres of gravity and the mass of each of these body segments as a percentage of the overall whole body mass.

For example, the position of the centre of mass of the forearm and hand segment would be found at 42% of the length of that segment. Once the position of the centre of mass is known, calculations can be performed to obtain moments about joints and the muscular effort required to perform a movement.

Equilibrium

A body is in equilibrium when the resultant of all forces acting upon it equals zero. Equilibrium may be static or dynamic. Static equilibrium may be either:

- stable
- neutral
- unstable.

Stable equilibrium

A body is in stable equilibrium when its potential energy is at a minimum and a force must be applied to it to cause a change in position. The forces acting upon the body at rest will tend to restore it to its original position after it has been displaced. The human body is in stable equilibrium when it is lying flat on a horizontal surface that totally supports it, as the centre of gravity is then at its lowest point. For the body to be in stable equilibrium, the centre of gravity must be held over the base of support and the line of gravity fall well within that base.

Neutral equilibrium

This occurs when a body comes to rest in a new position, with no change in the level of the centre of gravity when it is displaced, e.g. a ball rolling across a flat surface. If an individual is in an erect position such that his centre of gravity is some distance above his base of support, he has potential energy. When this potential energy remains constant the body is in neutral equilibrium. The larger the base, the greater will be the force needed to move the line of gravity outside of that base.

Unstable equilibrium

This occurs when potential energy is at a maximum and the base of support is small, so that very little force is required to move the line of gravity outside the base of support. An initial displacement and the forces acting on the object will increase this original displacement, making it unstable. Thus, there quickly becomes a position of dynamic equilibrium or else the object may return to a position of stable or neutral equilibrium, depending on the corrective action taken.

Dynamic equilibrium

Dynamic equilibrium varies between the neutral and unstable with activities such as walking, running and stooping. During all forms of locomotion the body strives to maintain its balance, and all of its postural reflexes are directed towards this goal. Equilibrium will be unstable when the base of support is small and become neutral as it becomes larger. Thus a patient with poor postural reflexes or one who is apprehensive of performing a task will adopt a wider base in order to maintain a position of neutral equilibrium for a greater length of time.

STABILITY

The stability of a body is determined by how easily its balance may be displaced by a force. The factors that will affect stability are:

- size of base of support
- mass of the body to be moved

- location of the centre of gravity within the body
- where the line of gravity falls within the base.

For a body to achieve maximum stability it must have a large base. The human body is at its most unstable when the feet are close together and parallel (Figure 1.20a). As the feet move apart, the size of the base increases and the person becomes more stable. This may be seen with toddlers learning to walk, who adopt a wide-stanced gait to achieve a large base and greater stability (Figure 1.20b). Similarly the use of walking aids such as crutches and sticks will act as artificial means to create a greater base of support (Figure 1.20c). It must be remembered that, depending on the method of using the walking aid, stability may be increased primarily in the anteroposterior or lateral directions. For further reading refer to Chapter 16.

There will be a point at which increasing the base ceases to increase stability. If a person continues to widen their stance so that their feet are positioned further apart than the width of the pelvis, then stability will be reliant on another factor, friction, to maintain contact between the feet and the ground. In addition to increasing the base of support, stability may be increased by lowering the centre of gravity within the

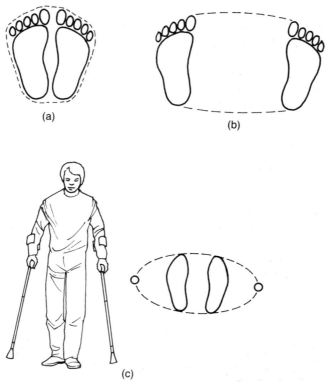

(a)

(b)

(c)

Fig. 1.20 Stability: (a) small base of support; (b) large base of support; (c) base of support increased by use of walking aid

body. For example, if stability is important a person would carry objects at arm's length by their sides as opposed to higher up against the chest wall.

A body is also more stable if its line of gravity falls within the base of support. In dynamic locomotion the line of gravity of the human body will frequently fall outside of the body's base of support. However, stability will be optimized if endeavours are made to minimize this effect, e.g. when lifting an object from the floor, a person may stoop or crouch down to pick it up. In a stooped position, the line of gravity will fall a long way in front of the feet and the centre of gravity will remain high and hence become unstable. In a crouched position, the line of gravity will stay closer to the base of support and the centre of gravity of the body will be lowered, making this a much more stable position as well as being metabolically less stressful.

FRICTION

Friction is the resistance that is felt when one object tends to move on another. Thus, the ability to walk is heavily dependent on the contact between the feet and the ground. Frictional force has the following qualities:

- it is not dependent on the area of contact between the two bodies
- it increases until a peak is reached, after which slipping will occur
- it varies with composition of the respective objects
- it increases with increased pressure.

The peak frictional force between two objects is called 'the limiting friction value' (Galley and Forster, 1987). Friction will also depend on the nature of the surfaces that are in contact with each other. For example, rubber has a high tendency to resistance or a high coefficient of friction. Thus, a walking stick with a rubber ferrule will have better frictional qualities than one with a wooden end.

Patients' footwear may also need to be chosen in order to maximize the effects of friction – they may best be served by undergoing walking training in shoes with good frictional force between the shoe and the floor to enhance stability. However, if the patient has difficulty in clearing the feet from the floor, one may wish them to wear shoes which offer little friction so that the feet can glide over the floor.

Friction will be increased further if the two surfaces are pressed firmly against each other. A patient walking with a stick will maximize friction if they lean heavily on it and press it firmly to the floor. Conversely, the patient who carries a stick and only passes minimal weight through it is more likely to lose contact between the stick and the ground and thus slip.

The therapist may wish to minimize the effects of friction by the use of slippery exercise boards or balanced slings to facilitate range of motion exercises in the leg, or by using oil or powder when performing soft-tissue massage to decrease the friction between the therapist's hands and the patient's skin.

PRESSURE

Pressure is an important aspect of force, indicating how that force is distributed over a given area. Pressure is defined as the total force per area of force application or $P = F/A$ (Rodgers and Cavanagh, 1984).

If one wishes to decrease the pressure on a surface, it will be necessary to increase the area that the force is applied to. In walking, if high-heeled shoes are worn all of the force is concentrated over a small area and high-pressure values will occur; increasing the size of the heel would spread the load and decrease the pressure. When considering the pressure problems of immobile patients, the area of force application may be maximized to try to combat problems with skin and circulatory breakdown. Thus pressure may be decreased by regular turning which alternates the areas under pressure or by the use of specialist bead- or fluid-filled mattresses, which adopt a much greater contact area with the patient's body and even out the pressure that is applied to a patient's skin. Similarly, in the fitting of prostheses, such as a below-knee artificial leg, the socket will be designed so that the contact force is distributed over as large a skin area as possible, usually by the use of anatomically conforming moulds.

ENERGY, WORK AND POWER

Energy

Energy is the ability to do work. Therefore, the energy that a body possesses is the measure of the work that it is capable of doing. While energy cannot be created or destroyed it can change its state; this is known as the principle of conservation of energy. Thus energy may be:

Potential
This is the energy of position or the potential to do work because of position or deformation. For example, a person standing on top of a wall has greater potential energy than one standing on the ground below; a ball that has been squeezed up has greater potential energy than one that is still and undeformed.

Kinetic
This is the energy of motion. The amount of kinetic energy will vary with the velocity of travel, so if more motor units are recruited, a greater velocity of movement will occur and the body will have greater kinetic energy.

Work

Work is the movement of a force through a distance, $W = F \times d$. For example, when biceps flexes, the arm work is being done as the arm is moved by shortening the muscle. If the work is done passively, e.g. eccentric muscle work when gravity returns the arm to its starting position, this is referred to as negative work.

Power

Power is the rate of doing work:

$$\text{Power} = \text{Work/Time} = Fd/t$$

Therefore, muscle power is the rate at which a muscle group can work irrespective of the force that it generates (muscle strength) – see Chapter 10.

STRESS

Stress occurs when a force is applied to a body that results in its deformation. It is expressed as force per unit area or Newtons per square metre (N/m^2). Stress may be linear or shear.

Linear stress

If an object is subject to equal and opposite forces applied perpendicular to its ends trying to pull it apart, it will create linear tensile stress. If the forces act to compress the body, it will create a linear compressive stress. Such a situation will often occur when muscles act as fixators.

Shear stress

If a force tends to produce a sliding effect between two planes, this will produce a shear stress. For example, a pair of scissor blades will apply equal and opposite forces to the material being cut tangentially, so creating shear stresses in the material.

Stress/strain diagrams

The amount of deformation produced in an object will depend on the material's mechanical properties. If a material is rigidly held at one end and progressively loaded at the other, the amount of stretching (strain) that occurs can be plotted against the force needed to produce it as a stress/strain diagram (Figure 1.21).

Hooke's law states that the strain is proportional to the stress producing it (this only applies to small strains). Therefore, there will be a linear relationship between the amount of deformation that occurs and the force applied. Eventually, as an increasing force is applied, a larger amount of deformation will occur for a proportionally smaller force. This is the elastic limit. Once the elastic limit has been reached, removing the force will not result in a return to the original state, as permanent deformation or distortion will occur (Biewener, 1992). Therefore, Hooke's law only applies until the elastic limit is reached.

Values for the ratio of stress:strain vary with different materials. This ratio for comparing different elasticities of materials is called the elastic modulus and is expressed for linear and shear stresses:

$$\text{Young's modulus} = \frac{\text{Linear stress}}{\text{Linear shear}}$$

$$= \frac{F/A}{(\text{Length2} - \text{Length1})/\text{Length1}}$$

$$\text{Shear modulus} = \frac{\text{Shear stress}}{\text{Shear strain}} = \frac{F/A}{\text{Angle of deformation}}$$

In the human body different structures such as tendons, cartilage and skin will respond differently to stress and this may be plotted (Barbenel *et al*., 1978).

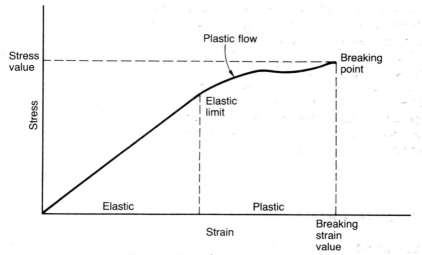

Fig. 1.21 Stress/strain diagram

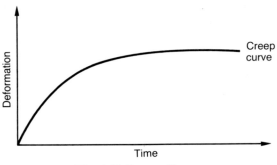

Fig. 1.22 Creep effect

Creep effect

If a constant stress or force is applied to a material, there is an immediate deformation or strain. Viscoelastic structures such as ligaments will continue to deform until they reach a steady state or exhibit a creep effect (Figure 1.22).

This effect may be utilized therapeutically, for example in serial splinting to minimize joint contractures. A stretch is applied to the limb manually and then held by a plaster of Paris cast. Upon removal, a further stretch is applied, more creep occurs and the tissues are thus gradually lengthened (see Chapter 12).

Springs

Springs are elastic and exhibit the characteristic properties described above. When they are stretched they offer resistance to muscle work and as they recoil, assistance. A spring will be at its most efficient when it is applied at 90° to the bone and parallel with the line of movement (see 'Angle of pull', earlier).

As the bones of the body describe an arc of movement, it will only be possible to achieve this 90° angle at one point on the arc. When setting up the spring apparatus, it is customary to use the centre of the arc so that the spring is working most effectively in the middle range of the muscle.

FLUID MECHANICS

An understanding of fluid mechanics will be necessary for the therapist intending to use movement therapy in the water, i.e. hydrotherapy.

The principal mechanical concepts that need to be considered when devising hydrotherapy programmes are:

- buoyancy
- centre of buoyancy
- specific gravity
- hydrostatic pressure.

Buoyancy

Buoyancy is the upward force that any fluid exerts on an object that is immersed in it. Archimedes' principle states that a body immersed in a fluid experiences a buoyant force equal to the weight of the fluid that it displaces. Therefore, the force of buoyancy will be equal to the weight of that volume of water which the object displaces.

Centre of buoyancy

In a manner that is analogous with gravity, buoyancy may be considered as acting at the centre of buoyancy of an object. If an object is of uniform density, the centres of buoyancy and gravity will coincide. However, this is not the case when humans are immersed. The centre of buoyancy is defined as the centre of gravity of the volume of the displaced fluid before its displacement, i.e. it is located in the region that displaces the largest volume of water. This is usually in the thorax, although a fatter person may have a centre of buoyancy closer to the hips.

Specific gravity

Specific gravity is a comparison of the density of an object with that of the fluid that it is immersed in. The specific gravity of the human body is between 0.95 and 0.97, depending on the physical characteristics of the individual (Davis and Harrison, 1988).

Hydrostatic pressure

At any point below the surface of the water, the pressure on the body will be greater than that at the surface by an amount directly proportional to the depth of that point and the density of the fluid. That pressure will be equal and opposite in all directions. Therapeutically, this may result in the movement of interstitial fluid from the lower extremities and a shift of around 700 ml of blood from the legs to the thorax (Arborelius *et al.*, 1972).

SUMMARY

- An understanding of mechanics provides the physiotherapist with the underlying principles to many of the treatment modalities that are used in everyday clinical practice such as sling suspension, balance exercises and strength training.
- Biomechanics considers the effect of forces on the body. Mechanics is further considered as kinetics, the study of forces and kinematics, the study of the effect of forces.
- Force is the effect that tries to change either the state or motion of a body.
- The effect of forces on the body may be analysed to visualize movement patterns by such means as free body diagrams.
- Forces may produce motion that is linear, angular or curvilinear and will obey Newtonian laws of motion.
- Levers transmit force and examples are found in many naturally occurring situations in the human body.
- Pulleys are means by which forces may be transmitted to change the direction of the force, make it produce a different movement or to obtain a mechanical advantage.
- The effects of gravity on the body need to be understood when studying movement or when designing exercise regimens.
- Stability, equilibrium and balance are all concepts that will need to be considered when developing exercise programmes.

APPENDIX: DEFINITIONS

Brief definitions of biomechanical terms and, where appropriate, their units of measurement are given below:

DYNAMICS	Concerned with bodies in motion.
VECTOR	A quantity that has both magnitude and direction.
FORCE	A vector quantity that describes the action of one body on another (newtons).
VELOCITY	Amount of displacement in a given time interval, e.g. metres per second (m/s).
ACCELERATION	Rate of change of velocity. Change in velocity divided by time gives the unit of acceleration, e.g. m/s^2.
MASS	Quantity of matter of an object.
CENTRE OF MASS	Point on a body that moves in the same way that a particle subject to the same external forces would move.
CENTRE OF GRAVITY	Point at which a force would be applied to balance the effects of gravity.

EQUILIBRIUM   Effect occurring when the resultant force and moment are zero.

FRICTION   Force acting between two bodies in contact that opposes motion.

WORK   Work is done when a force moves an object through a distance (joules).

TORQUE   Turning effect produced by a force (Nm).

POWER   Rate of doing work (watts).

LEVER   A system that changes the mechanical advantage and/or direction of a force.

MECHANICAL ADVANTAGE   Ratio of the effort-force lever arm to the resisting force lever arm.

STRESS   Force per unit that occurs as a result of externally applied loads.

STRAIN   Deformation that occurs as a result of loading.

REFERENCES

Arborelius, M., Balldin, U.W., Lilja, U.I and Lundgren, C. (1972) Haemodynamic changes in man during immersion with the head above the water. *Aerospace Medicine*, **43**, 592

Barbenel, J.C., Evans, J.H. and Jordan, M.M. (1978) Tissue mechanics. *Engineering in Medicine*, **7**, 5–9

Biewener, A.A. (1992) *Biomechanics: Structures and Systems*, Oxford University Press, Oxford

Brancazio, P.J. (1984) *Sports Science: Physical Laws and Optimum Performance*, Simon and Schuster, New York

Codman, E.A. (1934) *The Shoulder*, T. Todd, Boston

Contini, R., Drillis, R.J. and Bluestein, M. (1963) Determination of body segment parameters. *Human Biology*, **5**(5), 493–504

Davis, B.C. and Harrison, R.A. (1988) *Hydrotherapy In Practice*, Churchill Livingstone, Edinburgh

Dempster, W.T. (1955) Space requirements for the seated operator, *WADC Technical Report 55159*, Wright-Patterson Air Force Base, Ohio

Enoka, R.M. (1988) *Neuromechanical Basis of Kinesiology*, Human Kinetics Books, Illinois

Galley, P.M. and Forster, A.L. (1987) *Human Movement*, 2nd edn, Churchill Livingstone, Edinburgh

Gowitzke, B.M. and Milner, M. (1988) *Understanding the Scientific Bases of Human Movement*, 3rd edn, Williams and Wilkins, Baltimore

Inman, V.T. (1947) Functional aspects of the abductor muscles of the hip. *Journal of Bone and Joint Surgery*, **29**, 607

Le Veau, B. (1977) *Williams and Lissner: Biomechanics of Human Motion*, 2nd edn, W.B. Saunders, Philadelphia

Luttens, K. and Wells, K.F. (1982) *Kinesiology: Scientific Basis of Human Motion*, Saunders College Publishing, Philadelphia

MacDonald, F.A. (1973) *Mechanics for Movement*, G. Bell, London

O'Connell, A.L. and Gardner, E.B. (1972) *Understanding the Scientific Bases of Human Movement*, Williams and Wilkins, Baltimore

Rodgers, M.M. and Cavanagh P.R. (1984) Glossary of biomechanical terms, concepts and units. *Physical Therapy*, **64**, 1886–1902

2. *Skeletal basis of movement*

INTRODUCTION

The skeletal system is the bony framework of the body, which forms the centre of all body segments as well as acting to protect the body's organs. Articulation at the joints allows movement to take place, controlled by the nervous system. The motor force to perform the movement of the skeletal structure comes from the muscles via their bony attachments. The human skeleton is internal to the muscles with which it is closely functionally integrated, i.e. it is an endoskeletal system. There are over 200 bones in the human skeleton which articulate with each other to form joints and thus to perform movement. This chapter will outline:

- planes and axes of the skeletal system
- joints
- degrees of freedom
- link segment model
- range of motion of joints
- movement of joints.

PLANES AND AXES

Movement is described based from the anatomical position. In the anatomical position the body is standing erect and facing forwards, legs together and feet parallel, so that the toes point forwards. The arms hang by the sides of the trunk with the palms facing forwards and the thumb pointing laterally (Palastanga *et al.*, 1994). All movements are described from this basic position (Figure 2.1).

Planes of the body

The body is divided into three mutually perpendicular anatomical planes (Figure 2.1) which divide the body into equal parts and intersect

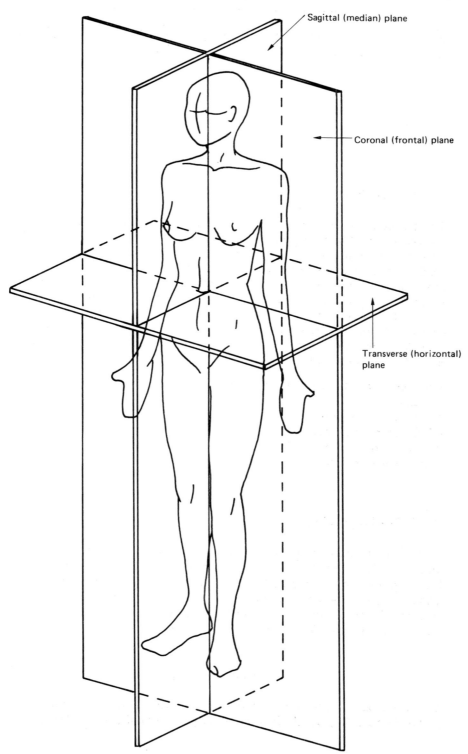

Sagittal (median) plane

Coronal (frontal) plane

Transverse (horizontal) plane

Fig. 2.1 Planes of the body (from Palastanga *et al.*, 1994)

Human Movement Explained

at the centre of gravity of the body. These are:

Sagittal (median) plane

This passes vertically through the body, dividing it into equal left and right halves. It is also sometimes referred to as the median plane. Any line parallel to this is called a sagittal plane.

Frontal (coronal) plane

This passes vertically through the body, dividing it into front and back halves. It lies at right angles to the sagittal plane. Any line parallel to this is called a frontal or coronal plane.

Transverse (horizontal) plane

This divides the body into equal upper and lower halves, passing at right angles to both the sagittal and frontal planes. Any plane parallel to this is called a transverse plane.

Axes of movement

With each plane there is a single axis about which movement takes place (Table 2.1):

- *Sagittal plane/coronal axis (M–L axis)* – this allows the movements flexion/extension and dorsiflexion and plantar flexion.
- *Coronal plane/sagittal axis* – this allows abduction and adduction, lateral flexion and ulnar and radial deviation.
- *Transverse plane/vertical axis* – this allows internal (medial) and external (lateral) rotation, and supination and pronation of the forearm.

All the various axes intersect at the centre of a joint, and the movement at any joint can thus be analysed into its component planes and axes.

Movement descriptors

Movements are described with respect to the anatomical position.

Table 2.1 Axes of movement

Plane	Axis	Movement
Sagittal	Coronal	Flexion/extension
Coronal	Sagittal	Abduction/adduction
Transverse	Vertical	Medial/lateral rotation

Sagittal plane movement

Flexion. The movement of adjacent body segments in a sagittal plane so that the angle between them decreases, e.g. bending the knee.

Extension. The moving apart of adjacent body segments in a sagittal plane so that the angle between them increases. It also refers to any movement beyond neutral that is the opposite of flexion.

Plantar flexion. Moving the foot downwards away from the leg.

Dorsiflexion. Moving the foot up towards the front of the leg.

Frontal plane movement

Abduction. The movement of a body segment away from the midline of the body in a frontal or coronal plane.

Abduction. The movement of a body segment towards the midline of the body in a frontal or coronal plane.

Radial deviation. A specific term for movement of the hand at the wrist joint in the direction of the thumb.

Ulnar deviation. Movement of the hand at the wrist joint in the direction of the fifth digit.

Lateral flexion. Movement of the trunk or head to one side in a frontal plane.

Transverse plane movement

Medial (internal) rotation. Rotation of a segment such that its anterior surface turns in to face the midline.

Lateral (external) rotation. Rotation of a segment such that its anterior surface turns outwards away from the midline.

Supination. Specific term associated with the forearm and foot. In the forearm, the palm of the hand is in the anatomical position in relation to the forearm. In the foot, the forefoot turns so that the sole faces medially.

Pronation. Movement of the palm of the hand away from the anatomical position to face posteriorly; movement of the forefoot so that the sole of the foot faces laterally.

JOINTS

A joint connects one component of a structure with one or more other components. The exact design of a joint will vary with its primary function, so that a joint that has a primary role of providing stability will be of a different design to one whose primary function is mobility. They may be broadly categorized as synovial joints (diarthroses) and non-synovial joints (synarthroses). Within each of these categories there will be further subclassifications.

Synovial joints (diarthroses)

In a synovial joint the bones are linked together by a fibrous capsule, together with accessory ligaments which may be inside or outside of this, so that the bone surfaces are in contact with each other but not in continuity (Williams and Warwick, 1980) (Figure 2.2). All synovial joints share certain characteristic features:

- joint capsule of fibrous tissue
- joint surfaces covered by hyaline cartilage
- capsule lined with a synovial membrane
- joint surfaces bathed in synovial fluid.

Intra-articular structures such as articular discs, menisci, fat pads and ligaments may be located within the joint. The role of the fat pads, discs and menisci, together with the synovial fluid, is to prevent excessive compression and wearing of the joint surfaces. Ligaments and tendons keep the joint surfaces in alignment and act as passive check reins to limit separation of the joint surfaces.

Synovial joints may further be divided based on shape into seven different varieties.

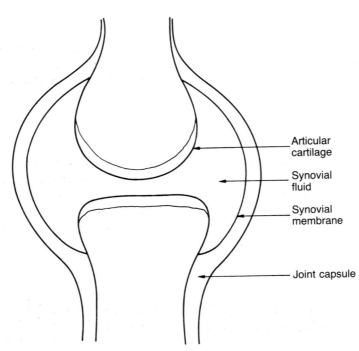

Articular cartilage

Synovial fluid

Synovial membrane

Joint capsule

Fig. 2.2 A synovial joint

Hinge joint

These are uniaxial, allowing movement in one plane only. They typically have strong collateral ligaments. Examples include the interphalangeal joints which allow movement (flexion and extension) in the sagittal plane around a coronal axis.

Pivot joint

These are uniaxial, comprising one component which acts as a central bony pivot and one that forms an osteoligamentous ring. Movement is restricted to a rotation around a longitudinal axis through the centre of the pivot. For example, the atlanto-axial joint has the odontoid peg formed by the dens of the axis and a ring by the anterior arch of the atlas and transverse ligament.

Bicondylar joint

The principal movement occurs in one plane, but a small amount of rotation can occur about a second axis perpendicular to the first. The joints have convex surfaces which are conventionally called male, formed by two distinct convex condyles which articulate with two concave (female) surfaces, e.g. the knee joint.

Saddle (sellar) joint

These are biaxial, in which opposing surfaces are concavoconvex, i.e. they are convex in one plane and concave in the other. Examples include the carpometacarpal joint of the thumb which allows flexion/extension and abduction/adduction.

Ellipsoid (condyloid) joint

These are biaxial and have joint surfaces shaped so that the concave surface of one bony component slides over the convex surface of the other component in two directions. Examples include the metacarpophalangeal joints, where movement occurs about two axes perpendicular to each other – flexion/extension around a coronal axis and abduction/adduction around an anteroposterior axis.

Plane joint

These are multiaxial, formed by the apposition of fairly flat articular surfaces, producing gliding movements between the bones. Examples include the carpal bones, where adjacent surfaces may slide on one another or rotate relative to each other in any plane.

Spheroidal (ball and socket) joint

A ball-like convex surface articulates with a concave socket, allowing movement in all three planes. For example, the hip joint allows flexion/extension in a sagittal plane around a coronal axis, abduction/adduction in a coronal plane around a sagittal axis, and rotation in a transverse plane around a vertical axis.

Non-synovial joints (synarthroses)

In these types of joints the bony components are connected by inter-osseous connective tissue. They are further divided, according to the type of tissue that forms the joint, into cartilaginous or fibrous.

Cartilaginous

The bony components are joined by either hyaline or fibrocartilage. They may be either symphyses or synchondroses.

Symphyses
These joints have the bony components joined by a fibrocartilaginous junction. They all occur in the median plane of the body. Examples include the symphysis pubis, where the pubic bones are joined by a fibrocartilaginous plate, allowing great strength but little or no movement except for during pregnancy when softening occurs under hormonal control.

Synchondroses
These are usually temporary cartilaginous joints where the bony components are joined by hyaline cartilage in the immature skeleton.

Fibrous

Fibrous tissue unites the two bones. There are three types of fibrous joint.

Suture joint
These are limited to the skull, occurring where the margins or broader surfaces of the bones meet and articulate, separated only by a zone of connective tissue. For example, coronal suture joint of the skull in infancy; when growth comes to an end, usually in the late twenties, osteogenic cells transform the joint to a bony fusion.

Gomphosis
This is a specialized type of peg and socket joint restricted to the fixing of the teeth in the mandible or maxillae.

Syndesmosis
This is a form of articulation in which closely apposed bone surfaces are bound together by an interosseous ligament, allowing a small amount of movement between the adjoining bones. For example, the shafts of the tibia and fibula are joined by the interosseous membrane.

DEGREES OF FREEDOM

Joints are described in relation to the number of planes that they can move in (Brunnstrom, 1983) (Table 2.2). Thus, uniaxial joints move around one axis and have one degree of freedom, i.e. the articulating bones only move in one plane.

Biaxial joints have two axes, two degrees of freedom and move in two planes.

Joints that have the ability to move in all three planes have three degrees of freedom, although they are termed multiaxial as they may move in oblique planes as well as the standard anatomical planes.

The maximum number of degrees of freedom that any joint can have is three. However, the effect of adjacent joints may be summated to express the total amount of freedom between one part of the body and an area more distal to it. The more distal a segment, the greater the degrees of freedom it will possess relative to the trunk. Gowitzke and Milner (1988) cite the example of the degrees of freedom between the distal phalanges of the hand and the trunk. These amount to 17, as shown in Table 2.3.

Table 2.2 Planes and degrees of freedom

Axis	Degrees of freedom	Planes
Uniaxial	1	1
Biaxial	2	2
Multiaxial	3	3

Table 2.3 Degrees of freedom between the distal phalanges of the hand and the thumb

Joint	Degrees of freedom
Distal interphalangeal joint	1
Proximal interphalangeal joint	1
Metacarpophalangeal joint	2
Wrist joint	2
Radio-ulnar joint	1
Elbow joint	1
Shoulder joint	3
Acromioclavicular joint	3
Sternoclavicular joint	3
Total	17

LINK SEGMENT MODEL

The concept of the body as a series of links was derived from engineering principles (Figure 2.3). A link is considered to be a straight line of constant length running from axis to axis. Each link will produce a reaction in relation to the next link in the system (Winter, 1979). This system of links is referred to as a kinetic chain. In engineering, the chain forms a closed system where the motion of one link will produce motion at all of the other joints in the system in a predictable way (Lehmkuhl and Smith, 1983). Similarly, the joints of the body may be linked together in such a way that motion at one joint must be accompanied by motion at an adjacent joint. For example, when performing a press up the proximal segment will be fixed at the hands by their contact with the floor and the feet by their contact with the ground. Brunnstrom (1983) considers that the pelvic girdle and the rib cage both constitute closed kinematic chains within the body.

It is more common to find open kinematic chains in the body, where the chain finishes with the distal segment free in space, e.g. waving the hand in the air. A chain is termed open if its end is free to move without causing motion at another more distal joint (Chapter 5).

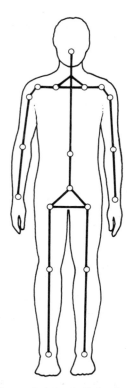

Fig. 2.3 Link segment model – each joint is linked to the next by a straight link

RANGE OF MOTION OF JOINTS

Galley and Forster (1987) define range of motion as the maximum amount of displacement possible at any one joint. The type of movement at a joint and its range of motion will depend on:

1. The shape of the articulating surfaces. Range of motion will be limited by bone-to-bone contact. For example, extension of the elbow is limited by approximation of the olecranon process of the radius in the olecranon fossa of the humerus.
2. Tightness of the joint capsule.
3. Restraining effect of the ligaments.
4. Restraining effect of muscles. For example, gastrocnemius crosses the ankle and knee joints, so that when the knee is in extension, ankle dorsiflexion will be limited by tension in gastrocnemius. A muscle may also act to restrain movement in isolation, for example tension in the hip adductors may limit hip abduction.
5. The bulk of the soft tissue in adjacent segments, e.g. knee flexion, may be limited by the bulk of the hamstrings and the calf musculature.

Measurement

Joint range is measured according to conventions laid down by the American Academy of Orthopaedic Surgeons (1966) (Figure 2.4). For

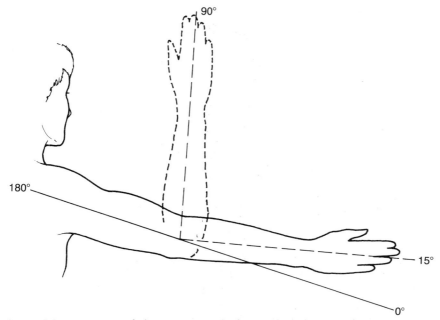

Fig. 2.4 Measurement of the joint range of motion – the arc of movement is 15°–90°

further details of the practical technique of joint measurement, the reader is referred to Chapter 18.

MOVEMENTS OF JOINTS

These can be divided into two categories:

* *arthrokinematics* – the movement of joint surfaces
* *osteokinematics* – the movements of bones.

Arthrokinematics

This subject is dealt with extensively by Williams and Warwick (1980), MacConnail and Basmajian (1977) and Norkin and Levangie (1992).

Arthrokinematics pressumes that one joint surface is more stable than the other and serves as a base for the motion, while the other surface moves on this relatively fixed base. By convention, the elements of the joint are referred to as male or female. The male component is either convex or, if the joint is a sellar joint, it is the larger articulating surface. The female component is concave or has a smaller articulating surface.

The types of motion that the moving part of the joint performs are described using the terms roll, slide and spin:

Roll. This describes the rolling of one joint surface on another. For example, the movement of the femoral condyles across the tibial plateau.

Slide. This is a pure translatory motion, where one component glides over the other. For example, the proximal phalanx slides over the fixed end of the metacarpal bone.

Spin. This describes the rotation of the movable component on the stationary component. For example, the radial head spins on the capitulum of the humerus in pronation and supination of the forearm.

The exact type of movement that occurs at a joint will depend on the shapes of the articulating surfaces of the bones. These may be ovoid or sellar.

In an ovoid joint, one surface is convex and one is concave, e.g. interphalangeal joints. When a convex (male) surface moves on a stable concave (female) surface, the convex articular surface will move in the opposite direction to the motion of the shaft of the bony level (Figure 2.5a). When a concave surface (female) moves on a stable convex (male) surface, the concave articulating surface will slide in the same direction as the bony lever (Figure 2.5b). So with the moving convex (male) surface, slide and roll will occur simultaneously in opposite directions, whereas with a moving concave (female) surface,

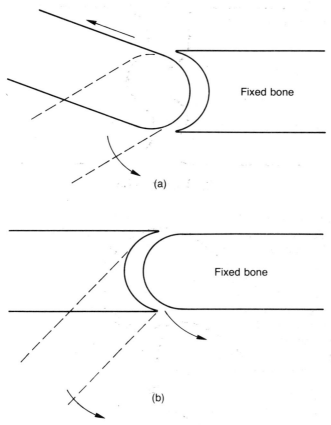

Fig. 2.5 Arthrokinematics: (a) convex surface moves on a concave surface – articular surface moves in opposite direction to the shaft of the bony lever; (b) concave surface moves on a convex surface – articular surface moves in the same direction as the shaft of the bony lever

they occur in the same direction. These combinations produce a curvilinear movement which increases the amount of angulation possible at a joint without increasing the size of the articulating surfaces.

In a sellar joint, each surface is both concave and convex (Figure 2.6).

Closed and loose packed joint positions

The two joint surfaces of an ovoid or sellar joint are only perfectly congruent in one position, usually at one extreme of the most habitual movement of the joint such as knee extension. In a state of full congruence, the joint surfaces have a maximal area of contact and are tightly compressed, held by tension in the capsule and ligaments, and the joint cannot be separated by force. This is referred to as the close packed position (Figure 2.7a).

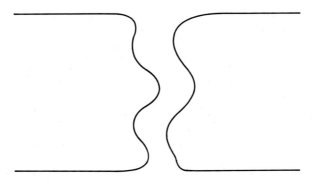

Fig. 2.6 Sellar joint

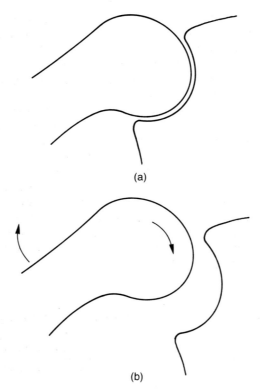

(a)

(b)

Fig. 2.7 Close packed (a) and loose packed (b) joint positions

In all other positions the articulating surfaces are not congruent and some parts of the capsule are lax. The joint is then referred to as being in a loose packed position (Figure 2.7b).

The principal differences between the close and loose packed positions of joints are summarized in Table 2.4.

Table 2.4 Differences between close and loose packing of joints

Close packed	Loose packed
Joint surfaces fully congruent	Joint surfaces non-congruent
Maximum contact between joint surfaces	Partial contact between joint surfaces
Joint surfaces tightly compressed	Joint surfaces ill-fitting
Capsule and ligaments spiralized and tense	Capsule and ligaments lax
Joint surfaces not distracted by force	Joint surfaces separate if a distraction force is applied
Joint surfaces locked together	Small and changing contact area
No further movement possible	Spin, slide and roll movement

Loose packing of the joint allows for combined elements of roll, slide and spin which characterize most normal joint movement. Joint surfaces are often of different sizes, e.g. the head of the humerus moving in the glenoid fossa has a larger moving (male) component than the stable (female) component. If only pure movements occurred, a roll would produce the effect whereby the larger moving component would roll off the smaller articulating surface before the movement was complete. With a loose packed joint, combined movements can occur where the moving component alternates rolling in one direction and sliding in the opposite direction to help increase the available range of motion and to keep the opposing joint surfaces in contact with each other. MacConnail and Basmajian (1977) cite the close and loose packed positions of some of the principal joints (Table 2.5).

Moving in and out of the close packed position is thought to have a beneficial effect on joint nutrition, as fluid is squeezed out during each compression and imbibed in when the compression is removed (Hertling and Kessler, 1990).

Table 2.5 Close and loose packed positions of some principal joints

Joint	Close packed	Loose packed
Gleno-humeral	Abduction and lateral rotation	Semi-abduction
Wrist	Dorsiflexion	Semiflexion
Metacarpophalangeal	Full flexion	Semiflexion and ulnar deviation
Interphalangeal	Extension	Semiflexion
Hip	Extension and medial rotation	Semiflexion
Knee	Extension	Semiflexion
Ankle	Dorsiflexion	Neutral
Metatarsophalangeal	Dorsiflexion	Neutral

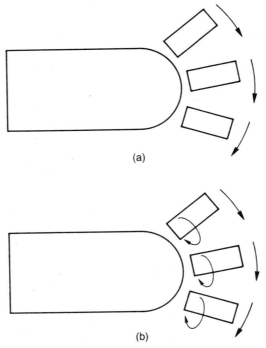

Fig. 2.8 Osteokinematics: (a) spin; (b) simultaneous spin and swing

Osteokinematics

This is the study of bone movements rather than of the joints (Figure 2.8). It divides movement into spin and swing:

Spin. This is defined as rotation around the long axis of the bone.

Swing. This is any movement other than pure spin.

SUMMARY

- The skeletal bases of movement have been described. It is important that accurate description of movement is given in relation to the anatomical reference position, to the plane of movement and to the axes about which the movement occurs.
- Different classifications of joint will have different possibilities for movement, but the most important are the synovial joints.
- The concept of degrees of freedom and the link segment model enables better understanding of joint movements in relation of one body segment to another.
- The concept of joint range of motion is described.

- Movements of joints are described according to the principle of arthrokinematics.
- The importance of the close and loose packed joint position is emphasized.
- An understanding of the above will help the reader to apply a scientific basis to their attempts to mobilize joints in clinical practice.

REFERENCES

American Academy of Orthopaedic Surgeons (1966) *Joint Motion Method of Measuring and Recording*, Churchill Livingstone, Edinburgh

Brunnstrom, S. (1983) *Brunnstrom's Clinical Kinesiology*, 4th edn, F.A. Davis, Philadelphia

Galley, P.M. and Forster, A.L. (1987) *Human Movement*, 2nd edn, Churchill Livingstone, Edinburgh

Gowitzke, B.A. and Milner, M. (1988) *Understanding the Scientific Bases of Human Movement*, 3rd edn, Williams and Wilkins, Baltimore

Hertling, D. and Kessler, R.M. (1990) *Management of Musculoskeletal Disorders*, 2nd edn, J.P. Lippincott, Philadelphia

Lehmkuhl, L. and Smith, L.K. (1983) *Brunnstrom's Clinical Kinesiology*, 4th edn, F.A. Davis, Philadelphia

MacConnail, M.A. and Basmajian, J.V. (1977) *Muscles and Movements*, 2nd edn, Robert E. Krieger, New York

Norkin, C.C. and Levangie, P.K. (1992) *Joint Structure and Function*, 2nd edn, F.A. Davis, Philadelphia

Palastanga, N., Field, D. and Soames, R. (1994) *Anatomy and Human Movement*, 2nd edn, Butterworth–Heinemann, Oxford

Williams, P.L. and Warwick, R. (1980) *Gray's Anatomy*, 36th edn, Churchill Livingstone, Edinburgh

Winter, D.A. (1979) *Biomechanics of Human Movement*, John Wiley, New York

3. *Musculoskeletal basis of movement*

INTRODUCTION

The human body contains around 640 skeletal muscles, varying from tiny vestigial muscles which perform such movements as moving the ears, to large muscles like the gluteus maximus (Gowitzke and Milner, 1988). Skeletal muscle constitutes over one-third of the human body's total mass (Palastanga *et al.*, 1994). Muscles, excited by the nervous system, act on the skeletal structure of the body to produce movement and to provide stability to the skeleton. Without these the body would be unable to attain its normal erect posture, as may be witnessed by the effect of paraplegia. The muscles generate forces that act on the joints of the body; generally these forces will be rotatory (providing mobility) or translatory (providing stability). The amount of mobilizing or stabilizing force that a muscle applies to a joint will vary with the muscles size, biomechanical characteristics and with the structure and position of the joint. This chapter will outline:

- skeletal muscle
- muscle structure
- mechanism of muscle contraction
- types of muscle fibre
- classification of muscle
- types of muscle contraction
- roles of muscles
- range of motion.

PROPERTIES OF SKELETAL MUSCLE

Skeletal, or striated muscle has four characteristic properties. These are:

Irritability, i.e. the ability to respond to a chemical, electrical or mechanical stimulus.

Contractility, i.e. the ability to develop tension.

50 ·

Distensibility, i.e. the ability to be stretched or lengthened by a force extrinsic to the muscle itself, such as antagonistic muscle force or gravitational forces.

Elasticity, i.e. the ability to recoil from a distended position.

Thus, skeletal muscle is viscoelastic, possessing both distensibility and elasticity which are antagonistic properties whose coexistence contributes to its unique qualities.

MUSCLE STRUCTURE

Skeletal muscle has two components – active contractile and inert compliant tissues. It is made up of non-branching muscle fibres bound together by connective tissue. The basic unit is the muscle fibre, each of which can be regarded as a single cell with many nuclei.

Muscle fibres

Each muscle fibre is enclosed in a cell membrane called the *sarcolemma* (Figure 3.1). The fibres are arranged in bundles of various sizes within each muscle called *fasciculi*. Within each fasciculus connective tissue called *endomysium* fills the spaces between the muscle fibres. Each fasciculus is bound by a strong connective sheath called the *perimysium* and all the fasciculi are bound together to form the whole muscle by a fibrous covering called the *epimysium* which is continuous with both the perimysial septa and the connective tissue of the external structures (Williams and Warwick, 1980). Muscle fibres are composed of cytoplasm called *sarcoplasm*. This contains the contractile structures of the muscle fibres, the myofibrils, as well as non-myofibrillar structures such as ribosomes, mitochondria and glycogen required for cell metabolism.

Ultrastructure of skeletal muscle

In longitudinal section, skeletal muscle appears to be striped or striated. These striations across the myofibrils can be divided into repeating regions of about 2.5 μm length in resting muscle called *sarcomeres* (Figure 3.2). The sarcomere has boundaries formed by Z lines to which actin filaments attach. The different areas of the sarcomere are called bands or zones and relate to different arrangements of the actin and myosin filaments:

A band. The portion of sarcomere that extends over the length of both myosin filaments and over a small portion of actin filament (anisotropic band).

Human Movement Explained

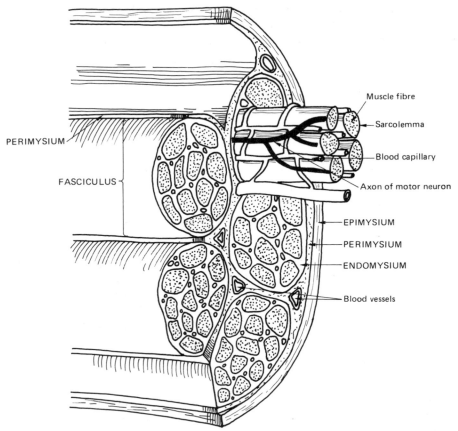

Fig. 3.1 Structure of muscle (from Palastanga *et al.*, 1994)

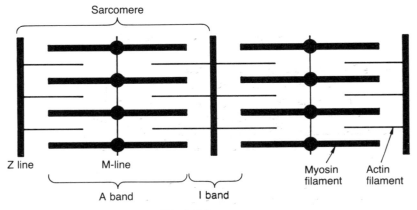

Fig. 3.2 Sarcomere

H zone. The central portion of myosin filament that bisects the A band where there is no overlap with actin.

M band. The central portion of the H zone formed by myosin myofilaments that connect one myosin myofilament with the filament above and below.

I band. Actin filaments only (isotropic band).

Microscopic structure of skeletal muscle

Each myofibril is about 1 μm in diameter and composed of small myofibrils which are made up of the proteins actin and myosin. The binding together of these two proteins results in a muscle contraction.

Actin

Actin is a thin myofilament formed by two chains of actin molecules wound around each other to form a helix. The globular proteins troponin and tropomyosin are attached to it, tropomyosin in the notches between the actin chains and troponin is bound to it at regular intervals (Figure 3.3a). These two proteins exert a regulatory effect on the binding of actin and myosin myofilaments.

Myosin

Myosin is comparatively thick, with the myosin molecules arranged to form long molecular filaments, each of which has a tail and a head, resembling a golf club in appearance. The myofilament is formed by

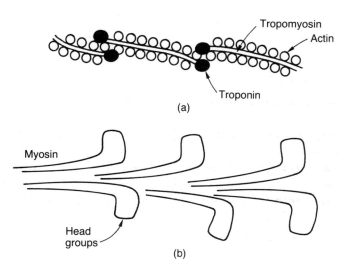

Fig. 3.3 Myofibril proteins: (a) actin filament; (b) myosin filament

the tails of the molecule which lie in parallel, their ends pointing towards the long axis. The heads of the molecule point laterally in pairs and can swivel to form a spiral pattern. These head groups are the binding sites for the attachment to the actin chains and play a vital role in muscle contraction and relaxation (Figure 3.3b).

Gross muscle structure

Gross muscle structure concerns the arrangement of the muscle fibres. Each bunch of fibres or fasciculus will be arranged in a manner that will affect the muscle's shape and function. There are three main groups – parallel, oblique and spiral – based on how the fasciculi are orientated relative to the final direction of pull (Figure 3.4).

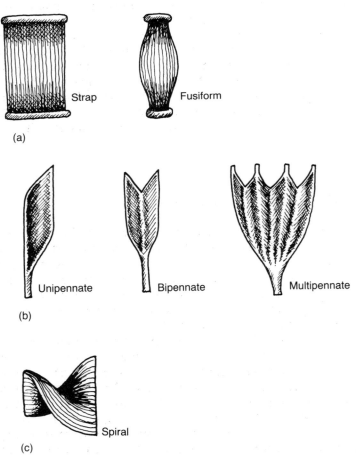

Fig. 3.4 Gross structure of muscle: (a) parallel; (b) oblique; (c) spiral

Parallel

The fasciculi are arranged parallel to the long axis of the muscle. They may be flat and quadrilateral, e.g. thyrohyoid, or strap like, e.g. sartorius. The fibres may be interrupted by transverse tendinous intersections, e.g. rectus abdominis, or form fusiform muscle bellies, e.g. biceps brachii.

Oblique

The fasciculi are oblique to the angle of pull. They may either be triangular, e.g. adductor longus, or pennate (feather like) in construction. For example, unipennate, flexor pollicis longus; bipennate, rectus femoris; multipennate, deltoid.

Spiral

The fasciculi are spiralled or twisted, e.g. latissimus dorsi undergos a 180° twist between its midline and lateral attachment.

Most muscles show a direct relationship in which they can shorten to approximately half their length at full contraction. In pennate muscles, this relationship is disrupted, as only a portion of the muscle force is directed at producing motion. Pennate muscles thus tend to be found where a limited range of motion is required coupled with considerable power. A parallel fibre arrangement tends to favour the production of motion and is found where a large range of motion is required (Guyton, 1981).

MECHANISM OF MUSCLE CONTRACTION

There are still some details of muscle contraction that remain hypothetical. It is generally accepted that it occurs by the sliding filament theory of Huxley (1974), where the two sets of filaments, actin and myosin, slide relative to each other. The filaments stay constant in length, as does the A band. The I band becomes narrower and will eventually disappear, the Z lines are drawn together and the sarcomere is shortened (Grinnell and Brazier, 1981).

Muscles are activated by nervous stimulation that evokes an action potential releasing calcium ions. The calcium binds with troponin to reposition the tropomyosin molecules, so that the receptor sites on the actin molecule are free to react with myosin. The heads of the myosin molecules move out perpendicular to their core and attach to the actin, forming 'cross-bridges'. After some cross-bridges have formed, the myosin heads undergo an energy-yielding conformational change, so that the heads swivel in an arc and pull the actin filament along the myosin breaking old cross-bridges and creating new ones.

This cycle of forming and breaking cross-bridges continues until there is a maximum overlap between the actin and myosin and the sarcomere is shortened. The formation of the cross-bridges generates muscle tension. Cross-bridges can only form and tension be generated when the actin and myosin overlap; both will be at a maximum when there is a maximal overlap of the myofilaments.

If enough muscle fibres actively shorten, the entire muscle will shorten (contract), active shortening producing a concentric contraction. If actin filaments are pulled away from the myosin, the cross-bridges are broken and reform in a lengthened position, i.e. an eccentric contraction. Tension will still be generated by the muscle as cross-bridges are still formed. An example is the eccentric contraction of a muscle resisting gravity (Astrand and Rodahl, 1986). Tension will not be generated if there is no overlap of filaments or if there is complete overlap and no binding sites for cross-bridge formation are available (Norkin and Levangie, 1992).

The motor unit

The motor unit is the basic functional unit of the muscle, consisting of an alpha motor neuron and all of the muscle fibres that it innervates (Figure 3.5). Contraction of a muscle may be as a result of many motor

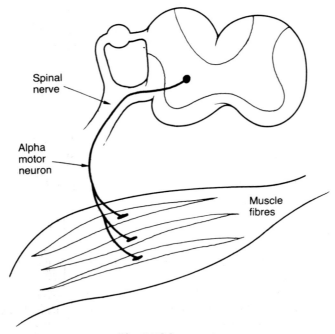

Spinal nerve

Alpha motor neuron

Muscle fibres

Fig. 3.5 Motor unit

units firing. The ratio of muscle fibres to motor units will vary with the demands of each particular muscle. For example, a muscle that is involved in gross movements, such as gastrocnemius, has up to 2000 muscle fibres per motor unit, whereas one that is involved with precision movement, such as the small muscles of the eye, may have as few as 6 muscle fibres per motor unit (Netter, 1987).

The way that the muscle will contract will vary with:

- the number of muscle fibres activated by each motor unit – this will affect the magnitude of the response to a given stimulus
- the number of motor units that are firing at any one time, which will affect the total response of the muscle
- the frequency at which the motor units fire
- the diameter of the axon, which will determine the velocity of the conduction of the impulse.

Muscle tension

Muscle tension may be active or passive.

Active tension

This is the tension that is developed by the contractile elements of the muscle. Tension will be increased by the formation of cross-bridges; the greater the number of cross-bridges that are formed, the greater the muscle tension that is generated. If a large number of motor units are recruited more cross-bridges will form and tension will increase. Muscle tension will also be greater in muscles that have a large cross-sectional area.

Passive tension

Tension will also be developed in the passive non-contractile elements of the muscle. Connective tissue structures may either assist with the muscle's action or become lax and not contribute. The total tension that develops will be a combination of the active contractile tension added to the passive, non-contractile tension.

Length/tension

The amount of tension in a muscle is also a function of the length of that muscle when it is stimulated to contract (Figure 3.6). There is a direct relationship between the tension in a muscle and the length of that muscle. Each muscle has an optimal length at which it is capable of generating maximal tension (Purslow, 1989). This optimal length is usually about 1.2 times the resting length of the muscle. It is at this

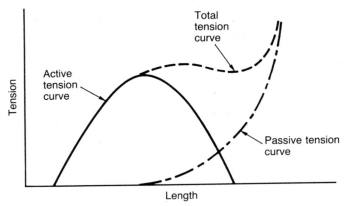

Fig. 3.6 Length/tension curve

optimal length that the most cross-bridges can form between the actin and myosin filaments and thus most tension can be generated. During a concentric contraction, muscles generate moderate tension in the lengthened (outer) range, maximum tension in the middle of the contractile range and minimal tension in the shortened (inner) range (Figure 3.7).

Length/tension curves

A length/tension curve can be produced by measuring the tension produced by the muscle at different lengths. Most work has been done *in vitro*.

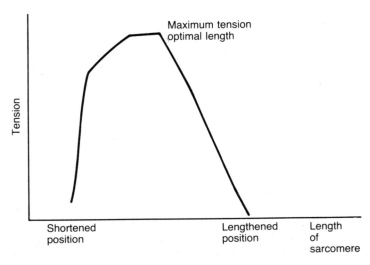

Fig. 3.7 Length and tension

Passive tension curves show the effect of stretching the muscle beyond its unattached resting length. Tension increases slowly at first until it has been stretched approximately 300% of its resting length when its ruptures (Ganong, 1979). In muscles that cross two joints, the passive tension developed in the muscle by stretching may limit joint range of motion. It tends to be higher in distal muscles than proximal ones and in those in the lower extremities.

The difference between the values for total tension (when the muscle contracts isometrically) and for passive tension values indicates the amount of tension generated by the contractile process (active tension).

The length of a muscle at which the active tension is maximal is its resting length. Many muscles are positioned in the body at this length in order to be able to perform strong muscle contractions.

Muscle insufficiency

Muscle insufficiency may be active or passive. Active insufficiency occurs when a muscle, usually crossing two or more joints, reaches a position where it is at a shortened length but can no longer generate muscle tension. For example, the hamstrings may become actively insufficient when they simultaneously extend the hip and flex the knee.

Passive insufficiency occurs when a muscle is unable to reach a sufficient length to permit completion of a full range of motion of the joints that it is crossing. For example, the hamstrings may limit the range of motion or the hip joint when the legs are held with the knees straight and an attempt is made to touch the floor with the hands.

TYPES OF MUSCLE FIBRE

Three main types of muscle fibre are found in skeletal muscle, each of which has different metabolic, mechanical and morphological characteristics. Different authors tend to use different classifications. The types of fibre are:

- fast twitch glycolytic/(type II)/(white) fibres
- fast twitch oxidative glycolytic/(type IIa) fibres
- slow oxidative/(type I)/(red) fibres.

Each muscle will contain all three types of fibres, but there are wide variations in their exact composition. To some extent this variation is determined genetically. The composition will also vary according to what the main function of an individual muscle is.

Muscles with a high proportion of fast glycolytic fibres (type II) tend to be those concerned with mobility and are also called phasic muscles. They frequently are concerned with producing large ranges of motion,

Table 3.1 Characteristics of muscle fibres

	Slow oxidative	Fast glycolytic	Fast oxidative glycolytic
Colour	Red	White	Red
Diameter	Small	Large	Intermediate
Speed of contraction	Slow	Fast	Fast
Rate of fatigue	Slow	Fast	Intermediate
Rate of recovery	Fast	Slow	Intermediate
Motor unit size	Small	Large	Intermediate

e.g. biceps brachii. They respond quickly but fatigue rapidly and have a relatively slow recovery rate (Colliander *et al.*, 1988).

Slow oxidative (type I) muscles are able to carry out sustained activity and so are called stabilizing or tonic muscles. They are efficient working isometrically and have a high aerobic capacity, e.g. soleus.

Muscles with a high proportion of fast oxidative glycolytic fibres tend to have intermediate characteristics. The main differences between the three muscle types are given in Table 3.1. Further details may be found in Williams and Warwick, 1980, or Gowitzke and Milner, 1988.

Most muscles will consist of a combination of fibres. Gastrocnemius and deltoid contain approximately equal proportions of fast twitch and slow oxidative fibres, while soleus contains twice as many slow oxidative fibres as fast twitch.

Muscles may be classified according to the composition of their fibres, e.g. tonic or phasic. Tonic muscles tend to have a high proportion of slow oxidative fibres, be penniform (feather shaped) in arrangement, be located deep and cross one joint to perform a stabilizing function. Working actively, they tend to produce extension, abduction and external rotation. They react readily to training and respond to stress by shortening.

Phasic muscles tend to have a high proportion of fast glycolytic fibres, be parallel in fibre arrangement and be positioned superficially. Their main function is to produce mobility and such movements as flexion, adduction and internal rotation (Norkin and Levangie, 1992).

CLASSIFICATION OF MUSCLES

Muscles may be classified in a number of different ways. They may be divided according to:

- shape
- location
- action
- role
- biomechanical action

Shape
Classification is based on the shape and appearance of the muscle, e.g. deltoid, rhomboid.

Location
Classification is made according to the attachments of the muscle, e.g. flexor pollicis longus (i.e. the long flexor of the thumb).

Action
Muscles performing similar actions may be grouped together, e.g. flexors, rotators.

Role
Muscles may be classified according to the role that they perform, e.g. agonist, synergist.

Biomechanical action (spurt or shunt)

Mathematical analysis led MacConnail (1966) to classify muscles as either spurt or shunt muscles. This theory was later revised by MacConnail and Basmajian (1977), However, other authors question its validity (Stern, 1971).

Spurt muscles have their origins at a distance from the joints on which they act and are inserted near them. They direct the greater part of their force across the bone rather than along it and provide the force which acts along the tangent to the curve transversed by the bone during movement (Chapter 1). Thus they tend to act to produce a larger range of movement.

Shunt muscles have their origins near the joints on which they act and their insertions at a distance from them, so the greater part of their contractile force is directed along the axis of the bones. Thus they tend to pull the joint together and act as stabilizers. They also provide the increase in centripedal force required in rapid movement. For example, pronator quadratus is the prime mover in pronation of the forearm and a spurt muscle, but when rapid movement is required pronator teres, a shunt muscle, assists.

If a muscle acts over two joints it will usually function as a spurt muscle for one and a shunt for the other. For example, biceps brachii acts as a spurt muscle at the elbow and a shunt muscle at the shoulder.

Types of muscle contraction

Muscle contraction implies a shortening of the muscle's contractile unit, and refers to the attempt of a muscle to shorten and develop tension. Rodgers and Cavanagh (1984) use the term 'muscle action' to

describe any type of tension development in the muscle, regardless of whether it is lengthening, shortening or maintaining the same length.
   Types of contraction include:

- dynamic concentric contraction
- dynamic eccentric contraction
- isotonic contraction
- isokinetic contraction
- isometric contraction.

Concentric contraction

The muscle actively alters in length by shortening, as the intrinsic forces created by the muscle contracting are greater than any extrinsic forces, e.g. biceps brachii flexing the elbow.

Eccentric contraction

The muscle alters in length by lengthening, as the extrinsic forces acting on the muscle are greater than those generated intrinsically. A muscle which is being lengthened while it is contracting can maintain greater tension then it can develop at any given equivalent static length, e.g. quadriceps femoris during knee flexion in standing.

Isotonic contraction

This involves the production of a constant force or tension as the muscle changes in length. However, this is an inaccurate description as tension never remains constant throughout range. During the isotonic contraction the effects of the load are maximal in only one small part of the range.

Isokinetic contraction

This involves a contraction where the velocity of contraction remains constant. There is no such thing as a naturally occurring isokinetic contraction in the body.

Isometric contraction

This is where the internal force of the muscle is such that the muscle stays at a constant length, the intrinsic force being equal to the extrinsic force applied.

ROLE OF MUSCLES

Muscles act to produce a given movement, often working together. Depending on the movement required, a muscle will act in a certain role. However, to produce a different movement it may act in a completely different way. The roles of muscles are:

- agonist/prime mover
- antagonist
- fixator/stabilizer
- synergist.

Agonist/prime mover

An agonist is the main muscle responsible for carrying out a given movement at a joint, e.g. quadriceps femoris when kicking a football is acting as a prime mover.

Antagonist

An antagonist produces the opposite effect to the agonist, or prime mover, relaxing reciprocally during the contraction of the prime mover, e.g. hamstrings are the antagonist in the above example.

Fixator/stabilizer

The muscle works to position a bone to provide a base from which the prime mover can act. They act to provide a fixed attachment for another muscle, e.g. when deltoid abducts the upper arm, the muscles of the shoulder girdle contract to stabilize the shoulder to allow deltoid to act from a fixed base.

Synergist

The muscle acts in conjunction with another, to produce a movement neither could produce on its own (conjoint synergist). They may act as a helping synergist, e.g. in flexion of the wrist the agonists will be flexor carpi radialis and ulnaris, but may be assisted by the flexors of the fingers acting as synergists.

As a true synergist (conjoint synergist), dorsiflexion at the ankle is produced by the synergistic action of tibialis anterior and extensor digitorum longus. Working alone, tibialis anterior would produce combined dorsiflexion and inversion and extensor digitorum, toe extension, dorsiflexion and eversion. Working together, they produce pure dorsiflexion.

Synergists will occur to neutralize or counteract the effect of a muscle. For example, the muscles that flex the fingers (flexor digitorum superficialis and flexor digitorum profundus) also pass over the wrist joint and act as wrist flexors. Where this simultaneous movement is not desired the wrist extensors contract to prevent wrist flexion whilst allowing finger flexion to occur. Thus muscles may act as a synergist to prevent unwanted movement from occurring, particularly where a muscle crosses more than one joint.

MUSCLE ATTACHMENTS

A muscle exerts a force on all of its attachments when it contracts. Until recently, by convention, the proximal attachment of the muscle, which is frequently the stationary component, was referred to as the origin and the distal moving attachment the insertion. However, as either the proximal or distal attachment may move, this can lead to confusion and the terms 'proximal' and 'distal' attachments are now considered to be more appropriate and have been adopted.

The muscle attaches to the bone by continuations from its outer sheath – the epimysium. These form tendons which attach each end of the muscle to the bone. The tendon is attached to the bone by Sharpey's fibres which are continuous with the bone periosteum.

RANGE OF MOTION

The range of motion of a muscle is the range it will move through when going from a position where it is fully stretched to one where it is fully contracted when it is working concentrically. Such a movement would describe full range. This can be divided into:

- *Outer range* – working concentrically, the muscle moves from a position of full stretch to the midpoint of full range.
- *Inner range* – working concentrically, the muscle moves from the midpoint of full range to a position of maximum contraction.
- *Middle range* – from the midpoint of the outer range to the midpoint of the inner range.

This classification will be used to describe the types of movement or exercise that a muscle is performing, e.g. inner range quadriceps exercises.

SUMMARY

- Human skeletal or striated muscle possesses characteristic properties of irritability, contractility, distensibility and elasticity that allow it to perform its function as the means of producing movement.
- Structurally, muscle consists of the basic unit, the muscle fibre, arranged in bundles to achieve different functions.
- The muscles contract due to linkages formed between actin and

myosin filaments under the stimulus of calcium and using energy from adenosine triphosphate (ATP) breakdown. The formation of such linkages results in muscle contraction and the production of muscle tension.

- There is a direct relationship between the length of a muscle and the tension that it develops.
- Muscle fibres may be fast or slow depending on the function they will perform; this will be either mobilizing or stabilizing.
- The way the fibres are laid down will vary with their intended function.
- Different classifications of muscles exist, the most important being phasic and tonic muscles or spurt and shunt muscles.
- Muscles can contract in different ways – concentric, eccentric, isometric, isokinetic or isotonic.
- Muscles can perform different roles – agonist, antagonist, fixator or synergist.

REFERENCES

Astrand, P.O. and Rodahl, K. (1986) *Textbook of Work Physiology*, 3rd edn, McGraw-Hill, New York

Colliander, E.B., Dudley, G.A. and Tesch, P.A. (1988) Skeletal muscle fibre type, composition and performance during repeated bouts of maximal concentric contractions. *European Journal of Applied Physiology*, **58**, 81–86

Ganong, W.F. (1979) *Review of Medical Physiology*, 9th edn, Lange, California

Ganong, W.F. (1993) *Review of Medical Physiology*, 16th edn, Lange, California

Gowitzke, B.A. and Milner, M. (1988) *Understanding The Scientific Bases of Human Movement*, 3rd edn, Williams and Wilkins, Baltimore

Grinnell, A.D. and Brazier, M.A.B. (1981) *The Regulation of Muscle Contraction: Excitation–Contraction Coupling*, Academic Press, New York

Guyton, A.C. (1981) *Textbook of Medical Physiology*, 6th edn, W.B. Saunders, Philadelphia

Huxley, A.F. (1974) Muscular contraction. *Journal of Physiology*, **243**, 1–43

MacConnail, M.A. (1966) The geometry and algebra of articular kinematics. *Biomedical Engineering*, **1**, 205–212

MacConnail, M.A. and Basmajian, J.V. (1977) *Muscles and Movement*, Krieger, New York

Netter, F.H. (1987) *The Ciba Collection of Medical Illustrations*, Vol. 8, Ciba Geigy, New Jersey

Norkin, C.C. and Levangie, P.K. (1992) *Joint Structure and Function*, 2nd edn, F.A. Davis, Philadelphia

Palastanga, N., Field, D. and Soames, R. (1994) *Anatomy and Human Movement*, 2nd edn, Butterworth–Heinemann, Oxford

Purslow, P.P. (1989) Strain induced reorientation of an intramuscular connective tissue network: implications for passive muscle elasticity. *Journal of Biomechanics*, **22**, 21–31

Rodgers, M.M. and Cavanagh, P.R. (1984) Glossary of biomechanical terms, concepts and units. *Physical Therapy*, **64**, 1886–1902

Stern, J.T. (1971) Investigations concerning the theory of spurt and shunt muscles. *Journal of Biomechanics*, **4**, 437–453

Williams, P.L. and Warwick, R (eds) (1980) *Gray's Anatomy*, 36th edn, Churchill Livingstone, Edinburgh

4. *Neurophysiological aspects of human movement*

This chapter will provide information about the nervous system which will be of use in the rehabilitation environment. It is not intended to provide a detailed account of the anatomy and physiology of the central and peripheral nervous systems because this information is readily available in many standard neurology texts.

This chapter will outline:

- an overview of the structure of the nervous system
- the simple stretch reflex
- the muscle spindle
- the Golgi tendon organ
- muscle tone
- higher centre influence on the control of movement
- the descending tracts
- the basal ganglia (basal nuclei)
- the cerebellum
- sensory aspects of motor control
- the ascending tracts
- role of the vestibular system in the control of movement
- the motor developmental sequence
- plastic adaptation of the nervous system.

OVERVIEW OF THE STRUCTURE OF THE NERVOUS SYSTEM

Without a nervous system a body might have form – due to the musculoskeletal system – but it certainly would not have function. One is seldom aware of the nervous system when it is functioning normally, except for the odd bout of 'pins and needles', but it would certainly be noticed if it was not working properly.

For example, one might have seen the young man who held on to his new motor cycle when its wheels suddenly left the road, only to find that he was left with a useless, limp arm as a result of a brachial plexus lesion. One might also have seen the stroke patient trying to walk again or the young rugby player who damaged his spinal cord when the scrum collapsed. So, too, could one recognize the patient with Parkinson's disease displaying the characteristic tremor, slowness of movement and shuffling gait.

These are just a few of many examples of nervous system dysfunction (Figure 4.1). Clinically the motor cyclist has a peripheral nervous system (PNS) injury damaging all the peripheral nerves which supply the arm, whereas the stroke patient and the rugby player have injured their central nervous system (CNS). The stroke victim has damage caused by haemorrhage into the brain or blockage of the blood vessels supplying it. The rugby player has damaged his spinal cord, the main connection between the environment, body periphery and brain. The patient with Parkinson's disease has a problem in one of the integrative centres in the brain known as the basal ganglia.

These examples are very different neurological conditions but they have one thing in common – they all have resulted in cell damage within the nervous system which has led to impairment of motor function.

For normal movement to occur, the musculoskeletal system must receive information from the sensory (afferent) part of the nervous

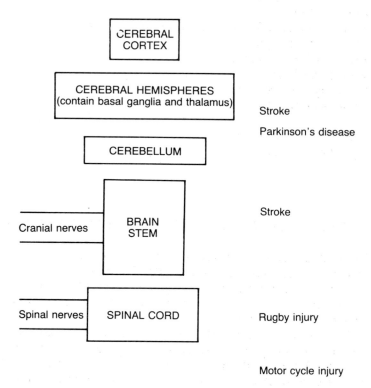

Fig. 4.1 The nervous system – see text for the examples listed on the right

system. It then processes this information and reacts accordingly. Provided that the motor (efferent) part of the nervous system is intact, the result is the production of normal movement.

If one steps on a drawing pin in bare feet, one does not think, 'Oh, I'm standing on a drawing pin, it's hurting and I must get off'. The body reacts very quickly to the situation and evasive action is taken without conscious decision. Likewise, the blink reflex has protected the eyes from many hazards since birth. The postural reflexes also affect head righting in the simplest of movements such as getting out of bed in the morning, and a new born baby would do badly without its rooting reflex!

All these responses do not require conscious thought, they are reflexes, which are involuntary, stereotyped responses to sensory inputs (Nolte, 1993). The reasons for some of these reflexes are not well understood. For example, why does the new born baby demonstrate the stepping reflex which leads some parents to pronounce proudly that their child will soon be walking? Along with many of the primitive reflexes, this reflex is suppressed soon after birth. Indeed the primitive reflexes of the newborn only remain if the infant is neurologically impaired, or they may reappear later following cerebral or brain stem lesions which are severe enough to result in a decerebrate state (see Table 4.5).

The notion that reflexes can be suppressed and need to be controlled for normal function leads to another important concept within the nervous system. Reflex activity is controlled from higher centres via nervous pathways which descend from the cerebral cortex, cerebellum, basal ganglia, brain stem and reticular formation. These pathways are known as tracts and it is their influence on the anterior horn cells located in the spinal cord which enables normal movement to take place.

There are also a number of ascending tracts in the white matter of the spinal cord which carry sensory information to all levels within the CNS, including the post-central gyrus of the parietal lobe where sensory discrimination occurs. Not all sensory information from body receptors is integrated at conscious level in the post-central gyrus; most is dealt with in the lower levels such as in the thalamus and brain stem. The initial 'pin' response would have taken place at spinal cord level. It is only when one has to concentrate to identify a particular sensation or to differentiate between the possible locations of a sensation, that the post-central gyrus assumes the principal role.

THE SIMPLE STRETCH REFLEX

The stretch reflex is the simplest of all human reflexes and its basic mechanism provides the key to human movement (Figure 4.2). Sensory or afferent information is brought into the CNS through the

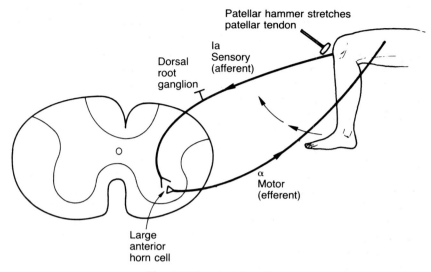

Fig. 4.2 The stretch reflex

posterior root of the spinal cord. The nerve cell bodies which relay the sensation are situated in a swelling on the posterior root known as the dorsal root ganglion (DRG). The sensory (afferent) fibres connect with the motor (efferent) fibres and their cell bodies, the anterior horn cells, which lie in the anterior horn of the spinal cord either directly or indirectly via connector neurons known as internuncial or inter-neurons.

Within the anterior horn there are two types of anterior horn cells (AHC): the large anterior horn cells, usually termed alpha motor neurons, which supply extrafusal (skeletal) muscle; and the small anterior horn cells, the gamma motor neurons, which provide the motor (efferent) supply to small sensors within muscles known as muscle spindles (intrafusal fibres).

In the example in Figure 4.2, the circuit is composed of a sensory (afferent) input connected to a motor (efferent) output by a single synapse in the spinal cord. Tapping the patellar tendon causes the quadriceps muscle to stretch, which is detected by Ia afferent neurons in the muscle spindles (intrafusal muscle fibres) of the quadriceps. The muscle spindles react to the stretch by causing excitation of the alpha (efferent) neurons in the anterior horn of the spinal cord which then fire to facilitate a contraction of the quadriceps.

THE MUSCLE SPINDLE

The muscle spindle (Figure 4.3) is capable of signalling fluctuations in length, tension, velocity of length change and acceleration of a muscle

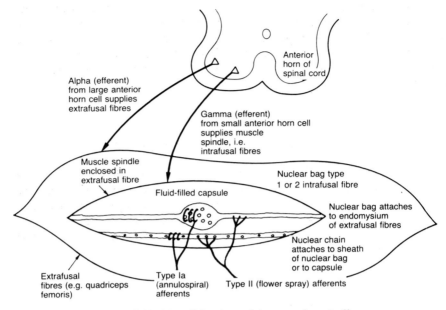

Fig. 4.3 Gross simplification of the muscle spindle

contraction. It also makes comparisons between the intended and the actual movement which takes place and is able to provide detailed input to spinal, cerebellar, extrapyramidal and cortical centres concerning the current state of muscular tension in the locomotor system.

Structure of the muscle spindle

Each muscle spindle consists of between six and 14 intrafusal muscle fibres which are enclosed in a connective tissue capsule containing fluid. The central portion of each intrafusal fibre is non-contractile.

There are two types of intrafusal fibres, the nuclear bag and the nuclear chain fibres. The ratio of nuclear bag to nuclear chain fibres in the human muscle spindle is about two to five. Nuclear bag fibres are larger in diameter and have a collection of nuclei at their centre. Two types of nuclear bag fibres have been identified and named dynamic bag 1 and static bag 2 fibres, their differences resulting from their histochemical and physiological properties (Warwick and Williams, 1973). The nuclear bag fibres are connected to the endomysium of extrafusal fibres.

Nuclear chain fibres are thinner and their nuclei are arranged in a single row throughout their length. They are attached to the capsule or to the sheaths of the nuclear bag fibres.

These intrafusal fibres of the muscle spindle make individual contributions to the control of human movement. The dynamic bag 1 fibres appear to signal the rapid changes in length which occur during movement. The static bag 2 fibres are less responsive to movement and the nuclear chain fibres have relatively slow adapting responses at all times.

Sensory innervation of muscle spindle

There are two types of afferent supply to the spindle which are wrapped around the centres of both intrafusal fibre types. The primary or annulospiral endings are the terminations of rapidly conducting group Ia afferent fibres and the secondary or flower spray endings, mainly located on the nuclear chains, are the terminations of group II afferent fibres.

Motor innervation of muscle spindle

Gamma motor neurons, also known as fusimotor neurons or small anterior horn cells (AHCs) provide motor innervation to the muscle spindle. Their neuronal cell bodies are found in the anterior horn of the spinal cord and they constitute 30% of all motor neurons in this location (Ganong, 1991). The cell bodies of alpha motor neurons or large anterior horn cells which innervate extrafusal fibres make up the majority of the remaining cells in the ventral root. Both large and small AHCs are influenced by the descending pathways of the spinal cord. In particular, the cerebellum, basal ganglia and reticular formation are known to exert strong influences over the gamma motor neurons.

Additionally, the muscle spindle receives sparse innervation from branches of fibres that innervate extrafusal fibres in the same muscle. These fibres are known as beta fibres but their function remains unknown.

Function of the muscle spindle

The principle of muscle spindle function is relatively simple in that when extrafusal fibres are stretched, the intrafusal fibres, including the non-contractile central portion, are also stretched. The increased tension of the central portion facilitates firing of both types Ia and II sensory endings. In response to this stimulation, the large AHC fires and causes the extrafusal fibres to contract, which removes the stretch from the muscle spindle. The tension in and therefore the sensitivity of the spindle is then reset via gamma efferent (small AHC) firing in response to higher centre stimulation. It is important to remember that higher centre influence greatly adds to the complexities associated with muscle spindle functioning.

THE GOLGI TENDON ORGAN

Neurotendinous endings, known as Golgi tendon organs (GTOs), were originally thought to initiate myotactic reflexes to inhibit the development of excessive tension during muscle contraction. Later, their role as a signaller of proprioceptive information, together with the muscle spindle during all forms of muscular activity, was recognized (Warwick and Williams, 1973).

Structure of the Golgi tendon organ

A GTO consists of a small bundle of tendon fibres enclosed in a delicate capsule. Each capsule is about 500 μm long and 100 μm in diameter and is related to a group of about 20 muscle fibres at the musculotendinous junction. It is possible for more than 50 GTOs to be found at a single musculotendinous junction.

Function of the Golgi tendon organ

The primary role of the GTO is to monitor the tension developed by a muscle contraction via its own Iβ (sensory) afferents (Kandel *et al.*, 1991). When there is an increase beyond the desired muscle tension, negative feedback from the GTOs inhibits further development of tension by increasing its firing on the alpha motor neuron. Conversely, if muscle tension decreases, due to fatigue for example, the GTO firing on the alpha motor neuron reduces, allowing muscle tension to increase.

MUSCLE TONE

When the nervous system is functioning properly, muscles demonstrate normal tone. The resistance of a muscle to stretch is often referred to as its tone or tonus, of which there are three states.

- normal tone
- high tone or hypertonic (spasticity or rigidity)
- low tone or hypotonic (flaccidity).

The motorcyclist's arm was described earlier as a 'useless limp arm'. Clinically this would be known as a flaccid or low tone arm. The patient with the stroke and the rugby player would probably within a few days or weeks develop increased tone in the affected limbs. The Parkinsonian patient would also have increased tone, but the characteristics of this hypertonicity would be different from that of the stroke patient and the rugby player.

Normal tone

Normal muscle tone is maintained by a balance of facilitatory and inhibitory activity on the gamma and alpha systems descending from higher centres to the anterior horn of the spinal cord. The basal ganglia are considered to have a particularly important role in the balance of normal tone (Walton, 1993).

High tone

Spasticity

Spasticity is characterized by hyperreflexia and an increased resistance to passive movement which tends to predominate in the antigravity muscles. It would probably be present in both the stroke patient and the patient with the spinal injury after cerebral and spinal shock have worn off (see Flaccidity, below).

The 'clasp-knife' phenomenon, a characteristic of spasticity, is distinguished by an initial strong resistance to passive movement followed by a sudden 'give', with the limb then moving easily into the new range. The velocity of passive movement will also determine the resistance to the movement – the faster the movement is performed, the greater the resistance.

Cause of spasticity

The cause of spasticity is not fully understood. Until recently it was attributed to an isolated lesion of the corticospinal (pyramidal) tract, the explanation being that the large anterior horn cells were freed from corticospinal higher centre inhibition which resulted in an increased firing of the stretch reflex.

Later evidence suggests that spasticity may result initially from an increased firing of gamma motor neurons which supply the intrafusal fibres of the muscle spindle. This is accompanied latterly by an increased firing of alpha motor neurons which supply the extrafusal fibres (Walton, 1993). Hypersensitivity of the gamma system will raise the level of excitability of the muscle spindle. This in turn will lead to a higher level of discharge and thus facilitate an increased response to muscle stretch. Hypersensitive alpha motor neurons react to this increased level of discharge by causing contraction of the extrafusal fibres.

Associations of spasticity (upper motor neuron lesions)
These include the following:

- heightened reflexes (hyperreflexia)
- clonus*
- predominantly unidirectional, mainly presenting in antigravity muscles
- occurs as mass pattern synergies (most usual: upper limb flexion, lower limb extension)
- clasp-knife phenomenon on passive movement
- extensor plantar response (Babinski/Hoffmann – see below)
- minimal muscle atrophy
- later onset of contractures.

Two simple clinical tests are available to enable immediate diagnosis of spasticity:

1. *Plantar response (Babinski sign).* This reflex is an abnormal response and is a strong indication that the corticospinal tract has been damaged. The reflex is elicited by firmly stroking the lateral plantar surface of the foot with a blunt object from the heel to the little toe. In response, the great toe will slowly dorsiflex and will be accompanied by fanning of the other toes. This is abnormal and would be recorded as a +ve Babinski or as a ↑ plantar response. The normal flexor response involves plantar flexion of all toes, particularly of the great toe (babies have an extensor plantar response until they are almost 18 months old).
2. *Hoffmann's sign.* A positive Hoffmann's sign is evidence of hyperreflexia in the upper limb. It is elicited by a rapid flexion of the distal interphalangeal joint of the middle finger. A positive result is observed if the thumb adducts and the index finger flexes. Usually no movement of these digits occurs if the response is normal; however, the response can occasionally be observed in the normal person.

Rigidity

Rigidity can be detected in both gravity-dependent and antigravity muscle groups and the resistance to passive movement is bidirectional. It is also relatively independent of velocity and is not characterized by a heightened stretch reflex.

---

* Clonus is a rapid intermittent contraction and relaxation of a muscle under stretch. It is best demonstrated in the calf following fairly rapid passive dorsiflexion of the patient's ankle.

Two types of rigidity can occur. Cog-wheel rigidity is characterized by an intermittent 'give' and return of resistance throughout the passive movement. Lead-pipe rigidity presents a uniform resistance throughout the range of passive movement, and is akin to the sensation experienced when bending a lead pipe.

This type of tone would typically be present in a patient with Parkinson's disease.

Cause of rigidity
Originally, rigidity was attributed purely to a lesion in a single system – the extrapyramidal system – but recent theories suggest that rigidity could also result from an altered corticospinal influence. In contrast to spasticity, however, increased alpha neuron discharge predominates over gamma system discharge (Walton, 1993).

Associations of rigidity
These include the following:

- reflexes usually normal
- no clonus
- bidirectional (detected in flexors and extensors during passive movement)
- no mass pattern synergies
- no clasp-knife phenomenon
- rigidity detected as either cog-wheel or lead-pipe
- minimal atrophy
- later onset of contractures.

'Decereberate rigidity'

Decereberate rigidity occurs in gross brain damage when higher centre control has been removed at midbrain level, thereby freeing the primitive spinal reflex systems from inhibitory influences (see Table 4.5). It is characterized by extension in the neck, trunk and all four limbs and medial rotation of the upper limbs.

Low tone

Flaccidity

Flaccidity is characterized by a diminished resistance to passive movement. Several states of flaccidity can be observed in clinical neurology and all inhibit normal movement:

1. Hypotonicity following either stroke or spinal injury is transient and usually followed by the onset of spasticity within days or weeks. It

occurs as a result of cerebral or spinal shock. The shock state involves the suppression of all motor reflexes despite the presence of intact spinal reflex arcs. Both the spinal cord injury and stroke victim would probably demonstrate this type of flaccidity in the early stage following injury.

2. Flaccidity also presents if the efferent or afferent pathways of the spinal reflex arc are interrupted. The lower motor neuron (LMN) constitutes the final common pathway or efferent pathway from the anterior horn of the spinal cord to the muscle. A lesion of the LMN will produce the characteristic symptoms of flaccidity, loss of reflexes and latterly muscle atrophy. This type of flaccidity is seen in peripheral nerve injuries such as the patient who sustained the brachial plexus lesion after the motorcycle accident or in disease of the AHC such as poliomyelitis.

3. A lesion of the cerebellum such as tumour or bleed which interrupts communication between the dentate and intermediate cerebellar nuclei, the motor cortex of the cerebrum and the brain stem will also cause hypotonicity. In this instance it is due to inhibition of gamma efferent discharge on the muscle spindle, with a resulting failure of the muscle spindle to reset in preparation for detecting changes in extrafusal fibre behaviour.

Associations of flaccidity
These include the following:

- loss of stretch reflexes
- loss of muscle tone – detected as diminished resistance to passive movement
- weakness or complete loss of muscle power
- eventual contracture formation due to replacement of muscle by fibrous tissue
- trophic changes in skin, partly due to loss of movement.

HIGHER CENTRE INFLUENCE ON THE CONTROL OF MOVEMENT

The 'higher centres' are the components of the central nervous system which are located above the spinal cord and include the brain stem, midbrain, cerebellum, basal ganglia, thalamus, cerebellum and cerebral cortex. Descending nervous pathways (tracts) originating in these centres are able to influence gamma and alpha motor neurons either in an inhibitory or facilitatory capacity to allow the generation of refined motor behaviour (Figure 4.4).

Higher centre involvement in motor control is extremely complex and although it is possible to identify the individual roles of the

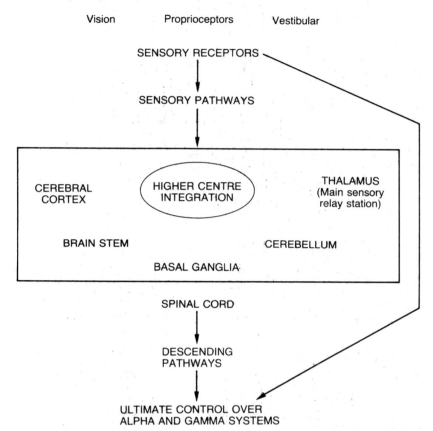

Fig. 4.4 Higher centre influence on motor control

components, their integrative actions and resulting influence on musculoskeletal functioning are far from clear.

Most information has been gleaned by studying patients with diseases affecting a particular component of the nervous system and from post-mortem studies. For example, more information is available about basal ganglia (nuclei) dysfunction than their normal function.

THE DESCENDING TRACTS

The descending pathways which influence motor control are the corticospinal, vestibulospinal, reticulospinal, tectospinal, olivospinal and rubrospinal tracts. They are located in either the anterior, medial, lateral or posterior funiculi (sections) of the white matter of the spinal cord together with the ascending tracts, as illustrated in Figure 4.5. A summary of their individual functions is outlined in Table 4.1.

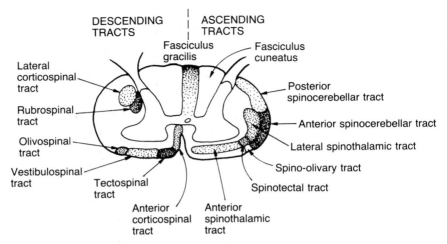

Fig. 4.5 Cross-section of motor and sensory tracts (after Snell, 1992, by permission)
Notes: (a) The reticulospinal tract is not shown because it is widely scattered in the anteromedial funiculus (Warwick and Williams, 1973). (b) The spinoreticular fibres are in the lateral funiculus but are not identifiable as a discrete tract (Warwick and Williams, 1973)

THE BASAL GANGLIA (BASAL NUCLEI)

Anatomical relations

Deep within the white matter of each cerebral hemisphere lie a group of nuclei which are collectively named the basal ganglia or basal nuclei. Individually they comprise the subthalamic nucleus, the substantia nigra, the caudate nucleus and the lentiform nucleus which has two parts, the globus pallidus and the putamen. The caudate and lentiform nuclei together are known as the corpus striatum.

Neural pathways have been identified between the basal ganglia, cerebral cortex, thalamus, and reticular formation of the brain stem. Associations with the cerebellum have also been documented (Ganong, 1991).

Normal function of the basal ganglia

The connections of the basal ganglia within the nervous system are highly complex and their function is far from certain. They are of particular importance in balancing alpha and gamma discharge in the control of normal tone (Walton, 1993) and a role in the planning and programming of normal movement is also assumed. However, much of the information about these nuclei has been derived from post-mortem study and from patients with diseased basal ganglia.

Table 4.1 The descending motor tracts

Tract	Origin	Termination	Function
Corticospinal (anterior and lateral)	Cerebral cortex: primary motor area, pre-motor area and parietal lobe	Via internuncial neurons to alpha and gamma motor neurons in the anterior horn (AH)	Possibly speed and agility of skilled movements
Vestibulospinal	Pons (vestibular nucleus)	Via internuncial neurons to alpha and gamma neurons in AH	Postural adjustments. Inhibition of flexor muscles, facilitation of extensor muscles
Reticulospinal	Reticular formation (brain stem)	Via internuncial neurons to alpha and gamma neurons in AH	Inhibition (lateral) and facilitation (medial) of voluntary movement and reflex activity
Tectospinal	Midbrain (superior colliculus)	Via internuncial neurons in the cervical region AH	Reflex postural adjustments in response to vision
Olivospinal	Possibly inferior olive	Unknown	Unknown
Rubrospinal	Midbrain (red nucleus)	Via internuncial neurons to alpha and gamma neurons in AH	Facilitation of flexor muscles and inhibition of extensor muscles

Motor disorders associated with basal ganglia dysfunction

These include the following:

- muscle tone alteration – usually presenting as 'extrapyramidal' rigidity
- loss of automatic associated movements such as facial expression and arm swing
- unwanted movements (chorea, athetosis, hemiballismus and dystonia)
- slowness and poverty of movement (bradykinesia)
- inability to initiate movement (akinesia)
- tremor.

Parkinson's disease

Parkinson's disease is the well-known disorder which occurs as a result of basal ganglia dysfunction. It is due to a steady loss of the neural transmitter dopamine together with a reduction in the number of dopamine receptors in the basal ganglia (Kandel *et al.*, 1991). The signs and symptoms of Parkinson's disease are outlined in Figure 4.6 and the motor disorders associated with basal ganglia dysfunction are summarized in Table 4.2.

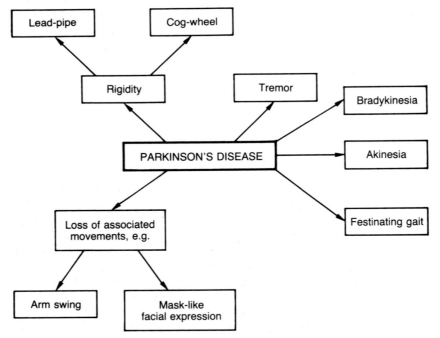

Fig. 4.6 Signs and symptoms of Parkinson's disease

Table 4.2 Other motor disorders associated with the basal ganglia

Disorder	Possible cause	Characteristics
Tremor	Imbalance in alpha and gamma systems	Present at rest. Decreases with voluntary movement. Frequency = 4–6 Hz. Usually asymmetrical. Pill rolling of thumb/index finger
Chorea	Lesion in the caudate nucleus	Rapid, dancing, involuntary movements
Athetosis	Lesion in the lentiform nucleus	Slow writhing movements
Ballismus	Lesion in the subthalamic nucleus	Flailing, intense, violent involuntary movements
Rigidity	Increased alpha and gamma activity in the flexors and extensors	Lead-pipe or cog-wheel; bidirectional resistance to passive movement

THE CEREBELLUM

Anatomical relations

The cerebellum is the largest part of the hind brain. In the adult male it weighs an average 150 g and is about one-eighth the size of the cerebrum. It is situated in the posterior cranial fossa. Above, it is separated from the cerebral hemispheres by a fold in the dura mater which is known as the tentorium cerebelli. It lies posterior to the pons and medulla oblongata and is joined to the brain stem by three pairs of peduncles which consist of bunches of afferent and efferent fibres. The superior peduncles connect the cerebellum to the midbrain, the middle peduncles connect it to the pons and the inferior peduncles form the connection between the cerebellum and the medulla oblongata.

Gross structure

The cerebellum consists of two cerebellar hemispheres which are joined in their median sagittal plane by the narrow vermis. It has an outer cortex of grey matter and an inner core of white matter. Deep within the white matter of each hemisphere lie the nuclei of the cerebellum – the dentate (lateral), the intermediate and the fastigial (medial). Their importance is illustrated in that, provided that they are intact, it is possible for an animal to function virtually normally if the movements are carried out slowly, even when a large portion of the cerebellar cortex has been removed (Guyton, 1986). The principal divisions of the cerebellum are the flocculonodular lobe and the corpus cerebelli.

Normal functions of the cerebellum

The cerebellum functions closely with the motor cortex, the thalamus, the basal ganglia, brain stem and spinal cord. It has a number of important roles, the combination of which lead to the ultimate production of smooth, controlled, co-ordinated and finely tuned movement.

It functions to:

- provide smooth co-ordinated movements of agonist and antagonist muscle groups
- provide controlled braking action during movement by exerting an inhibitory influence on the agonists via the motor cortex
- provide comparisons between movement intention initiated in the motor cortex and the performance itself
- provide correcting signals to modify the performance if necessary
- assess the rate of movement and calculate the length of time taken to reach the point of intention
- dampen momentum to allow precise, economic movement
- plan sequential movement, facilitating smooth progression through the sequence.

Dysfunction in the cerebellum

A lesion in one cerebellar hemisphere produces signs and symptoms in the ipsilateral side of the body. This is because each hemisphere receives information from the nervous pathways of the same side of the body.

The main signs and symptoms associated with a lesion of the cerebellum are summarized in Figure 4.7.

SENSORY ASPECTS OF MOTOR CONTROL

The correct motor response depends on the CNS receiving accurate information from the sensory receptors. The receptors provide the latest information about the internal and external environment.

Types of receptor

Receptors are usually classified as exteroceptors, proprioceptors and interoceptors.

Exteroceptors respond to external stimuli such as those involved in vision, hearing and smell. This group is sometimes subdivided into

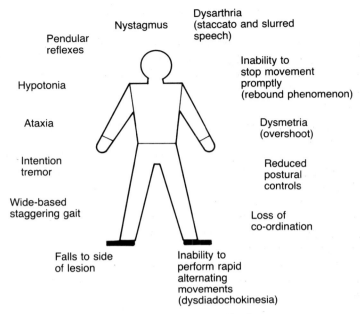

Dysarthria
(staccato and slurred speech)

Nystagmus

Pendular reflexes

Hypotonia

Ataxia

Intention tremor

Wide-based staggering gait

Inability to stop movement promptly (rebound phenomenon)

Dysmetria (overshoot)

Reduced postural controls

Loss of co-ordination

Falls to side of lesion

Inability to perform rapid alternating movements (dysdiadochokinesia)

Fig. 4.7 Clinical signs of a cerebellar lesion

telereceptors and contact receptors (Nolte, 1993). Telereceptors respond to stimuli at a distance from the body and include visual, olfactory and auditory receptors, while contact receptors are those which respond to stimuli applied directly to the body such as pain and cutaneous irritation.

Proprioceptors are the group of receptors concerned with aspects of locomotion such as mechanical stress and joint position. Two of the most important proprioceptors, the muscle spindle and Golgi tendon organ, have already been described.

Interoceptors constitute the group of receptors which are found in the walls of viscera, vessels and glands such as baroreceptors and chemoreceptors which control blood pressure and respiration, respectively.

Another popular classification subdivides receptors on the basis of the stimulus to which they are most responsive. For example, nociceptors are those receptors which respond to painful stimuli, chemoreceptors detect changes in body chemistry such as pH alterations and mechanoreceptors are concerned with mechanical deformation.

THE ASCENDING TRACTS

Sensory information is relayed in bundles of nerve fibres known as the sensory or ascending tracts, located in the white matter of the spinal cord. Some of the nerve fibres are in continuity with sensory neurons

Table 4.3 The ascending sensory tracts

Tract	Origin	Termination	Function
Anterior spinothalamic	Posterior grey columns of spinal cord (exact locations unknown)	Via ventral posterolateral nucleus of the thalamus to: Post-central gyrus of the cerebral cortex	Crude touch and pressure Conscious touch and pressure
Lateral spinothalamic	As above	As above	Pain and temperature
Posterior spinocerebellar	Posterior grey column of spinal cord at the level of L2 and L3	Passes into cerebellum via inferior cerebellar peduncle	Proprioception (muscle spindle and GTO); exteroception
Anterior spinocerebellar	Posterior grey columns of spinal cord	Passes into cerebellum via superior cerebellar peduncle	Proprioception; exteroception; information from the skin and superficial fascia
Posterior column (gracilis)	Posterior grey column at caudal end of spinal cord	Via nucleus gracilis and posterolateral nucleus of the thalamus to post-central gyrus of cerebral cortex	Tactile, vibration, proprioception
Posterior column (cuneatus)	Posterior grey column at mid-thoracic and cervical regions	Via nucleus cuneatus and posterolateral nucleus of the thalamus to post-cental gyrus of cerebral cortex	Tactile, vibration, proprioception
Spinotectal	Posterior grey column at cervical levels	Midbrain (superior colliculus)	Spinovisual reflexes; turning head and eyes to the side of stimulus
Spinoreticular	Posterior grey column	Brain stem reticular formation	Possibly to do with consciousness
Spino-olivary	Posterior grey column	Via medulla oblongata (inferior olivary nucleus) to cerebellum through inferior cerebellar peduncle	Proprioception; cutaneous sensation

in the peripheral nervous system and have their cell bodies in the sensory ganglia of the posterior roots. Other fibres originate in the posterior grey column of the spinal cord. Initially, all sensory information passes through the posterior grey columns via the dorsal root ganglion before entering the tracts in the white matter.

When learning the tracts for the first time, demarcation and function within the cord appear to be compartmentalized with definite borders, but this is not the case with most sensory information travelling in more than one pathway (Nolte, 1993).

A brief summary of the functions of the main ascending tracts is outlined in Table 4.3.

ROLE OF THE VESTIBULAR SYSTEM IN THE CONTROL OF MOVEMENT

A fairground is the ideal place to start talking about the role of the vestibular system in the control of movement, since many readers will have experienced nausea when on certain rides. Through its association with vision and proprioceptive inputs the vestibular system is concerned with the control and modulation of posture, balance and equilibrium. The immediate period after the ride will no doubt confirm that this is the case!

The vestibular system achieves its effect by providing a continuing flow of information into the CNS which relates to the effect of movement and gravitational force acting on the body. The components of the vestibular system interact as shown in the highly simplified diagram in Figure 4.8.

THE MOTOR DEVELOPMENT SEQUENCE

The physiotherapist is more likely to be successful when rehabilitating the brain-injured patient if the physiotherapist has knowledge of how human movement developed in the first instance, so that similar conditions of relearning can be facilitated and created during treatment the 'second-time around'.

The motor developmental sequence forms the basis of physiotherapy for patients with neurological disease. It is the process through which a normal baby develops into adulthood and is characterized by certain age-related motor milestones (Table 4.4). Failure to attain these milestones at the 'correct' stage of development may indicate neurological motor impairment, but there is great individual variation.

In addition to age-related motor milestones, there are also certain characteristics which can be observed if motor development is proceeding normally. These are summarized in Figure 4.9.

Table 4.4 Age-related milestones (first year only) of the motor developmental sequence

Age	Supine	Sitting	Standing
4 weeks	Turns head to side	Has a complete head lag when pulled to sit	
6 weeks	Follows with eyes to middle	Holds head up but it drops forwards	
12 weeks	Watches movements of hands in midline	Moderate head lag when pulled to sit	
16 weeks	Head and hands in midline	Slight head lag when pulled to sit	
20 weeks	Brings head forwards and reaches for feet	Head stable when body rocked	
24 weeks	Rolls from prone to supine	Lifts head and holds hand out in anticipation of being pulled to sit	When held, is able to put some weight on feet
28 weeks	Spontaneously lifts head from support; rolls from supine to prone	Sits with hands forward for support	
36 weeks	Tries to crawl	Sitting balance in anteroposterior direction, not sideways	Holds onto furniture
40 weeks		Pulls up to sit	Pulls up to standing
44 weeks		Leans over sideways	Lifts one foot off support while holding on
48 weeks	Walks on hands and feet	Turns around to pick up an object	Walks sideways around furniture
52 weeks		May shuffle on buttocks and hands	Walks with one hand held

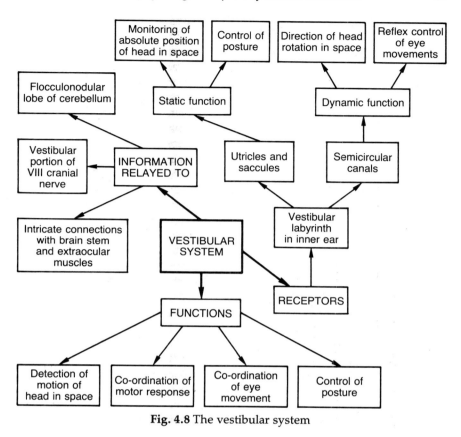

Fig. 4.8 The vestibular system

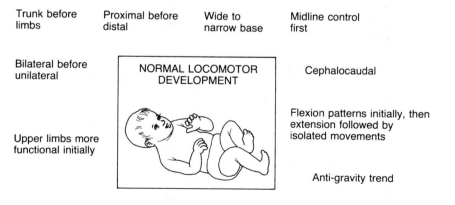

Fig. 4.9 Characteristics of normal locomotor development

Brief description of the locomotor developmental sequence

As soon as a baby is born it begins to show interest in its surroundings by using the eyes and lifting the head to look around. The need to explore is initially satisfied by rolling and then learning to crawl or to bottom-shuffle.

Increased mobility leads to a further need to explore. Of particular interest at this stage is any object out of reach. The baby begins to pull itself up to stand using furniture, mother's legs or anything that will be able to assist in the attainment of the goal – the desired object!

Standing becomes more frequent as the limbs gain strength and co-ordination. Objects are now reached by attempting to take a few steps, if feeling brave, or flopping down onto the floor to crawl or bottom-shuffle if not.

The 'brave' walk with a very wide base; they have minimal shoulder girdle and pelvic rotation and a very limited arm swing, if any. The arms tend to act as stabilizers, tending to be carried in abduction.

The base gradually decreases, the arm swing becomes more prominent and the gait becomes stable. Running is easy and jumping and hopping present no problem as the young child matures to adulthood.

Old age happens quickly, movement slows, balance becomes more difficult, the posture more flexed, with pelvis and shoulder girdle rotation reduced. Gait once more assumes a wide base and eventually a walking aid may become necessary.

The primitive reflexes and motor development

To enable smooth progression through the locomotor continuum, the new born is equipped with several primitive reflexes controlled at spinal or brain stem level. Some of the reflexes are clearly useful, e.g. the rooting reflex ensures an adequate supply of food, but some are poorly understood. For example, why does a newborn baby need a walking/stepping reflex? The majority of reflexes have either been suppressed or integrated into normal functional movement patterns by the time the child has reached the third birthday (Table 4.5).

Abnormal reflex activity in rehabilitation

Severe brain insult or injury can result in the re-emergence of some of these reflexes which have been released from higher centre control. Some rehabilitation methods utilize this abnormal reflex activity (Fay, 1948; Tannock, 1976) and one theory advocates that the brain-injured patient should pass through the abnormal reflex stage in order to gain maximal recovery (Brunnstrom, 1970). However, other theories tend to discourage their reinforcement by, for example, avoiding positions which are known to enhance their development (Rood, 1956; Bobath, 1990), by proprioceptive neuromuscular facilitation (Voss *et al.*, 1985) or by using language to achieve functional goals as in conductive education (Kinsman, 1989).

The classic example and best illustration of reflex re-emergence is that of the grasp reflex. Patients who have had a stroke often 'squeeze the ball to get the strength back in my fingers'. Explaining to relatives why this is not desirable can present problems for the inexperienced therapist. Inhibitory positioning is a usual method employed by therapists to control many of the unwanted reflexes in the brain-injured patient, and in the grasp reflex example, the patient is taught to keep the fingers in extension supported on a pillow or bedside table. A comprehensive critical review of rehabilitation theories may be obtained in Kidd *et al.* (1992).

PLASTIC ADAPTATION OF THE NERVOUS SYSTEM

Traditionally, rehabilitation after brain injury was aimed at teaching patients to compensate for the loss incurred as a result of the injury. The injured part of the brain was considered to be useless. This theory is clearly illustrated in gait re-education of stroke patients. These patients were provided with a tripod and instructed and assisted in its use to ambulate. The result was the 'typical hemiplegic gait'. The patient would hold the tripod with the unaffected upper limb, lean over onto it and then swing the hemiplegic leg forwards using circumduction of the hip and keeping the whole limb in extension. Evidence now suggests that the brain is capable of reorganization and adaptation following injury (Bishop, 1982a–d; Kidd *et al.*, 1992; Stephenson, 1993).

Recent theories suggest that the central nervous system can be manipulated and restructured and that this is now the key to successful therapy. To achieve success, 'intensive repeated stimulation is required to place demands on the reorganizing system which results in latent areas of the brain specializing to replace lost function and new pathways forming to by-pass the effects of a lesion' (Stephenson, 1993).

Plastic adaptation is encouraged by teaching the brain-injured patient to use the affected limbs and not to ignore them, as was the case in the 'compensation' situation identified above. It is facilitated by:

- commencing treatment as early as possible after brain injury
- allowing the patient to experience normal movement
- avoiding the use of compensatory treatment methods
- providing normal proprioceptive and sensory input
- eliciting active normal functional responses
- minimizing the development of abnormal tone and movement.

The mechanisms by which these objectives are achieved are beyond the scope of this chapter. The reader is referred to Goff (1972), Bishop

Table 4.5 Primitive reflexes (after Ganong, 1991)

Reflex	Stimulus	Response	Receptor
Stretch reflexes	Stretch	Contraction of muscle	Muscle spindles
Positive support-ing reaction	Contact with sole or palm	Foot plantar flexed to support body	Proprioceptors in distal flexors
Moro	Head drop into 10° extension while fully supporting body	Abduction of arms followed by the 'embrace'; legs extend and then flex	Proprioceptors in neck, vestibular apparatus
Walking	Supported stand-ing position	Walk	Proprioceptors and cutaneous in lower limbs
Grasp	Contact with palm	Strong flexion of the fingers	Cutaneous
Asymmetric tonic neck (ATNR)	(a) To side	Extension of limbs on side to which head is turned	Neck proprioceptors
Symmetric tonic neck (STNR)	(b) Up	Legs flex; arms extend	Neck proprioceptors
	(c) Down	Arms flex; legs extend	
Landau		Lifting of head and shoulders when in prone	Neck proprioceptors
Tonic labyrin-thine righting reflexes	Gravity	Head kept level	Otolithic organs in ear
Neck righting reflexes	Stretch on neck muscles	Righting of thorax and shoulders then pelvis	Muscle spindles
Body on head righting reflexes	Pressure on side of body	Righting of head	Exteroceptors
Body on body righting reflexes	Pressure on side of body	Righting of body even when head held sideways	Exteroceptors
Optical righting reflexes	Visual cues	Righting of head	Eyes
Placing reactions	Various visual, ex-teroceptive and proprioceptive cues	Foot placed on supporting surface in position to sup-port body	Various
Hopping reac-tions	Lateral displace-ment while stand-ing	Hops, maintaining limbs in a position to support body	Muscle spindles

Integrated in:	Present	Absent
Spinal cord, medulla	Birth onwards	
Spinal cord	Birth	Approximately 4 weeks
Spinal cord	Birth	36–48 weeks
Spinal cord	Birth	4 weeks
Spinal cord	Birth	12 weeks
Medulla	Birth	36 weeks
Medulla	6 weeks after birth; at 48 weeks is present as the Landau	3 years
Medulla	48 weeks	$2\frac{1}{2}$ years
Medulla	Birth	Integrated as balance reactions at 48 weeks
Midbrain	15 months	Integrated as balance and equilibrium reactions
Midbrain	24 weeks	Integrated as balance and equilibrium reactions
Midbrain	24 weeks	Integrated as balance and equilibrium reactions
Cerebral cortex		Integrated as balance and equilibrium reactions
Cerebral cortex		Integrated as balance and equilibrium reactions
Cerebral cortex		Integrated as balance and equilibrium reactions

(1982a–d), Downie (1986), Bobath (1990) and Kidd *et al*. (1992) for more detail on the theories of neural plastic adaptation and its implication for rehabilitation of the brain-injured patient.

SUMMARY

This chapter has summarized the function of the normal nervous system:

- the stretch reflex is the simplest of all human reflexes and provides the key to human movement
- the muscle spindle regulates length, tension and velocity of a muscle contraction. It is important for the maintenance of normal muscle tone
- the Golgi tendon organ's primary role is to monitor the tension developed when a muscle contracts
- muscle tone may be normal, hypotonic or hypertonic
- the descending tracts are motor tracts which convey impulses from the CNS to the periphery
- the ascending tracts are sensory tracts which convey information from the periphery to the CNS
- the functions and dysfunctions of the basal ganglia and the cerebellum are described
- the normal motor developmental sequence is outlined as it is the fundamental basis for the development of normal movement
- the brain is capable of reorganization and adaptation following injury. This process is known as neural plasticity or plastic adaptation.

REFERENCES

Bishop, B. (1982a) Neural plasticity. Part 1: Plasticity in the developing nervous system. *Physical Therapy*, **62**(8), 1122–1131

Bishop, B. (1982b) Neural plasticity. Part 2: Post-natal maturation and function induced plasticity. *Physical Therapy*, **62**(8), 1132–1142

Bishop, B. (1982c) Neural plasticity. Part 3: Responses to lesions in the peripheral nervous system. *Physical Therapy*, **62**(9), 1275–1282

Bishop, B. (1982d) Neural plasticity. Part 4: Lesion induced reorganisation of the CNS. *Physical Therapy*, **62**(10), 1442–1451

Bobath, B. (1990) *Adult Hemiplegia: Evaluation and Treatment*, 3rd edn, William Heinemann, London

Brunnstrom, S. (1970) *Movement Therapy in Hemiplegia*, Harper and Row, New York

Downie, P. (1986) *Cash's Textbook of Neurology for Physiotherapists*, 4th edn, Faber and Faber, London

Fay, T. (1948) The neurophysiological aspects of therapy in cerebral palsy. *Archives of Physical Medicine*, **29**, 327–334

Ganong, W.F. (1991) *Review of Medical Physiology*, 15th edn, Prentice-Hall, London

Goff, B. (1972) The application of recent advances in neurophysiology to Miss Rood's concept of neuromuscular facilitation. *Physiotherapy*, **58**(12), 409–415

Guyton, A. (1986) *Textbook of Medical Physiology*, 7th edn, W.B. Saunders, London

Kandel, E.R., Schwartz, J.H. and Jessell, T.M. (1991) *Principles of Neural Science*, Prentice-Hall, London

Kidd, G., Lawes, N., and Musa, I. (1992) *Understanding Neuromuscular Plasticity: A Basis for Clinical Rehabilitation*, Edward Arnold, London

Kinsman, R. (1989) A conductive education approach to stroke patients at Barnet General Hospital. *Physiotherapy*, **75**, 418–421

Nolte, J. (1993) *The Human Brain: An Introduction to its Functional Anatomy*, 3rd edn, Mosby Year Book, St. Louis

Rood, M. (1956) Neurophysiological mechanisms utilised in the treatment of neuromuscular dysfunction. *American Journal of Occupational Therapy*, **10**, 220–225

Snell, R.S. (1992) *Clinical Neuroanatomy for Medical Students*, 3rd edn, Little Brown, Boston

Stephenson, R. (1993) A review of neuroplasticity: some implications for physiotherapy in the treatment of lesions of the brain. *Physiotherapy*, **79**(10), 699–704

Tannock, R. (1976) Doman-Delacato method for treating brain-injured children: an assessment. *Physiotherapy Canada*, **28**, 203–209

Voss, D.E., Ionta, N.K. and Myers, B.J. (1985) *Proprioceptive Neuromuscular Facilitation*, 2nd edn, Harper and Row, Philadelphia

Walton, J. (ed.) (1993) *Brain's Diseases of the Nervous System*, 10th edn, Oxford University Press, Oxford

Warwick, R. and Williams, P.L. (1973) *Gray's Anatomy*, 35th edn, Longman, Edinburgh

Part II
Exercise in rehabilitation

5. *Classification of exercise*

The aim of this chapter is to provide a concise explanation and description of the terminology which is currently used to classify exercise. It will provide a classification of exercise according to:

- the movement performed
- the muscle contraction produced
- the muscle work undertaken
- the source of energy from muscle metabolic systems
- the concept of the kinetic chain.

TYPE OF MOVEMENT

Figure 5.1 gives a summary of the types of movement, which are described below.

Free active exercise

Free active exercise is carried out voluntarily by the patient. It requires no manual input from the physiotherapist and does not involve the use of apparatus. The exercising muscles are subjected only to the external assistance or resistance of the force of gravity.

Examples of free active exercise are gymnastic floor exercise routines, and running, walking, jumping and aerobic class regimens. More specific free active exercises used in rehabilitation include the circulatory exercises which are prescribed following surgery, the pendular arm swing exercises which feature prominently in shoulder rehabilitation programmes and the warm-up and cool-down routines of group exercise (Chapter 8).

Active assisted exercise

Active assisted exercise, as its name implies, is exercise carried out by the patient with some form of assistance. Assistance may be provided

Human Movement Explained

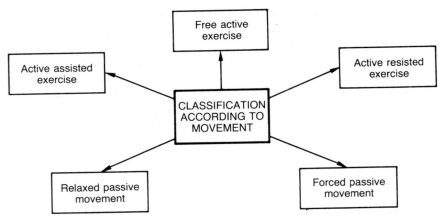

Fig. 5.1 Summary of types of movement

by patients themselves, such as when assisting movement of the affected limb with the unaffected limb. In this instance the exercise would be described as auto-assisted exercise. Assistance can also be applied mechanically using apparatus such as pulley circuits or sling suspension and by the physiotherapist who manually assists by possibly initiating a movement and then encouraging the patient to join in.

Active resisted exercise

This form of exercise is carried out against resistance. Patients can provide their own resistance, such as resisting knee extension with the other leg – auto-resisted exercise. Resistance can also be applied manually by the therapist – the proprioceptive neuromuscular facilitation techniques of Kabat and Knott (Voss *et al.*, 1985) provide a good example of manually resisted exercise (Chapter 9).

Apparatus used to provide resistance includes highly sophisticated computer-driven isokinetic dynamometers (Chapter 11), multi-gym apparatus and cycle-ergometers, through to the more basic sandbags, dumbbells and medicine balls which have long been part of the rehabilitation process.

Hydrotherapy is an additional and very effective medium for the application of resistance, assistance and support (Davis and Harrison, 1988).

Relaxed passive movement

Passive movements are carried out entirely by the therapist. They require no effort from the patient. They feature prominently in the treatment of the unconscious patient or the paralysed patient in the spinal injury unit, but they are also used in many other conditions. The patient who has recently had a cerebrovascular accident – the stroke patient – may initially require passive movements for the paralysed upper and lower limbs and trunk. The orthopaedic patient with a nerve injury or the patient who has been immobilized in plaster of Paris and the patient on traction are other examples of those who are likely to benefit from passive movements at some stage during their rehabilitation.

The uses of passive movement are:

- to maintain joint range of movement when (a) a muscle is paralysed and (b) muscle contraction is contraindicated
- to maintain soft-tissue length
- to promote and maintain circulation, although this is questionable since passive movements are performed for a relatively short time in a 24-hour period
- to reduce oedema when combined with passive accessory movements and elevation.

Performing passive movement

Passive movements involve moving each joint through its full anatomical range of movement. With practice, it is possible to perform passive movements smoothly and rhythmically while affording full support to the joints to which the movements are being performed. The movements are rarely performed in isolation, but are normally carried out in patterns of movement. For example, passive movements of the hip and knee are performed in one functional sequence.

The degrees of freedom of movement available at a joint will dictate the passive movements possible at this joint (Chapter 2). For example, the glenohumeral (shoulder) joint has three degrees of freedom and is therefore capable of being moved into flexion, extension, abduction, adduction, and medial and lateral rotation. As is the case in all joints which have three degrees of freedom, a combination of all of these movements will allow the shoulder to circumduct, as demonstrated when bowling a cricket ball, playing a racket sport or when stirring a large tin of paint!

To perform effective passive movements, the therapist should carry out approximately 8–10 repetitions, with at least one of the repetitions incorporating a gentle hold of approximately 3 seconds at the end of the range.

Precautions
When performing passive movements on patients with acute spinal injury, hip flexion should not exceed 60° for approximately 6 weeks for a patient who has a lumbar fracture, since there is a danger of moving

the fracture site. For the same reason, flexion of the upper limb should be limited to 90° during the first week after cervical injury. Ideally this time period should be longer, but loss of range in the shoulder region would be an obvious consequence.

Post-traumatic myositis ossificans

Care should be taken when performing passive movements, as force applied at the end of range by the inexperienced operator may result in the onset of a condition known as myositis ossificans. This is post-traumatic ossification of a subperiosteal haematoma (Apperley and Ross, 1992). The most common sites for myositis ossificans are in the muscles around the hip and elbow joints and in the quadriceps muscle*.

The signs and symptoms of myositis ossificans are:

- spongy end-feel on passive movement before range is lost
- loss of range of movement
- increase in pain on movement
- local oedema
- increase in temperature of the affected area
- pain on palpation
- palpation of a bony mass in the muscle in the later stages after onset
- an ultrasound scan will reveal abnormality before this is detected on X-ray
- X-ray will reveal bony deposits in the affected muscle once ossification has occurred.

Conflicting views are apparent concerning the treatment of myositis, with some schools advocating rest and others preferring vigorous passive movement to minimize the impending loss of joint range. Medication and surgery are also used, although the latter is only performed when the active stage has subsided and if movement is impeded and interferes with function, since there is a possibility of recurrence.

Forced passive movement

This technique should be carried out by physiotherapists who have had specialist postgraduate training in manipulation techniques. It involves the therapist taking the joint to the end of its existing range of movement and then, by providing an additional short burst of controlled force, pushing the joint into a new range.

Usually referred to as a manipulation, the forced passive movement technique can also be carried out under a general anaesthetic, in which

---

* Ossification which occurs abnormally around bone is known as heterotrophic ossificans and that which is seen around a joint is known as peri-articular ossificans.

case the procedure becomes known as a manipulation under anaesthetic (MUA).

Passive accessory movement

Active movements which occur at a joint do not necessarily include all of the movements which the joint structure allows. Some movements cannot be performed volitionally by the subject, but occur naturally during the performance of active movements of the joint. These are known as the accessory movements (Salter, 1955).

Examples of passive accessory movements occur when the therapist glides the tibia forwards and backwards in respect of the femur. Likewise, the interphalangeal joints can be moved in an anteroposterior and mediolateral direction. In both of these examples it is also possible to rotate the tibia and the proximal phalanx about their long axes. Passive accessory movements are important to the physiotherapist in that they form the basis of many joint mobilization techniques (Maitland, 1991).

Traction is an example of a passive accessory movement in a longitudinal direction that can be applied manually or mechanically; continuously or intermittently.

TYPE OF MUSCLE CONTRACTION

Isotonic

The term 'isotonic' is somewhat misleading since it implies that the exercising muscle maintains the same tension throughout range when contracting either eccentrically or concentrically. However, this is not the case when the muscle is contracting dynamically against a fixed load such as a sandbag. The tension varies as the muscle alters its length through the available range and the muscle develops its maximum tension at only one point in range (see Figure 3.6). This point usually equates with the habitual functional range for that specific muscle. Most muscles tend to develop maximal tension when approaching mid to inner range, since it is in this range that they tend to function in normal daily activities.

Isometric

During isometric exercise the exercising muscle contracts without shortening or lengthening. As the force of contraction increases, there is an increase in the tension generated by the muscle, but there is no change in muscle length and there is no visible joint movement. Isometric exercise is also called 'static' exercise.

Isokinetic

The concept of isokinetic exercise is relatively new (Hislop and Perrine, 1967). It involves training muscle strength under conditions of constant angular velocity. This is achieved using a machine called an isokinetic dynamometer which is able to control the speed of movement of the exercising muscle throughout its exercising range of movement.

In view of the increasing importance of isokinetic dynamometry as a research tool for providing objective and quantifiable information about dynamic muscle strength, together with its increasing contribution in clinical rehabilitation, the subject has been covered in detail in Chapter 11.

TYPE OF MUSCLE WORK

Eccentric exercise

During eccentric exercise the muscle contracts and lengthens. This type of contraction occurs when the external force applied to the muscle is greater than that developed internally by the muscle.

An example of a muscle working in this way is demonstrated when a person steps down from a stool with the right leg. The quadriceps femoris of the left leg contracts eccentrically to permit a controlled descent. Another example of a muscle group working eccentrically is observed when the subject bends forwards, as in touching the toes. The back extensors function eccentrically ('pay out') to enable a controlled lowering of the trunk.

Negative muscle work is carried out during eccentric exercise because the muscle moment acts in the opposite direction to the angular velocity of the joint. For example, as the elbow joint angle increases, biceps brachii, although contracting, is unable to prevent an increase in elbow joint extension. In this instance, biceps brachii is acting to control elbow extension by 'paying out'. It is also worth noting that the majority of non-contact soft-tissue injuries occur during the eccentric phase of the muscle contraction (Garrett, 1986).

Concentric exercise

When exercising concentrically, a muscle contracts and shortens. During a concentric contraction, the muscle is able to overcome an external force which is offering resistance. The force might be the weight of the forearm bones during elbow flexion, it may be manual resistance applied by the physiotherapist or a mechanical resistance supplied by a pulley circuit or free weight such as a dumbbell.

Positive muscle work is done during concentric exercise because when a concentric contraction occurs the muscle moment acts in the same direction as the angular velocity of the joint and by convention both are considered to be positive.

SOURCE OF ENERGY

Energy for any activity is derived aerobically and anaerobically and consequently these terms are sometimes used to describe exercise. Currently, this method of classification is very popular, with many would-be exercisers attending aerobic and step aerobic classes.

As a general guide, both types of energy are required for any physical activity but their contributions will differ. Energy provision for high-intensity, short-duration, fatiguing exercise predominantly relies on anaerobic sources, whereas that for exercise of mild to moderate intensity of long duration is derived aerobically. There are three energy systems. Two produce anaerobic energy and these are termed the adenosine triphosphate–creatine phosphate and the glycolysis–lactic acid systems. The aerobic system is the third source of energy provision.

These systems are described more fully, in relation to fitness testing, in Chapter 7.

KINETIC CHAIN EXERCISE

The concept of kinetic chain exercise is currently very popular in the United States of America. It was devised by Steindler (1973) from the closed kinematic and link concepts described by Reuleaux for the purposes of mechanical engineering (Gowitzke and Millner, 1988). Engineering link systems are described as a series of rigid segments connected by pin joints and are said to be 'closed' if fixed at both ends to an immovable framework. In a closed system, provided that there are a maximum of only four links, movement of one of these links will cause movement to occur in the others in a predictable way.

Using the engineering system model, Steindler proposed that the link system concept could be applied to and mimicked by the human body in different conditions of limb loading.

When the distal segment meets with 'considerable resistance' he noticed that muscle recruitment and joint motion differ from that produced when it is free to move in space. The latter scenario gave rise to the concept of 'open kinetic chain' exercise, whilst the relatively fixed distal segment system became known as 'closed kinetic chain' exercise (see Chapter 2).

It is important however to be aware that the above definition can be misleading in certain instances. Whilst the distal segment may be fixed in some exercises such as lower limb weight bearing activities, the proximal segments may not be in a similar state. For example, when standing erect and flexing the knees, which is often described as a closed kinetic chain exercise, the upper body is not in a true closed state. To promote this the patient should be instructed to touch a wall or other support with both hands.

Although currently described as closed kinetic chains some lower limb exercises could and perhaps should be described more accurately by placing the word 'pseudo' in front of the phrase 'closed kinetic chain exercise' (Major, 1994).

Examples of open kinetic chain exercises

In these exercises the limbs are free to move in space:

- straight leg raise
- circumduction of upper and lower limbs
- pendular exercises
- kicking a football
- upper limb weightlifting
- skipping.

Examples of closed kinetic chain exercises

In these exercises the limbs are relatively fixed:

- press-ups (upper limbs)
- ski sit (lower limbs)
- bench press (lower limbs)
- rowing machine (lower limbs)
- cycle-ergometer (lower limbs)

SUMMARY

- Exercise can be classified according to the movement being performed: free active, assisted or resisted and passive.
- It may be classified according to the type of muscle contraction: isotonic, isometric and isokinetic.
- Exercise can be classified according to the source of energy provided for muscle contraction: aerobic or anaerobic.
- Exercise may be classified as either open kinetic or closed kinetic chain.

REFERENCES

Apperley, C.E. and Ross, E.R.S. (1992) Fractures – physiotherapy and charts of fracture management. In *Cash's Textbook of Orthopaedics and Rheumatology for*

Physiotherapists, 2nd edn (ed. M.E. Tidswell), Mosby Year Book, London

Davis, B.C. and Harrison, R.A. (1988) *Hydrotherapy in Practice*, Churchill Livingstone, Edinburgh

Garrett, W.E. (1986) Basic science of musculotendinous injuries. In *The Lower Extremity and Spine in Sports Medicine* (eds J.A. Nicholson and E.B. Hershman), C.V. Mosby, St. Louis

Gowitzke, B.A. and Millner, S. (1988) *Scientific Bases of Human Movement*, Williams and Wilkins, Baltimore

Hislop, H. and Perrine, J.J. (1967) The isokinetic concept of exercise. *Physical Therapy*, **47**, 114–117

Maitland, G.D. (1991) *Peripheral Manipulation*, 3rd edn, Butterworth-Heinemann, Oxford

Major, R.E. (1994) Senior Bioengineer, Personal Communication, The Orthotic Research Locomotor Assessment Unit, Robert Jones and Agnes Hunt Orthopaedic Hospital, Oswestry, Shropshire

Salter, N. (1955) Methods of measurement of muscle and joint function. *Journal of Bone and Joint Surgery*, **37-B**, 474–491

Steindler, A. (1973) *Kinesiology of the Human Body Under Normal and Pathological Conditions*, Charles C. Thomas, Springfield, Illinois

Voss, D.E., Ionta, N.K. and Myers, B.J. (1985) *Proprioceptive Neuromuscular Facilitation* 2nd edn, Harper and Row, Philadelphia

6. *Exercise prescription*

Physiotherapists have in the past often been regarded as the 'exercise experts'. However, recent initiatives to promote a healthy population have produced a number of exercise prescribers who are not physiotherapists. For example, it is not unusual to attend a step aerobics class or an aquarobics class at the local sports centre run by a non-physiotherapist. Degrees in sports science are also now available at a number of universities. Indeed, it is not uncommon for a physiotherapist to possess this qualification as well as the professional qualification, and these physiotherapists are often highly regarded in the areas of health and fitness promotion.

This chapter will outline the:

- government initiative – 'Health of the Nation'
- planning of an exercise programme
- general principles of exercise prescription
- exercise prescription for patient groups
- exercise programme progression
- benefits of exercise
- dangers of exercise
- precautions of exercise
- motivational aspects of exercise
- critical analysis of the exercise programme.

GOVERNMENT INITIATIVE – 'HEALTH OF THE NATION'

The recent Government White Paper (1992) 'Health of the Nation' targets key areas for the reduction of risk factors and mortality from certain diseases such as stroke and coronary heart disease. Risk factors identified by the White Paper are:

- smoking
- diet and nutrition
- obesity
- raised blood pressure
- alcohol.

Table 6.1 Target reductions for the 'Health of the Nation' initiative

Risk factors	Aim	From (in 1990)	To (in 2005)
Smoking	↓ in M and F > 16 years	31% in M 28% in F	< 20% in M and F
Diet and nutrition	↓ Average % intake of saturated fatty acids	17%	< 11%
	↓ Average intake of fat	40%	< 35%
Obesity	↓ In M and F between ages of 16 and 64	*8% in M *12% in F	< 6% in M < 8% in F
Blood pressure	↓ Mean systolic		> 5 mm Hg
Alcohol	↓ % of population who consume weekly units: > 21 in M > 14 in F	28% in M 11% in F	18% in M 7% in F

M, male; F, female.
* 1987 figures quoted.

The target reductions for these risk factors are summarized in Table 6.1.

PLANNING AN EXERCISE PROGRAMME

The exercise programme should be tailored to the needs of the individual. The well-designed rehabilitative programme will contain exercises which retrain all of the following aspects:

- strength
- endurance
- flexibility
- balance, co-ordination and proprioception.

The amount of time apportioned to each will depend on assessment findings (e.g. age, sex, severity of disease or injury, occupation, etc.), on patient potential and on the final goal of rehabilitation.

This point is best illustrated by two examples:

1. Rehabilitation programmes will be very different for an international athlete and a middle-aged sedentary worker who have both ruptured their Achilles tendons. Although the injuries are the same, the patients' rehabilitation goals will be different because of levels of fitness, age, motivation and the requirements for their occupations.
2. A rehabilitation programme for an asthmatic child will differ substantially from that for a child who has cystic fibrosis. In this example, both are children with respiratory disease, but the diseases are dissimilar. Each condition can be controlled by medication, but the pathological processes and long-term prognosis can be very different. Commonly, asthmatics do not have additional pathological complications and live a normal life span, whereas the cystic fibrotic

may have cardiac or digestive problems and die of these complications in the late 20s or early 30s.

Planning exercise programmes can involve preparing for both healthy client and patient groups. To design an effective programme, the therapist requires a thorough knowledge of the musculoskeletal, neuromuscular, cardiovascular and respiratory systems in both the functional and dysfunctional state. By carrying out a full assessment, the levels of function can be ascertained and the findings used to plan a rehabilitation regimen.

GENERAL PRINCIPLES OF EXERCISE PRESCRIPTION

The questions below can be used as a guide to prescribing exercise:

- What is the purpose of each exercise?
- Is the exercise having an effect on the desired muscle group?
- Is the exercise intensity sufficient?
- Is the exercise biomechanically safe to perform?
- Is the exercise specifically designed to improve functional outcome?
- Does each exercise contain an objective measure of progress?
- Is the exercise programme realistic for the client group for which it has been prescribed?
- Does the programme consist of a variety of strengthening, mobilizing and co-ordination exercises?
- Does the exercise programme have a defined beginning and end?
- Does the exercise programme motivate the patient to work hard?

EXERCISE PRESCRIPTION FOR PATIENT GROUPS

The patient groups described in this section form the majority of patients for whom physiotherapists will at some stage during their careers provide advice on exercise. It is important to be aware that although patients are often grouped as conditions, they must be treated as individuals, because it is highly likely that personal needs will dictate exercise prescription; for example, sight and hearing may be affected in some elderly patients, whereas balance may be the main problem with others.

The elderly

Improved health care, diet and general lifestyle in the latter half of this century has led to greater life expectancy. Shephard (1990) cites the benefits of exercise for the elderly as improving health, providing increased opportunities for social contacts and gains in cerebral function.

Health

The risk of cardiovascular disease increases with age, with 10% of the over-70 age group presenting with clinically diagnosed coronary artery disease (Shephard, 1990). Regular exercise decreases this risk in all population groups including the elderly (Government White Paper, 1992).

Exercise also has a beneficial effect on the growing skeleton by acting to maximize peak bone mass, and it assists in the prevention of post-menopausal bone loss (Rutherford, 1990).

Perhaps foremost in the mind of the physiotherapist who works with the elderly is the role of exercise in the maintenance of function and independence. Flexibility of the sedentary individual has reduced by 20–30% by the age of 70 years (Adrian, 1981) and continues to decline, so that functionally the less active patient soon begins to present with mobility and self-care problems such as being unable to wash, dress, get into and out of bed or climb stairs.

Social contacts

Problems associated with isolation and lack of mental stimulation, possibly resulting from loss of partner, lack of mobility and diminished or lost sight and hearing, can be alleviated to an extent by encouraging participation in group activities. Attendance at day centres and community or sports centres allows ample opportunity to improve fitness while at the same time providing plenty of scope for social interaction.

Cerebral function

Shephard (1990) attributes improved cerebral function following prescribed aerobic conditioning in the elderly to a rise in blood pressure during exercise with a consequent increase in cerebral perfusion.

Exercise has also been shown to enhance body image, provide a greater sense of self-esteem (Emery *et al.*, 1989) and to reduce anger, depression and anxiety in patients where mood was initially disturbed (Ingebretsen, 1982).

Principles of exercise prescription for the elderly

1. Ensure medical clearance with GP or consultant prior to admitting patient to the group.

2. Encourage low to moderate intensity exercise – avoid extreme fatigue.
3. Incorporate low-impact or non-weight-bearing exercise, e.g. walking, chair exercises.
4. Target exercise towards functional improvement.
5. Avoid long sessions and complicated exercises.
6. Progress gradually following reassessment.
7. Closely monitor for warning symptoms, e.g. severe dyspnoea, dizziness, angina.
8. Encourage social interaction during the exercise class.
9. Encourage participation in group sports, e.g. dancing, rambling, swimming.
10. Never leave the class unsupervised.
11. Ensure ease of access to emergency equipment.
12. Tailor the programme to age, health and ability of the individual.

Children

Rehabilitative exercise for the young should be fun. Powers of concentration do not allow for complicated regimens. Success depends on building a trusting relationship with the child. Competition can provide additional motivation, but the therapist should be aware that enforcing competition can also have detrimental effects. The child who lacks confidence may be put off exercise if required to perform or demonstrate an exercise in front of the group. Alternatively, if handled empathetically, a shy child may gradually gain in confidence.

Principles of exercise prescription for the child

1. Aim to exercise in an attractive environment such as a brightly decorated room on the ward or in the physiotherapy department. A large, barren environment can be very intimidating to a small child.
2. Spend time establishing a rapport with the child. This may be achieved, for example, by asking questions about the teddy perched on the bed or by discussing a favourite football team.
3. Approach the child at his or her level by crouching down or sitting.
4. Encourage parents to remain with the child and involve them in the exercise session. However, should this result in bad behaviour by the child, the parents should be asked to leave.
5. Basing the exercise programme on a story with the child as the main character is often a successful and fun approach. For example, the story line could involve having to cross a large river on a narrow bridge (walking on a line painted on the floor). The bridge might then collapse and so it becomes necessary to row or swim back (upper limb general mobility). It must be said that a vivid imagina-

tion is a definite advantage when designing an exercise programme of this kind!

6. Avoid complicated and boring exercises because children do not have a long concentration span.
7. Use apparatus selectively and keep it under control at all times.
8. Children can be boisterous and impulsive, so it is essential to be in control.
9. Avoid long and regular periods of high-impact exercise, e.g. road running and prolonged jumping activities, if prescribing a training programme for the young sports person.
10. Avoid long periods of lifting heavy weights.
11. Ensure good technique. Demonstrate and correct frequently if teaching a new sport or a specific exercise.
12. Encourage the child to wear shoes which are appropriate for the exercise session.

Respiratory patients

Achieving physical independence using exercise is fundamentally important in pulmonary rehabilitation. Exercise tolerance can be increased by simple exercise regimens such as the 12 Minute Walking Test (Chapter 7) (McGavin *et al.*, 1970, 1976).

The severity and nature of respiratory disease will determine the type and amount of exercise these patients are able to carry out. Medical clearance is essential and assessment of exercise tolerance advisable before prescribing an exercise programme. It must also be recognized that respiratory patients desaturate on exercise and that if there is a tendency for this to occur, supplementary oxygen should be given throughout the exercise period.

Principles of exercise prescription for the respiratory patient

1. Initially prescribe low-intensity exercise of short duration.
2. Exercise at a level which makes the patient moderately breathless for a few minutes several times per day and increase the time spent exercising within the limits of patient tolerance (Cockcroft, 1988).
3. Teach posture correction and relaxation techniques (Chapters 13 and 14).
4. Correct and re-educate breathing patterns – diaphragmatic and lateral costal – at rest and during exercise.
5. Target different large muscle groups during the exercise session, e.g. trunk, then lower and upper limbs and then return to the trunk.
6. Provide real functional targets for the patient, e.g. timing a short walk to visit a friend or to the local shops.
7. Motivate the patient by making the exercises part of the daily routine

such as walking to meet the grandchildren from school or walking to the newsagent in the morning to collect the newspaper.

Cardiac patients

Coronary rehabilitation programmes are available in some of the larger hospitals for patients who are recovering from coronary illness or surgery. They aim to increase patient confidence in the ability to exercise by beginning in a controlled environment.

The following guidelines have been adapted from Bethell (1988).

Prior to commencing an exercise programme all patients will require a medical examination which includes a resting ECG and a submaximal exercise test performed on either a cycle-ergometer or treadmill (Chapter 7).

1. Start graduated exercise within (a) 3–4 weeks following a myo-cardial infarction; (b) 5–6 weeks following coronary artery surgery.
2. Avoid anaerobic exercise – aim to improve aerobic (endurance) fitness.
3. Circuit training is excellent for these patients since it is less boring and prepares them for many different functional activities. Examples of exercises which can be included in the circuit are: cycle-ergometry, step-ups, jogging on a trampette, jogging on the spot, upper and lower limb exercises using light resistance such as a dumbbell or weighted ankle or wrist band. A warm-up and cool-down period should also be included in the circuit programme. The circuit can be progressed by altering the number and speed of repetitions and/or the duration of each exercise.
4. Training should ideally involve 3 or 4 sessions which last between 20 and 30 minutes each. Twice weekly sessions are also acceptable. Sessions should be continued for 6–12 weeks.
5. Patients should exercise at 70% and 85% of maximum heart rate (for calculation equation see Chapter 7).
6. Encourage the partner to attend the exercise programme to gain an insight into the patient's exercise tolerance capacity.
7. Promote home exercises after 2–3 weeks to avoid dependence: brisk walking, step-ups and jogging on the spot are all suitable for inclusion.
8. Teach posture correction and relaxation techniques.
9. Encourage diaphragmatic and lateral costal breathing and discourage upper chest breathing.
10. Teach patients to take their pulse prior to and on completion of exercise and to keep a record of these values.
11. Ensure adequate access to emergency equipment.

NB. Exercise forms only a part of the pulmonary and cardiac rehabilitation programmes. Components of these programmes also deal with the

Table 6.2 Progression regimen for a total knee replacement

Stage	Exercise	Progression objective
1	Static quadriceps contractions	Palpable contraction
2	Inner range concentric quadriceps contraction	Straight leg raise with no lag (i.e. full extension)
3	Middle to inner quadriceps concentric contraction	Gait – partial weight-bearing with crutches
4	Functional range of quadriceps contraction	Climbing stairs
5	Functional goal oriented activities. Closed kinetic chain squats up and down from chair	Bending down to pick an object up from the floor

psychological implications of respiratory and cardiac dysfunction, medical care and educational aspects. Interested readers are referred to Cockcroft (1988), Bethell (1988) and Murray (1993).

EXERCISE PROGRAMME PROGRESSION

Progression of an exercise should be sequential and involve a series of graded stages. Ideally each stage should have a progression objective which is tailored to individual needs. For example, the patient who has undergone a total knee replacement might follow the typical exercise progression regimen for strengthening the quadriceps muscles, as outlined in Table 6.2.

Following initial assessment, ascertain the type and level of exercise required and thereafter reassess frequently to evaluate the level of progress and advance to the next stage as required.

There is no standard 'recipe' for progressing an exercise regimen. This will depend on the patient, on the condition and on the level of motivation. In the above example, for instance, the ability to reach the progression objectives will obviously depend on the ability to gain range of movement. The mode of progression will also depend on the experience of the therapist and on available apparatus. As a general guide, progression should be gauged by patient performance in each of the fundamental contents of the exercise programme: strength, endurance, flexibility, balance, co-ordination and function (Table 6.3).

THE BENEFITS OF EXERCISE

Patients often ask, 'Why do I need to exercise?' The beneficial reasons are outlined in Figure 6.1. Exercise may also provide benefits which are specifically related to a patient's condition such as a reduction in pain

Table 6.3 Guidelines for exercise progression

Progression	Strength	Endurance	Flexibility	Balance	Function
1	↑ length of lever arm	↑ number of repetitions	Alternate starting positions	↓ size of base of support	Voluntary activities with practice become involuntary
2	↑ resistance on distal aspect of lever arm	↑ duration of each exercise	↑ number of repetitions	↑ height of centre of gravity	Use closed kinetic chain concept
3	↓ speed on dry land	↑ number of exercises in the programme	Auto-assisted stretching	Fixed base to mobile base of support	Simple to complex activities
4	↑ speed in water	↑ duration of exercise programme	↑ length of time of stretch	Static to dynamic body activities	Normal functional activities
5	From gravity as an assistance to gravity as a resistance; add externally applied resistance		Partner stretching	Gross to fine controlled movements at low and high speeds	

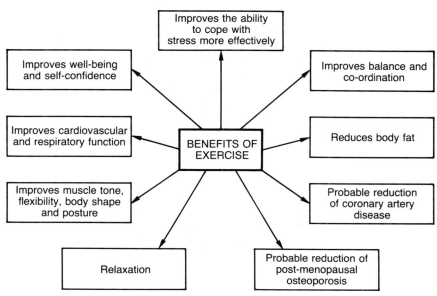

Fig. 6.1 Benefits of exercise

and swelling, an increase in joint range of motion and an improvement in circulation and skin condition.

THE DANGERS OF EXERCISE

Many people tend to assume that exercise is good for you. However, exercise can be dangerous if the following factors are not considered. The experienced and qualified therapist attending an aerobics class as a participant will immediately be able to assess whether the exercises are safe and effective.

The checklist below can be used when preparing and prescribing an exercise regimen, whether for an individual patient or a group of patients or clients. Failure to consider these aspects may result in injury or the delivery of an ineffective exercise class.

- Is the exercise environment safe? E.g. tidy, well lit, ventilated with non-slip floor.
- Is the equipment mechanically sound and functioning properly?
- Is adequate instruction and an effective demonstration given prior to the exercise?
- Are the exercises suited to patient capabilities? E.g. speed, range and intensity.
- Are the exercises biomechanically sound? E.g. avoiding stress on the lumbar spine, consider two joint muscles action and previous injury.

- Are physiological factors addressed? E.g. ballistic v. long slow stretch (Chapter 12).
- Is adequate correction given during the exercise programme?
- Is appropriate clothing and footwear worn?
- Have the existing level of fitness and state of health been assessed prior to exercise?
- Has an effective warm-up and cool-down period been included in the exercise programme?

THE PRECAUTIONS OF EXERCISE

Additionally, when designing an exercise programme it is important for the therapist to be aware of a number of precautions, lack of consideration of which can be a potential source of danger (Figure 6.2).

MOTIVATIONAL ASPECTS OF EXERCISE

The dictionary definition of motivation is 'the process which arouses, sustains and regulates behaviour'. A motivated patient will have the desire, the drive, interest or incentive to perform an exercise or task. Patients will generally perform better if motivation is maximized (Astrand and Rodahl, 1986), but along with compliance, motivation is difficult to control (Kisner and Colby, 1990).

Motivation may mean one thing to the patient and a different thing to the therapist. For example, patients may be happy to walk around

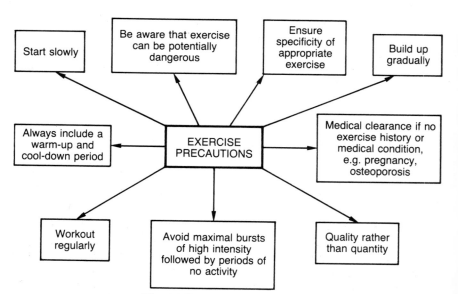

Fig. 6.2 Precautions of exercise

their homes but the therapist may consider that they are capable of walking to the shops. It is important to ensure that there is a degree of congruence between the goals of rehabilitation for the patient and therapist. Effective communication at the planning stage of the treatment programme will facilitate maximum compliance and motivation.

Basmajian and Wolfe (1990) describe two forms of motivation: external and internal. External motivation is provided by the therapist and internal motivation by patients themselves. The latter form of motivation can be either negative or positive in its orientation.

For example, a well-motivated and successful rugby player who has recently had a spinal cord injury resulting in paraplegia may remain highly motivated to succeed and end up as a top wheelchair athlete. Conversely, the reaction to injury may be such that the patient sees no future and consequently loses motivation to achieve. It is essential at this stage that the therapist, through an empathetic but firm approach, externally motivates the patient. The success of the therapist in this situation may facilitate a return of the patient's positive internal motivation and as a result he might still end up as a wheelchair athlete!

How to motivate

To facilitate motivation, patients should be involved in the planning and progression of their treatment regimens which should be realistically designed and within their capabilities. Boring long exercises should be avoided and encouragement and praise given where due. All exercises should have a set number of repetitions or a target to aim for, such as a distance or a height. The patient should always be able to see improvement which ideally should be measurable. Each exercise session should have a defined beginning and end and the entire programme should have a mutually agreed duration in terms of days, weeks or months.

CRITICAL ANALYSIS OF THE EXERCISE PROGRAMME

The ability to evaluate an exercise programme critically for effectiveness is often not considered to be a priority in the rehabilitation literature. Basmajian and Wolfe (1990) cite two main reasons for this: (a) therapists tend to have a fixation for doing rather than testing, and (b) there is a lack of training in measurement techniques.

Additionally, inexperienced therapists may lack the theoretical understanding of why they are prescribing a certain exercise or they may focus on just one small aspect of the patient's problem. For example, following a sprained ankle the patient will strengthen the ankle musculature but may fail to incorporate a series of proprioceptive exercises into the rehabilitation programme (Chapter 9).

The experienced physiotherapist will usually judge the success of an exercise protocol according to improvement in the patients' functional capacity. In other words, the ankle range of movement is restored and the patient is able to walk without the habitual limp. The less experienced therapist may just concentrate on the effect of the protocol directly on the impairment rather than on functional capacity, i.e. the range of ankle movement may be continuing to improve according to goniometric measurement and therefore the protocol is judged to be successful but the patient will still be walking with a limp.

Additionally, although difficult to assess, the probability of future injury should also be considered when examining the effectiveness of the exercise programme. In the above example, it is likely that the patient will resprain his ankle if proprioceptive rehabilitation is omitted (Irrgang *et al.*, 1994).

A common fault when evaluating the effectiveness of the exercise programme is failure to appreciate the patient as a whole but rather just treat as an injured part. Often therapists assume that, by treating the injury, functional improvement will automatically follow. This is certainly not the case.

The therapist who uses the holistic approach may fail to assess the contribution made by this approach to the functional outcome. For example, the amount of time spent localizing treatment to the ankle may be far less than the time spent correcting posture and re-educating gait. Psychologically the patient may not have the confidence to progress to full weight-bearing on the injured limb despite reassurance that this is possible.

Basmajian and Wolfe (1990) advocate the need to measure both functional outcome and changes in impairment (Chapter 18).

SUMMARY

- The 'Health of the Nation' is a Government initiative which aims to decrease risk and mortality from smoking, diet, obesity, raised blood pressure and alcohol consumption.
- A well planned exercise programme should consider strength, endurance, flexibility and balance.
- Exercises should be tailored to meet the needs of different patient groups.
- Progression of exercises should be sequential and include progression objectives for each stage.
- The benefits, dangers and precautions of exercise have been addressed and the need to develop skills of critical evaluation highlighted.

REFERENCES

Adrian, M.J. (1981) Flexibility in the ageing adult. In *Exercise and Ageing: The Scientific Basis* (eds E.L. Smith and R.C. Serfass), Enslow Publishing, New Jersey

Astrand, P.-O. and Rodahl, K. (1986) *Textbook of Work Physiology: Physiological Bases of Exercise*, 3rd edn, McGraw-Hill, New York

Basmajian, J.V. and Wolfe, S.L. (eds) (1990) *Therapeutic Exercise*, 5th edn, Williams and Wilkins, London

Bethell, H.J.N. (1988) Set up a coronary rehabilitation programme. *British Medical Journal*, **297**, 120–121

Cockcroft, A. (1988) Pulmonary rehabilitation. *British Journal of Diseases of the Chest*, **82**, 220–225

Emery, C.F., Pinder, S.L. and Blumenthal, J.A. (1989): Psychological effects of exercise among elderly cardiac patients. *Journal of Cardiopulmonary Rehabilitation*, **9**, 46–53

Government White Paper (1992) 'Health of the Nation'

Ingebretsen, R. (1982) The relationship between physical activity and mental factors in the elderly. *Scandinavian Journal of the Society of Medicine*, **29**, 153–159

Irrgang, J.J., Whitney, S.L. and Cox, E.D. (1994) Balance and proprioceptive training for rehabilitation of the lower extremity. *Journal of Sport Rehabilitation*, **3**, 68–83

Kisner, C. and Colby, L.A. (1990) *Therapeutic exercise. Foundations and Techniques*, 2nd edn, F.A. Davis, Philadelphia

McGavin, C.R., Gupta, S.P., Lloyd, E.L. *et al.* (1970) Physical rehabilitation for the chronic bronchitic: results of a controlled trial of exercises in the home. *Thorax*, **32**, 307–311

McGavin, C.R., Gupta, S.P., McHardy, G.J.R. (1976) Twelve Minute Walking Test for assessing disability in chronic bronchitis. *British Medical Journal*, **1**, 822–823

Murray, E. (1993) Anyone for pulmonary rehabilitation? Review paper. *Physiotherapy* **79**(10), 705–709

Rutherford, O.M. (1990) The role of exercise in prevention of osteoporosis. *Physiotherapy*, **76**(9), 522–525

Shephard, R.J. (1990) The scientific basis of prescribing exercise for the very old. *Journal of the American Geriatrics Society*, **38**, 62–70

7. *Fitness testing*

Involvement by the physiotherapist in the promotion of health and fitness is becoming increasingly important. It now ranges from basic assessment and prescription programmes in the physiotherapy department to working in highly sophisticated exercise laboratories testing patients with cardiac and respiratory disease.

This chapter primarily focuses on basic fitness assessments and provides information which will facilitate future progression to fitness testing using recent technological developments. It will outline:

- definitions of fitness
- reasons for performing an exercise test
- exercise and its energy systems
- test preparation and the Physical Activity Readiness Questionnaire
- exercise tests for the ATP-CP (anaerobic) power system
- exercise tests for the short term (glycolysis, anaerobic) endurance system
- exercise tests for the long term (aerobic) endurance system
- Maximal rate of oxygen uptake (the VO_2 max):

  (a) factors which affect the VO_2 max
  (b) factors which limit VO_2 max
  (c) the requirements for directly measuring VO_2 max
  (d) analysis of VO_2 max test results
  (e) test contra-indications

- Submaximal tests for predicting VO_2 max and/or aerobic endurance performance:

  (a) Astrand Submaximal Cycle-ergometer Test
  (b) Twelve Minute Field Performance Test
  (c) 'Rockport' One Mile Walk Test
  (d) Canadian Home Fitness Test
  (e) Harvard Step Test
  (f) Twelve Minute Walking Test
  (g) Self-paced Treadmill Test
  (h) Shuttle Walking Test

- prescribing aerobic exercise for health related fitness from test results.

DEFINITIONS OF FITNESS

Definitions of fitness are difficult because fitness tends to mean different things to different people. For example, an individual may consider him/herself fit but probably will take no exercise or perform activities which are considered to be beneficial for the cardiovascular system. In the Allied Dunbar National Fitness Survey (British Sports Council and the Health Education Authority, 1992) over 7 out of 10 men and 8 out of 10 women fell below the activity requirements to achieve health benefits.

For the endurance athlete, fitness may mean being able to run a sub $2\frac{1}{2}$-hour marathon; for the footballer, it may mean being able to play flat out for 90 minutes; for the middle-aged adult, it may mean being able to play in the local veterans tennis tournament or to take part in an evening ramble; and for the elderly, it may mean being able to function independently at home.

Perhaps to many sedentary individuals the concept of fitness tends to imply that one has to 'be sporty'. This is a misconception because to be fit does not require a sports outlook, a sports environment or sports equipment. Indeed, a perfectly acceptable and valid activity which provides many health benefits, especially for sedentary individuals, is brisk walking (Hardman and Hudson, 1989; Davison and Grant, 1993).

The contemporary definition of fitness falls into two general categories: (a) activity-specific fitness, and (b) health-related fitness. Individuals need a combination of both to allow them to achieve their optimum functional capacity.

REASONS FOR PERFORMING AN EXERCISE TEST

Testing exercise tolerance has become an integral part of medical rehabilitation and sports science. There are a number of general reasons why an exercise test might be performed. It might be required for the diagnosis/prognosis of sports performance and for designing training/rehabilitation programmes for the purposes of health related exercise.

Cardiovascular fitness testing is performed to ascertain the effect of a progressive and systematic increase in oxygen demand on the myocardium. Results of the evaluation provide information about the current status of cardiorespiratory function.

Musculoskeletal and neuromuscular fitness, i.e. muscle strength, endurance, flexibility, balance and co-ordination, are equally important if examining general fitness and these aspects are described in later chapters.

A fitness test can be used to:

- assess the maximal aerobic capacity (cardiovascular fitness) of normal subjects in preparation for exercise prescription or athletic training

- assess musculoskeletal and neuromuscular fitness prior to returning to sport/work
- assess maximal aerobic capacity of patients with cardiac disease
- detect coronary artery disease
- determine the prognosis and severity of disease
- evaluate the effect of medical and surgical intervention
- assess biomechanical function.

EXERCISE AND ITS ENERGY SYSTEMS

There are three systems which provide the metabolic capability to generate ATP for muscular contraction and these are classified by McArdle *et al.* (1994) as the:

1. Adenosine triphosphate–creatine phosphate (ATP-CP) system (the immediate energy system)
2. Glycolysis(–lactic acid) system (the short-term energy system)
3. Aerobic system (the long-term energy system)

All of the systems contribute energy during any activity, but the proportion of contribution depends on the type of exercise and intensity of activity being performed.

Two of the systems, the immediate and short-term, provide energy without using oxygen and these are known as the anaerobic systems. The third, the long-term system, requires the presence of oxygen to synthesize energy and is known as the aerobic system. This system is unable to provide immediate energy because even with available oxygen it is doubtful whether it could be utilized fast enough to produce ATP to meet the immediate demands imposed by certain high-intensity activities (McArdle *et al.*, 1994).

Immediate energy is required for activities which involve short sharp bursts of high-intensity exercise such as the first 6 seconds of the 100 metre sprint, the long jump, power lift, tennis serve or golf swing. Energy for these activities is provided by the ATP-CP system which is able to generate enough energy anaerobically for between 6 and 8 seconds of high-intensity exercise.

High-intensity activities lasting longer than 10 seconds but less than 3 minutes require energy from the glycolysis–lactic acid system. Activities which use energy from this system include the 400 metre sprint, a hard tennis or squash rally and the shorter swimming events.

The aerobic system provides energy for endurance activities such as jogging and hill walking. The threshold where aerobic metabolism provides the majority of energy is considered to be at distances of 800 metres and above (Astrand and Rodahl, 1986).

Table 7.1 gives a summary of the three energy systems.

Fitness tests are available to evaluate the individual contributions of the three energy systems to physical activity. Ideally, the tests used to

Table 7.1 Summary of the energy systems

Energy system	Supply	Time	Activity (e.g.)	Fitness Test (e.g.)
ATP-CP	Immediate (anaerobic)	6–8 sec of maximal exercise	100 m sprint/tennis serve/weight lifting	Power jumps Stair sprint 1st 6 sec of the Wingate Test
Glycolysis (lactic acid)	Short term (anaerobic)	Up to 3 min of maximal exercise	400 m sprint/ 100 m swim	Wingate Test
Aerobic	Long term (aerobic)	Longer duration, less in-tense periods of exercise	Marathon/hill walk	Walking tests Harvard Step Test Astrand Cycle Test Maximal treadmill/cycle tests

Human Movement Explained

assess fitness should be similar to the activity for which the energy capacity is being evaluated and in some instances it may be possible to use the activity as the test.

TEST PREPARATION

A full patient assessment should precede any fitness test. The Physical Activity Readiness Questionnaire (PAR-Q) (Shephard, 1991) provides a guide to the type of additional questions a patient should be asked before performing the test or before embarking on an exercise programme.

The Physical Activity Readiness Questionnaire (PAR-Q)

1. Has a doctor ever said that you have a heart condition   Y   N
   and recommended only medically supervised activity?
2. Do you have chest pain brought on by physical   Y   N
   activity?
3. Have you developed chest pain in the past month?   Y   N
4. Do you tend to lose consciousness or fall over as a   Y   N
   result of dizziness?
5. Do you have a bone or joint problem that could be   Y   N
   aggravated by the proposed physical activity?
6. Has a doctor ever recommended medication for your   Y   N
   blood pressure or a heart condition?
7. Are you aware, through your own experience or a   Y   N
   doctor's advice, of any other physical reason against
   your exercising without medical supervision?

NB. If you have a temporary illness, such as a common cold, or are not feeling well at this time, postpone the test until you have recovered.

Optimal testing conditions

1. The subject must feel well and be free from infection.
2. Several hours should have elapsed between the last meal and the test.
3. The subject should not have performed physical activity which is heavier than the work rate of the test within the last few hours.
4. Smoking within 2 hours of the test must be avoided.
5. The test room temperature should be between 18 °C and 20 °C and the room should be well ventilated.

EXERCISE TESTS FOR THE ATP-CP (ANAEROBIC) POWER SYSTEM

Power tests

Power tests are used to evaluate energy contribution from the short-term ATP-CP energy system. McArdle *et al*. (1994) refer to the following widely used power tests:

- Sargent jump-and-reach (vertical jump)
- standing broad jump (horizontal jump)
- stair sprint
- one repetition maximum.

Evaluation of these tests is possible using the equation:

$$P = F \times d \div T$$

where P = power expressed as watts (1 W = 6.12 kg-m^{-1}min^{-1}), F = force required to move an object or body, d = distance moved by the body as a result of the force, and T = time taken to move the object/body.

Sargent jump-and-reach (vertical jump)

This test is assessed by measuring the difference between a subject's standing reach and their maximum jumping reach from a semi-crouch position (Figure 7.1).

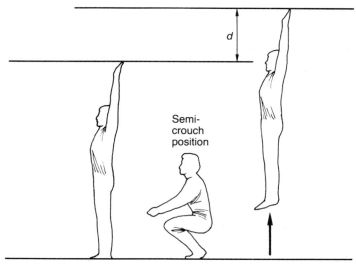

Fig. 7.1 Sargent jump-and-reach power test

The power will be equal to the force on the body mass due to gravity (e.g. 65 kg × 9.81 ms^{-2} Newtons)* multiplied by the distance travelled by the centre of gravity (CoG) of the body during the jump (the difference between finger positions pre and post jump) divided by the time it takes to do the jump.

Standing broad jump (horizontal jump)

In the standing broad jump, the subject jumps as far forwards as possible from a semi-crouched position and the horizontal distance covered is scored. Since this activity involves both vertical and horizontal components, the calculation of power will require vector analysis.

Criticism has been levelled at these jump tests because it is doubtful whether there is sufficient time to assess power generation. Clinically, however, both remain useful assessment tools for lower limb rehabilitation programmes because they provide an objective measurement of increased muscle strength and co-ordination for which a patient can be motivated to improve.

Stair sprinting

The subject is instructed to run up a flight of stairs, three at a time, as fast as possible. The power generated can then be calculated using the formula:

$$P = F \times d \div T$$

where F = vertical force due to gravity acting upon the subject's mass (Newtons), d = total vertical distance travelled by the subject in metres (vertical height of stairs), and T = time taken to cover the vertical distance in seconds (Figure 7.2).

For example, the power generated by a 70 kg man who climbs a vertical distance of 1.5 metres in 0.55 seconds would be calculated as follows:

$$P = F \times d \div T$$

$$\text{Force} = 70 \text{ kg} \times 9.81 \text{ ms}^2 = 686.7 \text{ N}$$

$$P = (686.7 \times 1.5) \div 0.55$$

$$P = 1873 \text{ W} = \text{absolute power}$$

It is important to be aware that the heavier subject covering the distance in the same time as the lighter individual will require a larger

---

* The Newton is that force which, when applied to a mass of one kilogram, produces an acceleration of one metre per second squared: 1 N = 1 kg.ms^{-2}.

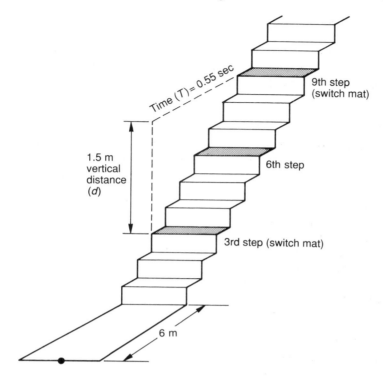

Fig. 7.2 Stair sprinting – see text

power output. If the test is to be suitable for comparing power generation between individuals of different body mass, power can be corrected by the following equation:

To correct for body mass calculate relative power (i.e. the power mass ratio) by dividing the absolute power score by the subject's body mass. In the above example, $1873 \div 70 = 26.76$ W/kg of body mass. If the subject was 65 kg, the power mass ratio would be 28.8 W/kg (i.e. $1873 \div 65$) and in this example the lighter subject would be able to generate more power per kilogram of body mass than the heavier subject.

The stair sprint can also be used to compare: (a) power output pre and post training for the same normal subject, and (b) power output pre and post rehabilitation progress for the same patient.

One repetition maximum

The one repetition maximum (1 RM) is the maximum load which can be lifted once by the subject and it is used to measure the maximal concentric force generated by a muscle group. The 1 RM is assessed by gradually increasing the load lifted (ensuring sufficient rest between lifts) until the subject is unable to lift any further load. It provides a good bench mark for designing a weight training programme (Chapter 10).

EXERCISE TESTS FOR THE SHORT-TERM (GLYCOLYSIS) ENDURANCE SYSTEM

Tests designed to evaluate the glycolytic contribution to exercise consist of maximal work performed for up to 3 minutes. Several tests have been used including shuttle runs, maximal sprints on a treadmill and cycle-ergometer, but it is generally difficult to evaluate the energy contribution from this system due to factors such as age, health status, motivation and body mass (McArdle *et al.*, 1994). Two tests are described in this section:

- The Wingate Test
- The Loughborough Non-motorized Treadmill Test.

The Wingate Test

The Wingate Test (Bar-Or *et al.*, 1980) is a test used to evaluate absolute leg power and the ability to sustain a high power output generated by the ATP-CP and glycolysis–lactic acid energy systems, respectively. To perform the test, the subject is required to pedal flat out for 30 seconds on a mechanically braked cycle-ergometer.

Calculation of power output per minute at any point during the test is usually performed by a computer interfaced with the ergometer, but it can also be performed manually (see below). Typical values of between 500 and 1000 W are obtained in the untrained, whereas values in excess of 1000 W are usual for a 100 m sprinter or sprint cyclist.

During the test recordings are made of:

(a) The resistance on the flywheel (braking power) measured in kg or kiloponds (kp = force acting on 1 kg mass at the normal acceleration of gravity) (Anderton and Bigg, 1972).[*]
(b) The flywheel radius and circumference ($2\pi r$).
(c) The number of pedal revolutions.

Clinical example to show the calculation of power output from the Wingate Test

A young female patient who is recovering from a partial rupture of the left quadriceps produces the following data using a Monark cycle-ergometer:

Force (resistance) on flywheel $= 5.0 \text{ kg} \times 9.81 \text{ ms}^{-2} = 49.05 \text{ N}$

[*] The kilo-pound is sometimes seen as the unit of measurement on cycle-ergometers.

Flywheel radius = 0.25 m (Monark cycle-ergometer only)

Flywheel circumference = 1.57 m (calculated using the formula: circumference = $2\pi r$)

1 pedal revolution = 4 flywheel revolutions (Monark cycle-ergometer only)

For every pedal revolution, the distance travelled is therefore equal to 1.57 × 4 = 6.28 m

Pedal revolutions per minute (rpm) = 120

The power output per minute is calculated using the formula:

$$P = \frac{F \times d}{t}$$

$F = 5.0 \text{ kg} \times 9.81 \text{ ms}^{-2} = 49.05 \text{ N}$

$d = 120 \text{ pedal revolutions} \times 6.28 = 753.6 \text{ m in 1 min}$

$t = 60 \text{ sec}$

$$P = \frac{49.05 \text{ N} \times 753.6 \text{ m}}{60 \text{ s}}$$

$$P = \frac{36964.08 \text{ Nm}}{60 \text{ s}}$$

$P = 616 \text{ W}$

NB. Forces required to overcome inertia and maintain flywheel revolutions have not been taken into account in the above calculation. These are minimized, however, by beginning the test from a rolling rather than a stationary start.

The Non-motorized Treadmill Test (Lakomy, 1984)

The protocol for this test is very similar to that of the Wingate Test except that the subject runs on a non-motorized treadmill restrained by a belt around the waist (CoG of body). The belt is tethered to a force transducer.

Power output is calculated by:

$$P = \text{Force} \times \text{Velocity}$$

where:

$$\text{Velocity} = \frac{\text{Distance}}{\text{Time}}$$

$F = \text{Force measured by the force transducer}$

For example: if a subject propels the treadmill belt up to a peak speed of 8 m/s (approximately 18 miles per hour) by applying a horizontal force of 150 N, the power output will be 1200 W in the horizontal direction.

EXERCISE TESTS FOR THE LONG-TERM (AEROBIC) ENDURANCE SYSTEM

Test may be maximal or submaximal. Maximal tests are more appropriate for use in sports testing or in the final stages of rehabilitation when the assessment is for a specific activity or sport and where it is safe to ask an individual to exercise maximally. In most instances, the physiotherapist will be dealing with submaximal tests.

The ability to take in and use oxygen is the aerobic capacity of an individual. It is well established that subjects who perform sporting activities which involve long-duration high-intensity effort possess a large aerobic capacity, e.g. cross-country skiers, marathon and middle distance runners, road-race cyclists and rowers.

MAXIMAL RATE OF OXYGEN UPTAKE (THE VO_2 MAX)

During exercise there is a ceiling to the oxygen uptake by each individual which can be determined experimentally and which is very reproducible. This point is known as the VO_2 max (Holly, 1988).

VO_2 max can be expressed as an absolute value in litres of oxygen per minute, e.g. VO_2 max = 2.6 litres.min^{-1}, or in relation to body mass, so that relative comparisons between individuals of varying body mass can be made, e.g. VO_2 max = 42.0 ml kg^{-1} min^{-1}. The conversion equation which allows these values to be interchanged is given as follows:

$$VO_2 \, max \, (ml \, kg^{-1} \, min^{-1}) = \frac{VO_2 (l \, min^{-1}) \times {}^*1000 \, ml \, l^{-1}}{Body \, mass \, (kg)}$$

Factors which affect the VO_2 max

The ability to take up and utilize oxygen varies between individuals. Factors which will affect this include (Figure 7.3):

(a) Genetics (Klissouras, 1971). This contributes between 25% and 40% of the VO_2 max by influencing such factors as an individual's

---

* Millilitres is used in the calculation, otherwise the value in litres would end up being a very small fraction.

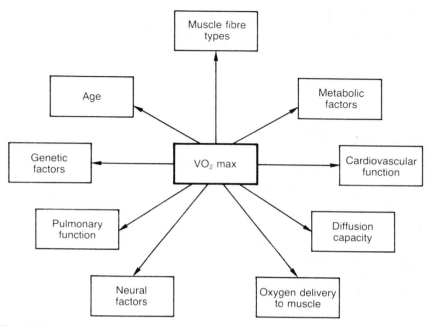

Fig. 7.3 Factors which affect maximal oxygen uptake

potential maximal cardiac output and the proportion of slow twitch muscle fibres.

(b) Training – this may improve oxygen uptake by between 6% and 20% (Davies and Sargeant, 1975).

(c) Sex differences in body composition and haemoglobin content are some of the factors why VO_2 max values for females are 15–30% less than those for males (McArdle *et al.*, 1994).

(d) Modes of exercise testing. Maximal oxygen uptake may vary with different modes of testing, e.g. VO_2 max is less for untrained subjects when measured on a cycle-ergometer than on a treadmill (Jones and Campbell, 1982).

Factors which limit VO₂ max

There is a continuing debate on which factors have the largest influence on limiting maximal oxygen uptake. These factors are usually divided into two groups – the central cardiovascular and the peripheral muscular and metabolic factors (Davies and Sargeant, 1975). All of these factors can be measured objectively in the laboratory using cardiopulmonary and radiographic techniques, blood sampling and muscle biopsy.

Normal pulmonary function

Healthy subjects utilize only about two-thirds of their pulmonary capacity during maximal exercise and therefore pulmonary function at sea level is not considered to be a limiting factor (Astrand and Rodahl, 1986). However, respiratory disease can compromise pulmonary function and this may be reflected in lower VO_2 max values.

Cardiovascular limitations

For large muscle group exercise, evidence suggests that the central factor, i.e. the cardiovascular system, may limit oxygen transport, as demonstrated in the single-leg training studies of Davies and Sargeant (1975). VO_2 max values for single-leg cycling were significantly higher per limb when compared with two-leg cycling.

Oxygen utilization

This has been investigated as another limiting factor. The question was asked – if the muscles were able to receive more oxygen could they actually use it? Two main techniques have featured in the literature. The first involves elevating oxygen delivery by a process of increasing blood volume by reinfusion. Ekblom *et al*. (1972, 1976) used autologous transfusion (removal and storage of blood from the same subject and then reinfusion after 3 weeks), i.e. 'blood doping'.

The second method of increasing oxygen to the muscles involved breathing hyperoxic gas (Ekblom *et al*., 1975). Rises in VO_2 max of between 11% and 13% were recorded. The muscles seem to be able to use as much oxygen as the cardiovascular system can deliver.

Nervous system control

Sympathetic nervous system control mechanisms effecting vasoconstriction of the arterioles of skeletal muscle are considered another source of limitation to maximal oxygen uptake (Secher *et al*., 1977).

Diffusion capacity

During normal pulmonary blood flow, blood becomes almost saturated with oxygen by the time it has passed through a third of the pulmonary capillary, and so even with a shortened time of exposure in exercise the blood can still become fully oxygenated or nearly so.

Metabolic factors

Conflicting evidence has been put forward concerning the link between metabolic factors and VO_2 max. The consensus suggests that increased muscle enzyme activity improves rather than limits VO_2 max (Gollnick *et al.*, 1973; Ivy *et al.*, 1980).

The requirements for directly measuring the VO_2 max

A test for direct measurement of the VO_2 max is usually carried out on a static cycle-ergometer or on a motorized treadmill, but other ergometers such as rowing machines and swimming flumes have been used.

Several protocols are available and are usually designed in one of two ways. There is either one short, sharp burst of hard exercise or a series of incremental increases in exercise intensity.

Treadmill protocols usually involve a constant pace, with the exercise intensity progressed by increasing the gradient of the treadmill so that the subject gradually walks/runs up a steeper incline.

During the test, expired air is collected either in a Douglas bag (Figure 7.4) for subsequent analysis, or is analysed through a metabolic cart (Figure 7.5) which may be capable of carrying out breath-by-breath analysis during the test. In general, the tests should be completed in 6 minutes where the subject exercises to volitional fatigue.

Initially, in the well-equipped laboratory, it may be necessary to decide whether to use the cycle-ergometer or the treadmill. The relative merits of both are outlined in Table 7.2.

Protocols commonly in use today include the Balke, Naughton, Astrand and Bruce tests (McArdle *et al.*, 1994). All of the tests show a progression by increasing the exercise speed, duration and the grade of the treadmill slope or the resistance applied to the cycle-ergometer.

To illustrate how a treadmill test is carried out, the Bruce test (Bruce, 1971) is shown in Table 7.3.

During the test, heart rate, blood pressure and ECG are monitored for diagnostic purposes (i.e. assessing for heart disease). Respiratory gases are collected and analysed, especially when testing for athletic performance. In addition, a visual analogue scale – the Borg scale (Borg, 1982) – is often used so that the testers are aware of how the subject is feeling during the test (see later).

Analysis of VO_2 max test results

Expired air

The amount of oxygen in a given volume of expired air is compared with the known constant of the same volume of atmospheric air using the Haldane transformation equations (McArdle *et al.*, 1994). Aerobic fitness is measured by the largest consumption of oxygen an individual achieves during the test.

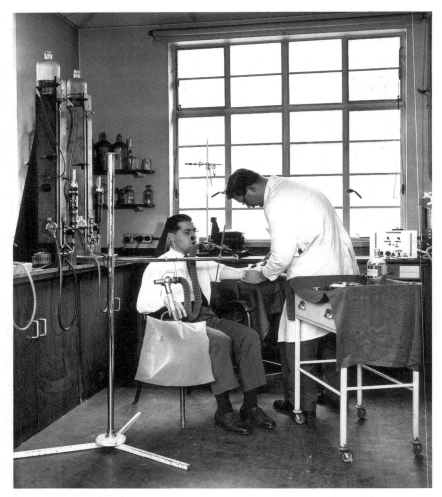

Fig. 7.4 An early version of the Douglas bag, the principle of which is un-changed although the technical read-outs, etc., are now far more advanced (photograph by courtesy of Mr J. May, Plysu Protection Systems Ltd)

Anaerobic (ventilatory) threshold

The anaerobic threshold is the point at which aerobic energy produc-tion is supplemented by anaerobic mechanisms (Wasserman *et al.*, 1973). This is determined either by carbon dioxide ventilatory thresholds (Wasserman *et al.*, 1973) or by coinciding measurements of the onset of blood lactate (OBLA). It is believed that a ventilatory threshold occurs when an excess of carbon dioxide is produced as an end product of buffering accumulated lactic acid in the blood (Rhodes and Loat, 1993).

The threshold can be used both in rehabilitation and the sports environment to prescribe training levels for improving aerobic fitness.

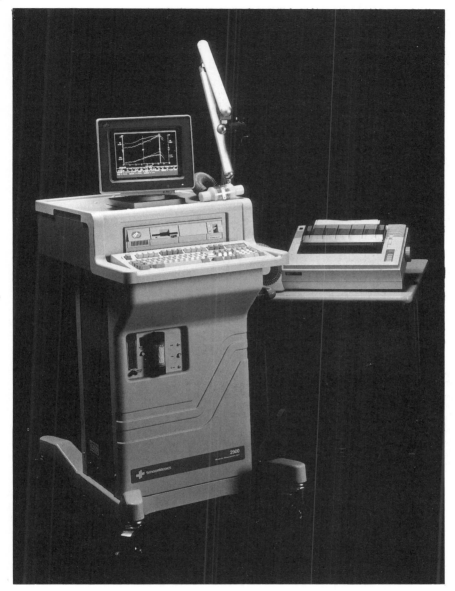

Fig. 7.5 A metabolic cart (photograph by courtesy of SensorMedics BV, Rugby)

As this improves, the onset of the threshold during a retest will occur at a later point in the test at the same exercise intensity.

Blood lactate

When exercise becomes harder, aerobic systems are unable to supply sufficient energy and anaerobic energy sources are required to supplement muscle metabolism to generate adenosine triphosphate. This,

Table 7.2 Treadmill v. cycle-ergometer

Treadmill		Cycle-ergometer	
Advantages	*Disadvantages*	*Advantages*	*Disadvantages*
1. Functional – patient uses same muscles as for walking 2. Standard settings for speed and grade can be used because body weight is a determinant of power at a given setting 3. The obtained VO_2 max will be higher with the treadmill than on the cycle-ergometer 4. Generates higher power and maximal heart rates in cardiac patients	1. Large and noisy 2. Expensive 3. Difficult to take BP and ECG 4. Walking on a moving belt is not easy for some patients 5. Breathing apparatus awkward to control	1. Takes up less room 2. Portable and quieter 3. Relatively inexpensive 4. Easier to take BP and ECG 5. Power output is independent of subject body mass which is easily calculated and regulated	1. Not considered to be a daily functional activity – muscular fatigue may develop faster 2. Older people with weak quadriceps may find it difficult to reach desired heart rate before the onset of muscle fatigue

Table 7.3 The Bruce protocol

Stage	Duration (min)	Speed (kph)	Grade (°)	Grade (%)	Mets (metabolic unit)*
1	3	2.7	5.7	0	4
2	3	4.0	6.9	12	6–7
3	3	5.5	8.0	14	8–9
4	3	6.8	9.1	16	15–16
5	3	8.0	10.3	18	21

*MET (metabolic unit) describes energy expenditure and exercise intensity, where 1 MET = 3.5 ml O_2 uptake $kg^{-1}min^{-1}$.

however, results in the accumulation of blood lactic acid, with a concomitant increase in muscle acidity; the process of muscle fatigue becomes inevitable.

Heart rate (HR)

1. Heart rate normally increases linearly with oxygen uptake (VO_2) during incremental exercise (Donald *et al.*, 1955).
2. Patients with heart disease demonstrate steep increases in HR during exercise because of a low stroke volume. During a stress test, the exercise becomes more difficult. As the patient approaches maximum work rate, the relationship between VO_2 and HR will usually become even steeper because oxygen uptake tends to slow down while the heart rate continues to increase (Wasserman *et al.*, 1986).

Blood pressure

During exercise, the systolic pressure shows large increases but the diastolic pressure remains essentially unchanged. Physiologically, if cardiac disease is present, at some point during the test, the systolic blood pressure may drop due to cardiac dysfunction. This may require emergency medical attention.

VO_2 max test contraindications

Maximal exercise is defined as the point at which the patient is unable to continue because of symptoms or because safety limits have been reached (Jones and Campbell, 1982).

Hillegass (1989) cites a number of absolute and relative contraindications to maximal cardiovascular fitness testing.

Absolute contraindications

- recent myocardial infarction
- acute pericarditis and myocarditis
- resting or unstable angina
- serious ventricular or rapid atrial arrhythmias
- second- or third-degree heart block
- congestive cardiac failure
- any acute illness displaying mental and or physical manifestations.

Relative contraindications

- aortic stenosis
- left main coronary artery disease
- hypertension (systolic greater than 220 mm Hg and diastolic greater than 150 mm Hg at rest)
- idiopathic hypertrophic subaortic stenosis
- severe depression of the ST segment on a resting ECG
- compensated heart failure.

All patients should be thoroughly assessed before performing a maximal exercise test. The test is generally not suitable for patients who are undergoing a hospital-based rehabilitation programme. Submaximal tests are preferable in this situation.

SUBMAXIMAL TESTS FOR PREDICTING VO₂ MAX AND/OR AEROBIC ENDURANCE PERFORMANCE

Submaximal tests are less expensive and are safer to use for many patients. In the following section the tests have been divided into those which allow exercise prescription from a predicted VO_2 max and those which measure endurance performance (Table 7.4).

Tests which predict VO₂ max

Astrand Submaximal Cycle-Ergometer Test

Described by Astrand and Rodahl (1986), this is a widely used test. It requires a cycle-ergometer and heart monitor or the ability of the tester to take accurate pulse readings. A gradually decreasing HR at a standard work rate during repeated tests is indicative of increasing cardiovascular fitness.

Method
The test is performed for 6 minutes. The aim during the test is to raise the subject's HR above 120 beats per minute (bpm). This should be achieved within the first 2 minutes. If this is not achieved, the work

Table 7.4 Submaximal exercise tests

Tests which predict VO₂ max	Tests for measuring endurance performance
Astrand Submaximal Cycle-Ergometer Test	Harvard Step Test
Cooper's 12 Minute Run	Twelve Minute Walking Test
'Rockport' One Mile Walk Test	Self-Paced Treadmill Walking Test
Canadian Home Fitness Test	Shuttle Walking Test

rate should be increased by 50% for unfit subjects or by 100% for younger and/or fitter subjects (Table 7.5).

When the work rate is increased, at least 3 minutes should be allowed for the HR to level off for recording purposes. For example, if at 4 minutes into the test the work rate is increased, then the test should be extended to 7 minutes.

The pulse rate is counted during the last 15–20 seconds of each minute and provided that the work rate is not too heavy, a steady state is usually reached after the first 4–5 minutes.

The averaged pulse rate for the fifth and sixth minute is taken as the HR response to the exercise demand and can be used for calculation purposes. However, if the difference between these two values is greater than 5 bpm, the exercise should be continued until the steady state is reached.

The HR should not exceed 200 minus the subject's age, especially for the older and less fit subjects. If this does occur, then the test should be stopped immediately and the subject's HR recorded and used in the regression equations and nomogram for calculating the predicted VO₂ max (Tables 7.6 and 7.7).

Using the Astrand nomogram to predict VO₂ max
The Astrand nomogram can be used to predict VO₂ max (Astrand and Rodahl, 1986). Tables 7.6 and 7.7 for males and females have been adapted (Buckley, 1994) to simplify the use of the nomogram.

Table 7.5 Guidelines for selecting the pedal speed and starting work rate in the Astrand Submaximal Cycle-Ergometer Test

	Males		Females	
	Pedal speed (RPM)	Work rate (Watts)	Pedal speed (RPM)	Work rate (Watts)
Sedentary or age 50+	60	60	50	30
Moderately active and under the age of 50	70	70	60	50
Highly active/trained	75–85	75–85	60–75	60–75

Table 7.6 Astrand nomogram for predicting $VO_2\,max\,(l\,min^{-1})$ in males (WR = work rate in Watts; HR = beats per minute)

WR	50	55	60	65	70	75	80	90	100	105	110	120	130	140	150
HR															
120	2.2	2.3	2.4	2.6	2.7	2.8	3.0	3.2	3.5	3.6	3.7	4.0	4.2	4.5	4.8
122	2.1	2.2	2.3	2.5	2.6	2.7	2.8	3.1	3.4	3.5	3.6	3.9	4.1	4.4	4.7
124	2.0	2.15	2.35	2.4	2.5	2.65	2.75	3.0	3.3	3.4	3.5	3.7	4.0	4.2	4.5
126	1.95	2.10	2.25	2.35	2.45	2.6	2.7	2.9	3.2	3.3	3.4	3.65	3.85	4.1	4.4
128	1.90	2.05	2.15	2.3	2.4	2.5	2.65	2.8	3.15	3.2	3.3	3.5	3.75	4.0	4.2
130	1.85	2.0	2.10	2.2	2.35	2.4	2.6	2.75	3.0	3.1	3.2	3.4	3.6	3.9	4.1
132	1.80	1.95	2.05	2.15	2.3	2.35	2.5	2.7	2.9	3.0	3.1	3.3	3.5	3.75	4.0
134	1.75	1.90	2.0	2.10	2.2	2.3	2.4	2.65	2.8	2.9	3.0	3.2	3.4	3.65	3.9
136	1.70	1.85	1.95	2.05	2.15	2.2	2.35	2.6	2.75	2.8	2.95	3.15	3.35	3.6	3.8
138	1.65	1.80	1.90	2.0	2.10	2.15	2.3	2.5	2.7	2.75	2.9	3.1	3.25	3.5	3.7
140	1.60	1.75	1.80	1.95	2.05	2.1	2.2	2.4	2.6	2.7	2.8	3.0	3.2	3.4	3.6
142	1.58	1.70	1.75	1.90	2.0	2.05	2.15	2.35	2.55	2.65	2.75	2.9	3.1	3.3	3.5
144	1.50	1.65	1.70	1.85	1.95	2.0	2.1	2.3	2.5	2.6	2.7	2.85	3.0	3.25	3.4
146		1.60	1.65	1.80	1.90	1.95	2.05	2.25	2.4	2.55	2.6	2.8	2.95	3.15	3.3
148			1.60	1.75	1.85	1.90	2.0	2.2	2.35	2.5	2.55	2.75	2.9	3.1	3.25
150				1.70	1.80	1.85	1.95	2.15	2.3	2.4	2.5	2.7	2.8	3.0	3.2
152				1.65	1.75	1.8	1.90	2.1	2.25	2.35	2.45	2.65	2.75	2.95	3.1
154				1.6	1.70	1.8	1.85	2.05	2.2	2.3	2.4	2.6	2.7	2.9	3.0
156					1.65	1.75	1.80	2.0	2.15	2.25	2.3	2.6	2.65	2.8	2.95
158					1.60	1.7	1.75	1.95	2.1	2.2	2.25	2.5	2.6	2.75	2.9
160					1.55	1.65	1.7	1.9	2.1	2.15	2.2	2.45	2.55	2.7	2.85
162						1.60	1.7	1.85	2.05	2.10	2.2	2.35	2.5	2.65	2.8
164							1.65	1.8	2.0	2.05	2.15	2.3	2.45	2.6	2.75
166							1.6	1.8	1.95	2.0	2.1	2.25	2.4	2.55	2.7
168								1.75	1.9	2.0	2.05	2.2	2.35	2.5	2.65
170								1.7	1.85	1.95	2.0	2.15	2.3	2.45	2.6

Regression equation for age correction for males:
Males ($^{*}VO_2\,max \times 0.348) - (0.035 \times Age) + 3.011$
where $^{*}VO_2\,max$ is the uncorrected value in litres/min ($l\,min^{-1}$).

Table 7.7 Astrand nomogram for predicting VO_2 max ($l\,min^{-1}$) in females (WR = work rate in Watts; HR = beats per minute)

WR	50	55	60	65	70	75	80	90	100	105	110	120	130	140	150
HR															
120	2.5	2.6	2.8	2.9	3.1	3.2	3.4	3.7	4.1	4.2	4.4	4.6	5.0	5.2	5.6
122	2.4	2.5	2.7	2.8	3.0	3.1	3.3	3.6	3.9	4.1	4.2	4.4	4.8	5.0	5.4
124	2.3	2.4	2.6	2.7	2.9	3.0	3.2	3.4	3.8	3.9	4.1	4.2	4.6	4.9	5.2
126	2.2	2.3	2.5	2.6	2.8	2.9	3.0	3.3	3.6	3.7	3.9	4.1	4.4	4.7	5.0
128	2.15	2.25	2.4	2.5	2.7	2.75	2.9	3.2	3.5	3.6	3.7	4.0	4.3	4.6	4.8
130	2.05	2.2	2.3	2.4	2.6	2.7	2.8	3.1	3.3	3.5	3.6	3.8	4.2	4.4	4.7
132	2.0	2.1	2.2	2.35	2.5	2.65	2.7	3.0	3.2	3.4	3.5	3.7	4.0	4.2	4.5
134	1.95	2.05	2.15	2.3	2.4	2.6	2.65	2.9	3.1	3.3	3.4	3.6	3.9	4.1	4.4
136	1.90	2.0	2.1	2.2	2.35	2.5	2.6	2.8	3.0	3.2	3.3	3.5	3.8	4.0	4.2
138	1.80	1.95	2.0	2.15	2.3	2.4	2.5	2.75	2.9	3.1	3.2	3.4	3.6	3.9	4.1
140	1.75	1.9	1.95	2.1	2.2	2.3	2.4	2.65	2.8	3.0	3.1	3.3	3.5	3.8	4.0
142	1.7	1.8	1.9	2.05	2.15	2.25	2.35	2.6	2.75	2.9	3.0	3.2	3.45	3.6	3.9
144	1.65	1.75	1.85	2.0	2.1	2.2	2.3	2.5	2.7	2.8	2.9	3.1	3.35	3.55	3.8
146	1.6	1.7	1.8	1.95	2.05	2.15	2.2	2.4	2.65	2.75	2.8	3.0	3.25	3.45	3.7
148	1.55	1.65	1.75	1.9	2.0	2.1	2.15	2.35	2.6	2.7	2.75	2.95	3.2	3.35	3.6
150	1.5	1.6	1.7	1.85	1.95	2.05	2.1	2.3	2.55	2.6	2.7	2.9	3.1	3.3	3.5
152		1.55	1.65	1.8	1.9	2.0	2.05	2.25	2.45	2.55	2.65	2.8	3.0	3.2	3.4
154		1.5	1.6	1.75	1.85	1.95	2.0	2.2	2.4	2.5	2.6	2.75	2.95	3.1	3.3
156			1.55	1.7	1.8	1.9	1.95	2.15	2.35	2.4	2.55	2.7	2.9	3.0	3.2
158			1.5	1.65	1.75	1.85	1.9	2.1	2.3	2.35	2.45	2.65	2.8	2.95	3.15
160				1.6	1.7	1.8	1.85	2.05	2.25	2.3	2.4	2.6	2.75	2.9	3.1
162					1.65	1.75	1.8	2.0	2.2	2.25	2.35	2.5	2.7	2.85	3.0
164					1.6	1.7	1.75	1.95	2.15	2.2	2.3	2.45	2.65	2.8	2.95
166						1.65	1.7	1.9	2.1	2.15	2.25	2.4	2.6	2.75	2.9
168						1.6	1.7	1.85	2.05	2.1	2.2	2.35	2.55	2.7	2.8
170						1.55	1.65	1.8	2.0	2.05	2.15	2.3	2.5	2.65	2.75

Regression equation for age correction for females:
Females ($^*VO_2$ max $\times$ 0.302) $-$ (0.019 $\times$ Age) $+$ 1.593
where $^*VO_2$ max is the uncorrected value in litres/min ($l\,min^{-1}$).

To calculate the predicted VO₂ max from the tables

1. Find the end test HR (i.e. the averaged value for 5th and 6th minute) in the left-hand column
2. Read across to the end test work rate (WR)
3. Obtain the predicted VO_2 max value at the intersection of HR and WR. This value is in litres/min ($1\ min^{-1}$)
4. Enter the value into the appropriate regression equation to correct for age*.

The Twelve Minute Field Performance Test – 'Cooper's 12 Minute Run' (Cooper, 1968)

The subject is allowed to run or walk during this test. The objective is to cover as much distance as possible in a 12-minute period. This distance has been shown to correlate reasonably well ($r = 0.897$, where r = the correlation coefficient) with the VO_2 max measured during an incremental treadmill test (Spiro *et al.*, 1974) (Table 7.8).

The 'Rockport' One Mile Walk Test (Kline *et al.*, 1987)

The Rockport protocol is a submaximal field test which is used to estimate VO_2 max using a one-mile walk. Subjects are required to perform the walk as quickly as possible. During the walk, HR is monitored continuously and recorded for 1 minute at every one-quarter of a mile. The VO_2 max can then be predicted using the standard equation:

$$VO_2\, max = 6.9652 + (0.0091 \times WT) - (0.0257 \times Age)$$

$$+ (0.5955 \times Sex) - (0.2240 \times T_1)$$

$$- (0.0115 \times Average\ of\ HR\ 1-4)$$

where: WT = body weight in kg; T_1 = track walk time in minutes; HR = average heart rate calculated by adding HR readings obtained at every quarter of a mile and dividing by 4; Sex = male = 1, female = 0.

In the absence of a HR monitor, immediately on finishing the walk the HR should be taken for 15 seconds and multiplied by 4. This value is not an entirely accurate representation of the average HR, but may be used in the above equation.

Table 7.9 gives the 'Rockport' fitness charts for men and women of various ages. To use the charts:

1. Locate correct chart for the age and sex of subject.

---

* The standard regression equations which are located below Tables 7.6 and 7.7 for males and females should be used in conjunction with the nomogram, since they take into consideration the gradual decrease in maximal heart rate with age.

Table 7.8 Predicted maximal oxygen consumption on the basis of 12-minute performance (From Cooper, 1968, by permission)

Distance (miles)	Laps (1/4 mile track)	Maximal oxygen consumption (ml/kg/min)
< 1.0	< 4	< 25.0*
1.000	4	25.0*
1.030		26.0*
1.065	4.25	27.0*
1.090		28.2
1.125	4.5	29
1.150		30.2
1.187	4.75	31.6
1.220		32.8
1.250	5.0	33.8
1.280		34.8
1.317	5.25	36.2
1.340		37.0
1.375	5.5	38.2
1.400		39.2
1.437	5.75	40.4
1.470		41.6
1.500	6.0	42.6
1.530		43.8
1.565	6.25	45.0
1.590		46.0
1.625	6.5	47.2
1.650		48.0
1.687	6.75	49.2
1.720		50.2
1.750	7.0	51.6
1.780		52.6
1.817	7.25	53.8
1.840		54.8
1.875	7.5	56.0
1.900		57.0
1.937	7.75	58.2
1.970		59.2
2.000	8	60.2

2. On the horizontal axis draw a vertical line from the time taken to do the walk.
3. On the vertical axis draw a horizontal line from the averaged heart rate at the end of the test.
4. The bisection of the two lines is indicative of the cardiovascular fitness level.

The Canadian Home Fitness Test

Developed in the mid-1970s by Bailey *et al*. (1974, 1976), the Canadian Home Fitness Test (CHFT) is widely used in Canada to screen for

Table 7.9 Fitness charts for men and women at various ages (from American College of Sports Medicine, 1992, by permission)

Age   Men

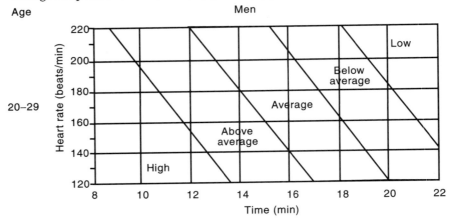

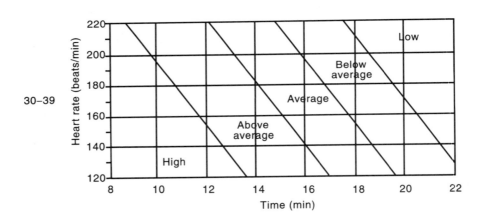

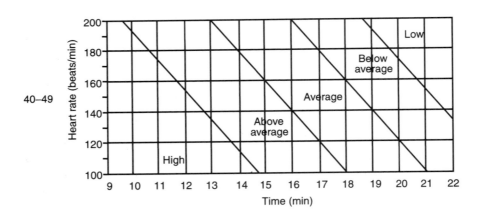

Table 7.9 *Cont*

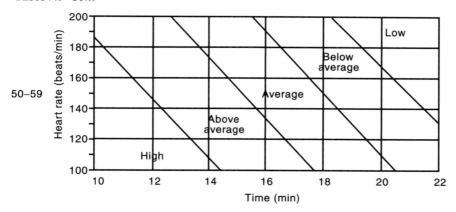

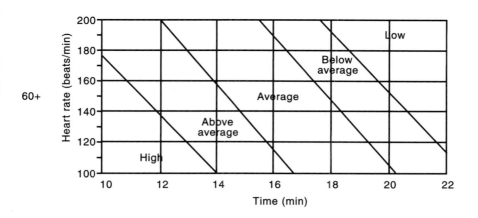

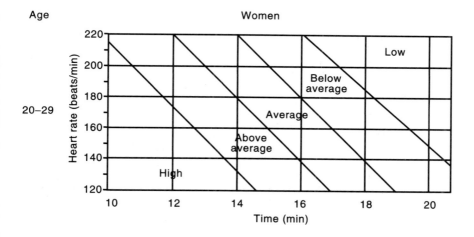

Table 7.9 *Cont*

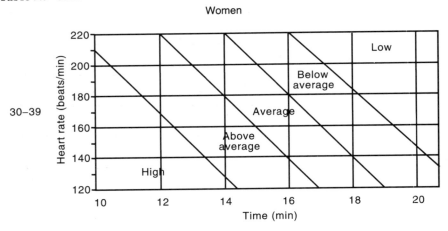

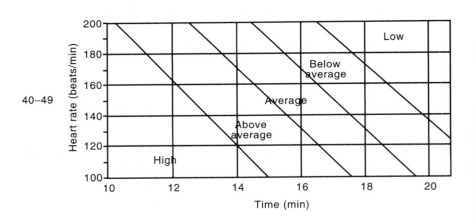

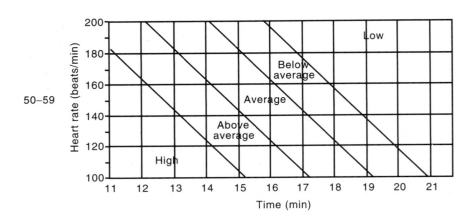

Table 7.9 *Cont*

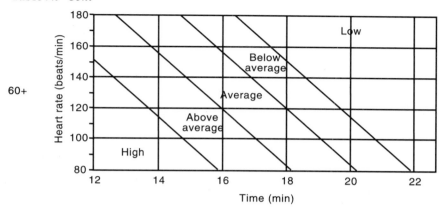

fitness*. It consists of stepping up and down two steps for a prescribed period of time, after which the pulse is taken. If the pulse rate falls within the charted age group/sex limits, the subject can progress to the next stage. If not he/she stops and their 'fitness score' is read from the chart from which VO_2 max can be predicted.

Tests which measure endurance performance

The Harvard Step Test

Step tests provide a simple method of using the HR to evaluate cardiovascular fitness. They are easy to perform and can provide a 'bench mark' for future evaluation following an aerobic training pro-gramme. The Harvard Step Test was originally designed to select physically fit army recruits (Clarke, 1976).

Method
The subject is required to step up onto a stool first with one foot and then with the other so that both feet are on the stool, and then to step down so that both feet contact the ground. The height of the stool should be 20 in (51 cm). Thirty step ups are performed per minute and continued for 5 minutes. A metronome can be used to assist the subject to maintain the required stepping rate.

The test can also be performed using a stool height of 18 in (46 cm), a 30 steps per minute rate and a test duration of 4 minutes.

The pulse rate is counted from $1-1\frac{1}{2}$, $2-2\frac{1}{2}$ and $3-3\frac{1}{2}$ minutes after stopping the test. Using the formula below, the physical efficiency

---

* Full details of this test can be obtained from: Fitness and Amateur Sport Canada, 365 Laurier Avenue West, Ottawa, Ontario, Canada K1A 0X6.

index (PEI) can be calculated (Brouha *et al.*, 1944) (Table 7.10). This remains the same for both protocols.

$$PEI = \frac{\text{Duration of exercise (sec)} \times 100}{2 \times \text{Sum of pulse counts in recovery period}}$$

This test can also be used by members of a 'keep fit' or step aerobics class as a progress assessment. A gradual decrease in the resting pulse over a period of weeks is indicative of an increasing cardiovascular fitness.

The Twelve Minute Walking Test (McGavin *et al.*, 1976)

This test is often used in the hospital situation (corridor test) to assess level of fitness of respiratory patients. It is similar in design to the field performance test described above, in that it involves instructing the subject to walk as far as possible in a 12-minute period. The patient is told to try to keep going, but not to worry if they have to slow down or stop for a rest. The main aim is for the patient to feel that at the end of the test they could not have walked any further in the available time. A note of the final distance walked is recorded by the therapist and used to form a baseline measure against which subsequent attempts can be measured. A weak but significant correlation ($r = 0.52$) has been found between walking distance and VO_2 max for a group of patients with lung disease (Bradley *et al.*, 1976).

Disadvantages of the corridor walking test
- it does not provide information about physiological and symptomatic changes during exercise
- it is difficult to standardize the test
- the test result may be influenced by motivation and encouragement (Guyatt *et al.*, 1984)
- the test does not have an incremental facility
- obstacles in the corridor may hinder the test.

Table 7.10 Classification of fitness from PEI scores (Brouha *et al.*, 1944)

PEI scores	Physical condition
Below 55	Poor (P)
55–64	Low (L)
65–79	High average (HA)
80–89	Good (G)
Above 90	Excellent (E)

The Self-paced Treadmill Walking Test

Beaumont *et al*. (1985) describe a self-paced treadmill walking test in which patients are required to perform the corridor test by walking on the treadmill. Distances covered, stride length and walking speed are similar for both tests. These authors suggest that the treadmill may be better for assessing respiratory patients because it enables physiological evaluation to take place during the test. However, there are a number of disadvantages associated with its use and these have been outlined earlier in Table 7.2.

The Shuttle Walking Test

The Shuttle Walking Test, modified by Singh *et al*. (1992), is currently a popular method used for assessing exercise tolerance in respiratory patients: 'It is a standardized incremental field walking test which provokes a symptom-limited maximal performance. It provides an objective measurement of disability and allows direct comparison of patients' performance' (Singh *et al*., 1992).

The test is progressive and externally paced and is based on the 20 m Shuttle Running Test, described by Leger and Lambert (1982), which is used to evaluate functional capacity in athletes.

Method
1. The patient is required to walk up and down a 10 m course (Figure 7.6).
2. The speed of walking is dictated by a computer-generated audio signal played on a tape recorder (tape available from Dr Singh, Dept of Respiratory Medicine, The Glenfield Hospital, Groby Rd, Leicester LE3 9QP).
3. The patient is instructed to walk at a steady pace and to aim to turn around on hearing the signal.

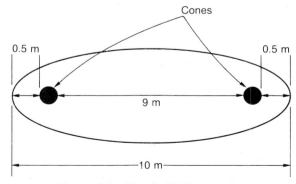

Fig. 7.6 The Shuttle Walking Test

Table 7.11 The modified shuttle walk protocol (From Singh *et al.*, 1992, by permission)

Level	No. of shuttles per level	Speed (m/sec)	Speed (mph)
1	3	0.50	1.12
2	4	0.67	1.50
3	5	0.84	1.88
4	6	1.01	2.26
5	7	1.18	2.64
6	8	1.35	3.02
7	9	1.52	3.40
8	10	1.69	3.78
9	11	1.86	4.16
10	12	2.03	4.54
11	13	2.20	4.92
12	14	2.37	5.30

4. The speed of walking is increased by a small increment at 1-minute intervals.
5. If the patient arrives at the end of the shuttle early, they are instructed to wait until they hear the audio signal which signifies that they can continue.
6. The tester observes the test but gives no verbal encouragement except to instruct the patient at each minute to increase walking speed slightly.
7. The test ends if: (a) the patient is too breathless to maintain the required speed; (b) the subject fails to complete a shuttle in the set time (i.e. is more than 0.5 m away from the cone when the bleep sounds); (c) 85% of the predicted maximum HR is reached. This is calculated according to the formula $[210 - (0.65 \times age)]$.

The methodology of the shuttle test is supported by the underlying philosophy that most day-to-day activities are of an irregular nature and rarely reach a steady state (Spiro, 1977). Consequently, patients' symptoms are more likely to be revealed during an incremental test than in a self-paced exercise test of a single work rate (Table 7.11) (Singh *et al.*, 1992).

PRESCRIBING AEROBIC EXERCISE FOR HEALTH RELATED FITNESS FROM TEST RESULTS

This section will explain how to use the predicted VO_2 max values obtained in the above tests to prescribe exercise.

The validity of submaximal tests has been questioned (Astrand and Rodahl, 1986), but assuming that they do give an indication of a person's exercising capability they can generate reliable test–retest data which can be used for prescribing exercise.

The protocols can also be used effectively to assess subjective responses, particularly of sedentary individuals or patients to incremental exercise.

Based on the fact that treadmills and exercise cycles are becoming increasingly available within physiotherapy departments, the following charts have been adapted by Buckley (1994) from the relationships described by Astrand and Rodahl (1986) and Williams (1985). Both Williams (1985) and Astrand and Rodahl (1986) described a direct relationship with oxygen consumption (VO_2) and running speed, and Astrand and Rodahl (1986) showed a similar relationship between exercise cycling and oxygen consumption (VO_2).

The charts can be used with any protocol that either measures or predicts VO_2 max, based on the relationship (Williams, 1985; Astrand and Rodahl, 1986) that for every 1 kilometre per hour ($km\ h^{-1}$) of walking or running 3–3.5 $ml\ kg^{-1}\ min^{-1}$ O_2 are required, and for exercising on a cycle-ergometer at 70 W, 1 $l\ min^{-1}$ of oxygen is required. The above holds true if 7–7.5 $km\ h^{-1}$ is used as the transition pace for walking to running (Table 7.12).

Astrand and Rodahl (1986) state that the sedentary individual will be able to sustain prolonged activity (more than 10 min) at around 55–65% of VO_2 max, moderately active individuals between 65% and 75% of their VO_2 max and highly trained individuals up to 90% of their VO_2 max. Therefore, once the VO_2 max is determined, Tables 7.13 and 7.14 can be used for prescribing walking/running/exercise cycling pace.

Using the heart rate to prescribe exercise

Exercise prescription for HR and end performance measurements tend to rely on trial and error and therefore may require more time for assessment and prescription.

Submaximal exercise tests can be stopped when a predetermined percentage of the maximal predicted HR is reached. This assumes, however, that HR increases linearly with oxygen uptake or work rate, but this may be inaccurate for some subjects since there are many individual variations of maximal HR within a given age group.

Table 7.12 Subjective descriptions of pacing

Subjective pace rating	Actual pace ($km\ h^{-1}$)
Good strolling walk	3.0–5.0
Brisk walk	5.5–7.0
Slow jog	7.5–8.0
Moderate jog	8.5–10
Running	10.5 +

Table 7.13 Prescribed walking, jogging and running paces in relation to VO_2 max

VO_2 max ($ml\,kg^{-1}\,min^{-1}$)	Fitness index (pace I– sedentary) (55–65% VO_2 max) ($km\,h^{-1}$; min/mile)	Fitness index (pace II– moderately active) (65–75% VO_2 max) ($km\,h^{-1}$; min/mile)	Fitness index (pace III– trained) (75–90% VO_2 max) ($km\,h^{-1}$; min/mile)
20–25	4.2 (23.00)	5.0 (19.24)	5.4 (17.54)
26–30	5.5 (17.36)	6.5 (14.54)	7.0 (13.48)
31–35	6.5 (14.53)	7.7 (12.34)	8.3 (11.40)
36–40	7.5 (12.54)	9.0 (10.45)	9.7 (9.58)
41–45	8.6 (11.15)	10.2 (9.29)	11.1 (8.43)
46–50	9.6 (10.05)	11.5 (8.25)	12.4 (7.48)
51–55	10.7 (9.2)	12.3 (7.37)	13.7 (7.04)
56–60	11.7 (8.16)	14.0 (6.55)	15.0 (6.27)
61–65	12.8 (7.34)	15.2 (6.22)	16.5 (5.52)
66 +	13.8 (7.01)	16.5 (5.52)	17.8 (5.26)

Table 7.14 Prescribed exercise cycling pace in relation to VO_2 max

VO_2 max ($l\,min^{-1}$)	Fitness index (pace I– sedentary) (55–65% VO_2 max) (Watts)	Fitness index (pace II– moderately active) (65–75% VO_2 max) (Watts)	Fitness index (pace III– trained) (75–90% VO_2 max) (Watts)
0.9–1.1	36	41	44
1.2–1.4	43	50	53
1.5–1.7	65	75	80
1.8–2.0	84	96	102
2.1–2.3	97	112	119
2.4–2.7	111	127	136
2.8–3.0	129	149	159
3.1–3.3	143	165	176
3.4–3.6	157	181	192
3.7–3.9	171	196	210
4.0–4.2	185	213	227
4.3–4.5	195	226	240
4.6–4.8	209	241	257
4.9–5.1	223	257	274
5.2+	236	273	291

Target heart rate

The target training HR is defined by the lower and upper HR limits for aerobic exercise and can be calculated as follows (American College of Sports Medicine, 1992):

1. Calculate predicted maximum HR using (220 − Age).
2. Multiply by 0.6 to give the *lower* limit for aerobic exercise.
3. Multiply by 0.8 to give the *upper* limit for aerobic exercise.

By reaching and maintaining the target HR for 20–60 minutes, 3–5 days per week, physical and mental benefits from exercise can be obtained.

The calculations below are commonly used in conjunction with the HR target zone chart (Figure 7.7) to enable individuals to achieve a training HR.

Predicting the maximal heart rate

There are two recognized calculations for predicting maximal HR. These are described below:

1. $HR_{max} = 210 - [0.65 \times Age (yr)]$ (Lang-Anderson *et al.*, 1971)
2. $HR_{max} = 220 -$ Subject age (McArdle *et al.*, 1994)

The Borg scale

The Borg scale of perceived exertion (Borg, 1982) is a visual analogue scale which has been shown to correlate highly with HR and other cardiorespiratory and metabolic values such as VO_2 max and blood

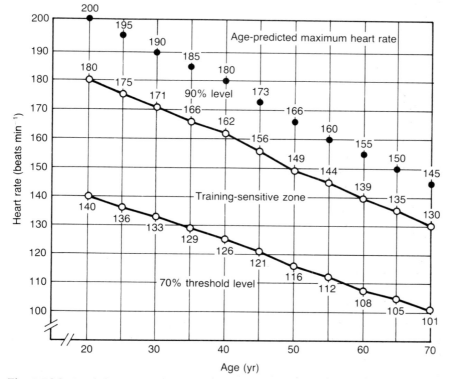

Fig. 7.7 Maximal heart rates and training-sensitive zone for use in aerobic training programmes for men and women of different ages (from McArdle *et al.*, 1994, by permission)

Human Movement Explained

lactate concentration (American College of Sports Medicine, ACSM, 1986).

It is a simple guide to progress during testing and is used by the subject to indicate their perceived level of exertion, i.e. how hard they are finding the exercise. It is a numerical scale ranging from 6 to 20, with verbal descriptors at every odd number.

When working at a perceived rate of exertion (PRE) of 12, the heart is functioning at 60% of maximum. At a PRE of 15 this increases to a functioning level of 90% of maximum.

The Borg scale (Figure 7.8) is often used to prescribe exercise intensity for endurance training which should ideally by performed at a PRE of between 12 and 15 (ACSM, 1986).

Predicting VO$_2$ max from age

The VO$_2$ max can also be predicted from age for subjects of 20 years and above using the equations below, but they serve only as a guide because they do not take into account individual responses to exercise (Jones and Campbell, 1982). This can often be a useful calculation prior to assessing an individual for the purposes of determining test workload.

Males $VO_2 max = 4.2 - 0.032 \times Age (yr) = l\,min^{-1}\ (SD \pm 0.4)$

$VO_2 max = 60 - 0.55 \times Age (yr) = ml\,kg^{-1}\,min^{-1}\ (SD \pm 7.5)$

Females $VO_2 max = 2.6 - 0.014 \times Age (yr) = l\,min^{-1}\ (SD \pm 0.4)$

$VO_2 max = 48 - 0.37 \times Age (yr) = ml\,kg^{-1}\,min^{-1}\ (SD \pm 7.0)$

6	
7	Very, very light
8	
9	Very light
10	
11	Fairly light
12	
13	Somewhat hard
14	
15	Hard
16	
17	Very hard
18	
19	Very, very hard
20	

Fig. 7.8 The Borg scale of perceived exertion (from Borg, 1982, by permission)

SUMMARY

- Fitness testing can be performed in relation to the three energy producing systems: the anaerobic power, anaerobic endurance and aerobic systems.
- Maximal oxygen uptake (VO_2 max) is a widely used parameter for predicting aerobic fitness.
- Sub-maximal tests are more commonly used in a non-athletic population to ascertain fitness levels.
- Exercise prescription can be performed based on the data obtained from fitness test results.

REFERENCES

American College of Sports Medicine (1986) *Guidelines for Exercise Testing and Prescription*, 3rd edn, Lea and Febiger, Philadelphia

American College of Sports Medicine (1992) *Fitness Book*, Leisure Press, Champaign, Illinois

Anderton, P. and Bigg. P.H. (1972) *Changing to the Metric System*, HMSO, London

Astrand, P.-O. and Rodahl, K. (1986) *Textbook of Work Physiology: Physiological Bases of Exercise*, 3rd edn, McGraw-Hill, New York

Bailey, D.A., Shephard, R.J. and Mirwald, R.L. (1976) Validation of a self-administered home test of cardio-respiratory fitness. *Canadian Journal of Applied Sports Sciences*, 1, 67–78

Bailey, D.A., Shephard, R.J., Mirwald, R.L. and McBride, G.A. (1974) Current levels of Canadian cardio-respiratory fitness. *Canadian Medical Association Journal*, 111, 25–30

Bar-Or, O., Dotan, R., Inbar, O. et al. (1980) Anaerobic capacity and muscle fibre type distribution in man. *International Journal of Sports Medicine*, 1, 89–92

Beaumont, A., Cockcroft, A. and Guz, A. (1985) A self paced treadmill walking test for breathless patients. *Thorax*, 40, 459–464

Borg, G.A.V. (1982) Psychological basis of perceived exertion. *Medicine and Science in Sports and Exercise*, 14(5), 377–387

Bradley, C.A., Harris, E.A., Seelye, E.R. and Whitlock, R.M.L. (1976) Gas exchange during exercise in healthy people. 1. The physiological dead-space volume. *Clinical Science and Molecular Medicine*, 51, 323–333

British Sports Council and the Health Education Authority (1992) Allied Dunbar National Survey

Brouha, L., Fradd, W. and Savage, M. (1944) Studies in physical efficiency of college students. *Research Quarterly*, 15(3), 211–224

Bruce, R.A. (1971) Exercise testing of patients with coronary heart disease. *Annals of Clinical Research*, 3, 323

Buckley, J. (1994) Personal communication, *Lifestyle Fitness*, Shrewsbury, Shropshire

Clarke, H.H. (1976) *Application of Measurement – To Health and Physical Education*, Prentice-Hall, New York

Cooper, K.H. (1968) A means of assessing maximal oxygen uptake. *Journal of the American Medical Association*, 203, 201–204

Davies, C.T.M. and Sargeant, A.J. (1975) Effects of training on the physiological responses to one and two leg work. *Journal of Applied Physiology*, 38(3), 377–385

Davison, R.C.R. and Grant, S. (1993) Is walking sufficient exercise for health? *Sports Medicine*, **16**(6), 369–373

Donald, K.W., Bishop, J.M., Cumming, C. and Wade, O.L. (1955) The effect of exercise on the cardiac output and central dynamics of normal subjects. *Clinical Science*, **14**, 37–73

Ekblom, B. (1972) Response to exercise after blood loss and reinfusion. *Journal of Applied Physiology*, **33**, 175–180

Ekblom, B. (1976) Central circulation during exercise after venesection and reinfusion of red blood cells. *Journal of Applied Physiology*, **40**, 379–383

Ekblom, B., Huot, R., Stein, E. and Thorstensson, A. (1975) Effect of changes in arterial O_2 content on circulation and physical performance. *Journal of Applied Physiology*, **39**, 71–75

Gollnick, P., Armstrong, R., Saltin, B., Saubert, C., Sembrowich, W. and Shepherd, R. (1973) Effect of training on enzyme activity and fibre composition of human muscle. *Journal of Applied Physiology*, **34**, 107–111

Guyatt, G.H., Berman, L.B., Townsend, M., Pugsley, S.O. and Chambers, L.W. (1984) A measure of the quality of life for clinical trials in chronic lung disease. *Thorax*, **42**, 773–778

Hardman, A.E. and Hudson, A. (1989) Walking for health – a closer look at exercise. *Health Trends* (Department of Health), **21**(3), 91–92

Hillegass, E. (1989) In Scully, R.M. and Barnes, M.R. (1989) *Physical Therapy*, J.B. Lippincott, Philadelphia

Holly, R.G. (1988) *American College of Sports Medicine Resource Manual for Guidelines for Exercise Testing and Prescription*, Lea and Febiger, Philadelphia

Ivy, J.L., Withers, R.T., Van Handel, P.J., Elger, D.H. and Costill, D.L. (1980) Muscle respiratory capacity and fibre type as determinants of the lactate threshold. *Journal of Applied Physiology*, **48**(3), 523–527

Jones, N.L. and Campbell, E.J. (1982) *Clinical Exercise Testing*, W.B. Saunders, Philadelphia

Kline, G.M., Porcari, J.P., Hintermeister, R. *et al*. (1987) Estimation of VO_2 max from a one-mile track walk, gender, age and body weight. *Medicine and Science in Sports and Exercise*, **19**(3), 253–259

Klissouras, V. (1971) Genetic aspects of physical fitness. In *Physical Fitness*, Proceedings of a Satellite Symposium of the XXV International Congress of Physiological Sciences, Prague, pp. 217–223

Lakomy, H.K.A (1984) An ergometer for measuring the power generated during sprinting. *Journal of Physiology*, **33**, 354

Lang-Anderson, K., Shephard, R.J., Denolin, H. *et al*. (1971) *Fundamentals of Exercise Testing*, World Health Organisation, Geneva

Leger, L.A. and Lambert, J.A. (1982) A maximal multistage 20-m shuttle run test to predict VO_2 max. *European Journal of Applied Physiology*, **49**, 1–12

McArdle, W.D., Katch, F.I. and Katch, V.L. (1994) *Essentials of Exercise Physiology*, Lea and Febiger, Philadelphia

McGavin, C.R., Gupta, S.P. and McHardy, G.J.R. (1976) Twelve minute walking test for assessing disability in chronic bronchitics. *British Medical Journal*, **1**, 822–823

Rhodes, E.C. and Loat, C.E.R. (1993) Relationship between the lactate and ventilatory threshold during prolonged exercise. *Sports Medicine*, **15**(2), 104–115

Secher, N.H., Clausen, J.P., Klausen, K., Noer, I. and Trap-Jensen, J. (1977) Central and regional circulatory effects of adding arm exercise to leg exercise. *Acta Physiologica Scandinavica*, **100**(3), 288–297

Shephard, R.J. (1991) The Canadian Home Fitness Test. *Sports Medicine*, **11**(6), 358–366

Singh, S.J., Morgan, M.D.L., Scott, S., Walters, D. and Hardmann, A.E. (1992) Development of a shuttle walking test of disability in patients with chronic airways obstruction. *Thorax*, **47**, 1019–1024

Spiro, S.G. (1977) Exercise testing in clinical medicine. *British Journal of Diseases of the Chest*, **71**, 145–172

Spiro, S.G., Juniper, E., Bowman, P. and Edwards, R.H.T. (1974) An increasing work rate test for assessing the physiological strain of submaximal exercise. *Clinical Science and Molecular Medicine*, **46**, 191–206

Wasserman, K., Hansen, J.E., Sue, D.Y. and Whipp, B.J. (1986) *Principles of Exercise Testing and Interpretation*, Lea and Febiger, Philadelphia

Wasserman, K., Whipp, B.J., Koyal, S.N. and Beaver, W.L. (1973) Anaerobic threshold and respiratory gas exchange during exercise. *Journal of Applied Physiology*, **35**(2), 236–243

Williams, C. (1985) Nutritional aspects of exercise induced fatigue. *Proceedings of the Nutritional Society*, **44**, 245–256

8.   *Group exercise*

Patients at similar stages of rehabilitation and with a common disability often benefit from working together in an exercise class. It is usual to attend the physiotherapy department for either a preliminary session of individual treatment or for assessment of suitability prior to joining a specified class. During the class patients will undergo an exercise programme which has been specifically designed to meet their needs of continued rehabilitation.

This chapter will outline:

- advantages of group exercise
- disadvantages of group exercise
- role of the physiotherapist in group exercise
- planning an exercise class
- safety aspects in group exercise
- examples of common exercise classes: lower limb, back and wrist (Colles' fracture)
- examples of specific exercise which can occur in group situations: postoperative, antenatal, postnatal
- design of the home exercise programme
- circuit training.

ADVANTAGES OF GROUP EXERCISE

1. Encourages patients to take responsibility for their own rehabilitation, i.e. to take a more active rather than passive role. Patients tend to develop a greater self-reliance when no longer the subject of individual attention.
2. Encourages patients to work with others and not to set themselves apart because of their disability.
3. Introduces an element of controlled competition, as patients compare their progress to that of others with similar disabilities.
4. Restores patient confidence in the injured body part.
5. Enables patients to forget their disability temporarily, especially

when joining in with team games which are often a part of the well-designed class.

6. Allows emphasis to be placed on achievement rather than on specific movements or muscle strengthening.
7. Facilitates patient independence by providing a home exercise programme.
8. Improves patient motivation and self-esteem. Patients who leave a successful class are likely to be motivated to continue with a home exercise programme.
9. Motivation to achieve promotion to a more advanced group.
10. Time is saved by the physiotherapist if several patients can be treated together effectively.

DISADVANTAGES OF GROUP EXERCISE

There are certain instances when group exercise is not appropriate. A group exercise programme will never suit all patients equally. A standard programme can be used initially but it should be tailored to individual needs as soon as possible. This can be achieved by arranging a shortened group exercise programme at the start of the session and then allowing patients to work on their own routines.

The main disadvantages of group exercise are:

1. Incorrect selection of patients, e.g. too wide an age range or patients at very different stages of rehabilitation.
2. The mental attitude of the patient may be unsuitable for inclusion in the group.
3. Loss of control by the therapist, e.g. if a group is too large, the therapist may fail to lead the group.
4. Competition may reduce the confidence of some patients.
5. Failure to consider the needs of individual patients.
6. There can be a tendency to use the exercise class as a replacement for individual treatment because it takes less time and frees the physiotherapist to treat more patients.

ROLE OF THE PHYSIOTHERAPIST IN GROUP EXERCISE

The success of an exercise class depends upon the ability of the physiotherapist to control and motivate the group. Some therapists will never feel comfortable taking a class, whereas others will thrive on the challenge. By following the simple guidelines below, it should be possible to successfully master at least one group exercise session!

The physiotherapist should:

- ensure that he/she introduces him/herself to the group
- plan and prepare the class in advance

- conduct him/herself in a professional manner when taking the class
- give clear, concise instructions for each exercise
- tell the patient what to do rather than ask them
- demonstrate difficult exercises
- set a specific achievable number of repetitions for each exercise
- give clear, concise commands such as 'begin', 'stop'
- correct individuals who are performing exercise regimens incorrectly
- recognize patients who are failing to keep up with the group
- give encouragement and praise where due
- encourage a happy and yet hard-working environment
- sort and test all apparatus before beginning the class
- ensure that he/she moves around the room to supervise all of the class, not just those who are closest.

PLANNING AN EXERCISE CLASS

Careful planning, together with a professional but inspired rendition, will usually result in the delivery of a successful class. The well designed class should last between 20 and 30 minutes and contain the following components:

1. Initial welcome and explanation of class aims and objectives.
2. A warm-up period (5 minutes).
3. Gentle exercises which are specific to the injured body part (approximately 4 different exercises).
4. More difficult exercises which are graded according to the level of rehabilitation (approximately 4 different exercises).
5. General functional exercises (e.g. team game or gait re-education).
6. A cool down period (5 minutes).
7. Home exercise reminder.
8. Announcement of date and time of next class.

A variety of exercises should be used to re-educate strength, endurance, balance and co-ordination. The failed group exercise programme is the programme which does not work the patient sufficiently hard and incorporates too many enjoyable but ineffective activities.

SAFETY ASPECTS IN GROUP EXERCISE

Class discipline is essential, particularly during advanced classes. Classes should ideally be limited to between 8 and 10 participants. However this will obviously depend on factors such as room size, patient availability and type of class.

Patients selected for a specific exercise class should be medically cleared prior to participation. If in doubt following initial assessment, medical advice should be sought.

All apparatus should be tested before the class begins. Apparatus which is not being used during the class should be placed well away from the exercise area. When finished with, it should not be left on the floor where patients can trip over it.

Before joining the class, some patients will benefit from seeing a similar class so that they will know what to expect. They should also be encouraged to bring suitable clothing and footwear for the class.

TYPES OF EXERCISE CLASS

Classes are usually named according to their function. The most common general exercise classes led by the physiotherapist on the ward or in the physiotherapy department are the lower limb classes, the back class and the Colles' fracture class. Examples of exercises which are suitable for each of these classes are provided below. However, it should be remembered that the list is by no means exhaustive and that there are plenty of other exercises which are very suitable for inclusion in group exercise sessions.

The lower limb exercise class

Rehabilitation following lower limb injury frequently involves the patient attending a lower limb exercise class. These classes are designed to accommodate patients at various stages of recovery and are referred to as early, intermediate or late leg classes.

The precise definition of early, intermediate and late classes will differ between physiotherapy departments. For example, an early leg class in one centre will imply that the patient will be non weight-bearing, whereas, at another the early class will contain patients who are partially weight-bearing (PWB).

The main aim of these classes is to restore function. Therefore exercise should, where possible, be functionally oriented and contain a selection of open and closed kinetic chain exercises which target strength, flexibility, balance, proprioception and co-ordination.

Patients may join the early leg and progress through to the late leg class before being discharged or they might just join the advanced (late) rehabilitation class. Allocation to a class will depend on diagnosis, stage of rehabilitation and patient suitability as judged by the physiotherapist. Examples of exercises which are deemed suitable for the three classes are described below.

All classes should be preceded by a warm up period which includes total body range of movement and muscle stretches.

Early leg class

A patient who would be suitable for the early leg class outlined below could have had a recent minor lower limb surgical procedure and the post-operative instruction is to mobilize partially weight-bearing (PWB) using crutches.

Following assessment, treatment objectives will be defined for the patient by the physiotherapist and these might be to:

(a) achieve full range of motion at all lower limb joints
(b) ensure that muscle control around the joints is regained
(c) achieve balance and stability when using crutches
(d) learn how to achieve a normal PWB gait pattern

These objectives could be achieved using the following exercises:

1. Lying – ankle plantar flexion, dorsiflexion and circumduction.
2. Lying – static quadriceps and gluteal contractions.
3. Lying – straight leg raise (SLR).
4. Sitting – raising buttocks from bed by pushing down onto bed with arms.
5. Sitting in bed – SLR.
6. Sitting in bed – SLR with small weight with/without using a thigh wedge/quads block.
7. Sitting in bed – lower limb circumduction in a clockwise and anticlockwise direction.
8. Gym-based – gait re-education.
9. Standing by wall bars – alternate hip hitching.
10. Standing by wall bars – both knees bend.
11. Standing by wall bars – toe standing alternating with heel standing.
12. Team game – walking with crutches around an obstacle course.

Intermediate class

When the patient is ready to progress from the early class he may attend the intermediate class. Progression to this class will be gauged by having achieved the objectives set on entry to the early leg class.

The objectives of the intermediate class will be similar to those for the early class but the emphasis this time will be on restoring normal function, increasing weight bearing and gait re-education. The exercise class described below therefore contains a selection of exercises which are designed to restore flexibility, balance, proprioception, strength and co-ordination in the lower limb, since all of these properties are needed for the restoration of normal function.

1. Standing – marking time.
2. Standing – rock walking forwards onto toes and backwards onto heels.
3. Standing – walking in a circle with regular change in clockwise and anticlockwise direction.

4. Standing – walking in a small defined area with random change in direction.
5. Standing – bent knee walking.
6. Standing – ski sit against a wall (making sure the heels remain in contact with the floor).
7. Standing – controlled step-ups using a low bench.
8. Standing – balance on right leg then on left leg.
9. Sitting – static quadriceps progressing to straight leg raise – hold for 10 sec and lower slowly.
10. Sitting – straight leg raising using small weight.
11. Prone lying – stretch affected quadriceps by flexing knee with un-affected leg.
12. Team game – sitting astride a small blanket on the floor – blanket walking race forwards and backwards.

Late leg class

The final stage of rehabilitation should be designed to ensure that the patient is able to return to and cope with the demands of their job and lifestyle. The exercises must therefore be designed with these object-ives in mind.

The exercises described below are typical examples for a young patient who needs to get back to an active job and sporting lifestyle.

1. Standing – gentle jogging on the spot – progress speed.
2. Standing – alternate fast high knee jogging with gentle normal jog.
3. Standing – jumping on the spot – high jump alternating with normal jump.
4. Standing – hopping on the spot – progress to hopping around the room.
5. Standing – zigzag by jumping to right and left of a straight line.
6. Standing – skipping.
7. Standing – star jump.
8. Standing – step-ups at speed.
9. Standing – unilateral stance on tip toe.
10. Sitting – straight leg raise with moderate weight – raise, hold for 10 sec and lower slowly.
11. Crouch walking.
12. Remedial team games, e.g. crab football, bunny hopping relay race.

Exercise and rehabilitation programmes for the back

Patients with acute low back pain are often referred to the physio-therapy department by their GP or orthopaedic consultant. Here they undergo various treatment regimes which, depending on thorough assessment, may include manual therapy, electrotherapy, education and exercise.

When the acute pain is over, emphasis tends to shift from 'hands on' modalities towards strengthening, mobilizing and education by teaching, for example, graded exercise, correct lifting techniques and postural awareness. At this stage patients often attend back school where emphasis is placed on back care education, so that the risks of further injury are minimized.

Back school

The back school approach initially began in Sweden and has been in existence since 1969 (Hayne, 1984). Its main purpose is to educate back care. In a controlled trial, Klaber Moffett found the back school to be better than just a traditional exercise class (Klaber Moffett *et al.*, 1986).

Components of schools vary, but generally they include exercises (see below) and biomechanical and ergonomic advice. A variety of patients can attend the back school, e.g. housewives, young mothers, office professionals and manual workers. Some will have had surgery, others will have had back pain for many years, whilst for many this will be their first episode of pain.

Traditionally those attending the school were referred as outpatients with chronic low back pain by their GP or consultant, but now patients in the acute stage can also attend schools which are specifically designed to meet their needs and are often run in the ward setting (White, 1983).

Exercise
Specific exercises may be prescribed for each patient according to assessment findings. For example, the patient with a disc problem may initially require exercises which emphasize extension, because lumbar flexion can increase intra-discal pressure posteriorly (McKenzie, 1981). Spinal discs are also weakest postero-laterally and it is in this direction that a prolapse tends to occur (McKenzie, 1981).

Later on, controlled flexion exercises may be incorporated into the programme so that the spine is provided with its natural corset of balanced muscles – the abdominals and lumbar extensors.

Examples of exercises:
1. Prone lying – arms by sides. Raise right leg off floor by contracting hip extensors, hold for 5 sec and then lower slowly. Repeat with left leg.
2. Supine lying – knees bent, feet flat on the floor. Roll knees to the right, return to the vertical and repeat to the left.
3. Supine lying – knees bent, feet flat on the floor. Tilt pelvis posteriorly by flattening lumbar spine onto the floor.
4. Supine lying – knees bent, feet flat on the floor, hands resting on anterior aspect of thighs. Tuck chin in and slide hands up thighs towards knees. Avoid full sit-up.
5. Prone kneeling – hollow and arch (hump) spine.

Functional restoration programmes

Functional restoration programmes cater mainly for patients with chronic low back pain. Many centres run these or work-hardening programmes which are aimed at decreasing illness behaviour and reliance on hospital staff and increasing the patients' ability to manage their own pain. Examples of such programmes are INPUT (de C. Williams, 1993) at St. Thomas's Hospital and the Oxford Rehabilitation Programme (Frost and Klaber Moffett, 1995).

There is no standard programme. All are different in design, but their philosophies are similar, i.e. to involve the patient in a complete rehabilitation programme. The main contents of this type of programme are:

- promotion of independence by increasing confidence and encouraging patients to lead an active life in spite of their pain
- general fitness, e.g. circuits, flexibility, strength, balance and co-ordination
- hydrotherapy
- education and advice, e.g. back care, relaxation, employment
- psychological support, e.g. professional and from other patients.

Wrist (Colles' fracture) class

Patients attending a Colles' fracture class will have just had the plaster removed from their wrist. They will have been assessed and probably treated on an individual basis prior to joining the class.

The class will be designed to assist the patient to regain range of movement, to improve strength and to restore normal function. Typically it is often made up of middle-aged and elderly females who have sustained their fracture by falling on the outstretched hand.

Examples of exercises which are suitable for inclusion in a wrist class:

1. Sitting – elbows flexed to 90° and stabilized against the trunk. Flex, extend and circumduct both wrists.
2. Sitting as above. Palms together forwards. Alternately point fingers to the ceiling and then to the floor.
3. Sitting as above. Palms together forwards. Slowly abduct elbows to produce stretch on anterior apsect of wrist.
4. Standing – hands placed palms down onto a table. Gradually and with care apply an increased force through the upper limbs by leaning forwards onto hands.
5. Limb bath with warm water – place lower arm in the bath so that the wrist is fully submerged. Encourage circumduction and flexion, extension, abduction and adduction.
6. Wax therapy – apply wax to wrist and hand for 10 minutes and then use wax to provide increasing resistance to exercise as it cools.
7. Encourage functional exercises such as window cleaning and dusting as part of the home exercise programme.

SPECIALIST EXERCISE

The following patient groups are frequently attended by the physiotherapist and therefore a brief overview is provided in this chapter,
although it should be recognized that these instances do not constitute
routine group exercise.

Postoperative exercise class

Until recently, postoperative exercise classes were performed routinely
on surgical wards. However, physiotherapists are now tending only
to see those patients who have postoperative complications. Such
patients may be seen on an individual basis or as a member of a small
ward-based group.

The main purpose of a postoperative visit is to assist the patient to
regain breathing control and to prevent circulatory complications by
encouraging exercise whilst on bed rest. The exercises described below
are suitable for patients who have had surgical intervention, e.g.
general abdominal, thoracic or gynaecological surgery.

The patient should be told to practise the exercises regularly, i.e. 4–5
repetitions of each exercise every hour. Some of the exercises can be
taught on the first day postoperatively, whereas others will be dependent upon patient progress. All can be continued at home.

Deep breathing
The patient is instructed to breathe in through the nose and out
through the mouth. Encouragement is given to ensure that the patient
achieves normal breathing control, i.e. that he breathes using the lower
rather than upper chest. This can be facilitated by placing the patient's
right hand on the upper chest in the region of the xiphisternum and
the left hand over the diaphragm. By telling the patient to avoid
movement under the right hand when breathing, but to allow the left
hand to 'move' in time with the breathing cycle the required control is
usually achieved with practice.

Foot and ankle exercises
With the legs straight and flat on the bed, the patient is instructed to
move the feet up and down briskly (alternate dorsiflexion and plantar
flexion) and to circumduct in both directions.

Isometric (static) buttock and thigh exercises
With the legs straight and flat on the bed, the patient is instructed
alternately to tighten and relax the buttock and thigh muscles. The
mechanism of this circulatory exercise can be likened to squeezing a
sponge. When the patient tightens the buttock muscles, blood is
squeezed out, and it is then returned when the buttocks are relaxed.

Alternate hip and knee bends
In crook lying, i.e. lying flat on the back with the knees bent and feet flat on the bed, the patient is instructed to straighten and bend one leg and then straighten and bend the other leg.

Pelvic floor exercise (Laycock, 1987)
This exercise is important to many patients. Weakness of the pelvic floor can lead to faecal and urinary incontinence, both of which will lead to social embarrassment (see later in this chapter for more detail).

Pelvic tilting
In crook lying, the patient is instructed to tilt the pelvis by pulling up the abdominal muscles and squeezing the buttocks. This can be difficult to teach, but is made easier by placing a hand under the patient's lumbar spine and instructing them to flatten their back against the hand.

*Knee rolling**
Lying flat on the back with the knees bent and feet flat on the bed, the patient is instructed to tilt the pelvis as above, keep the shoulders flat on the bed and then slowly roll the knees to the left, return to the middle and stop. The process is repeated and the knees are rolled to the right.

Bridging
This exercise is necessary to enable the patient to sit on a bed pan and for nursing care.
   Lying flat on the bed, the patient is instructed to bend the knees, tilt the pelvis as described above and then to lift the buttocks no more than four inches off the bed. There should be no 'arching' of the back during this exercise.

*Hip hitching**
Lying flat on the bed with one knee bent, foot flat and one leg straight and supported, the patient is instructed to pull in the abdominal muscles, to draw the straight leg up from the waist as if shortening it without bending the knee, and then to stretch it down as though lengthening it.

* The majority of exercises described under the heading of 'Postoperative exercise class' are suitable for inclusion in the antenatal exercise class; however, those asterisked (*) are not suitable for inclusion in a late antenatal class. They should also be avoided where mothers have a diastasis (i.e. division of the rectus abdominis) of more than 2 fingers' breadth and should not be performed in the early post partum period. Knee rolling can be used to relieve flatus 2 days post partum for mothers who have had a Caesarian section. Movement, however, should be limited to between 4 and 6 inches either side of the midline.

Straight partial sit-up*

In crook lying, the patient is instructed to rest the hands on the thighs and adopt a pelvic tilt as described above. The chin is then tucked in as the patient attempts to slide the hands down the thighs towards the knees, flexing the trunk. The head and shoulders should only lift a short distance from the bed. After a 5-second hold the trunk is lowered slowly back onto the bed. This exercise is a strengthening exercise principally aimed at the rectus abdominis muscle. If the exercise is not performed as described and the trunk not 'curled', the hip flexors become the prime movers rather than the abdominal muscles (Norris, 1993).

Diagonal partial sit-up*

In crook lying, and after adopting a pelvic tilt, the patient is instructed to curl to either the right or to the left by sliding the hands together up the right or left thigh, respectively. This exercise is designed to strengthen the internal and external oblique abdominal muscles.

The antenatal exercise class

Antenatal exercises form part of antenatal classes attended by mothers-to-be and their partners. The purpose of these classes is to minimize the stresses and strains on the musculoskeletal system which occur during pregnancy and in the post-partum period. The physiotherapist may also provide advice on breathing and pain control in labour. Full details of these roles can be found in Polden and Mantle (1990).

Additionally, the physiotherapist can encourage the mother-to-be to maintain her existing levels of fitness safely. Obviously, vigorous contact sports which may put mother or baby at risk should be avoided. Generally, competition athletes should be able to maintain their present fitness/training level but they should be advised against progressing their routine.

The postnatal exercise class

This can be a ward-based group exercise and may involve women who are newly delivered or those several days post partum. Therefore, it is necessary to grade the exercise programme according to patient ability.

Rarely is it necessary to design a postnatal exercise class from scratch because most maternity units have a standard exercise sheet which has been designed by the obstetric physiotherapist. The exercise sheet should highlight the importance of exercise and provide suitable examples.

Postnatal exercises can be conveniently grouped under five categories:

1. Circulatory exercises (described above)
2. Abdominal exercises (described above)
3. The pelvic floor
4. Posture and back care advice
5. Relaxation techniques (see Chapter 13).

The pelvic floor

The importance of the pelvic floor in the prevention of incontinence
The role of childbirth, especially vaginal delivery, in the pathogenesis of incontinence is well documented. A number of authors have described damage to the nerves supplying the levator ani muscle during childbirth, causing denervation and subsequent reinnervation (Snooks *et al.*, 1984a,b; Kiff and Swash, 1984, Allen *et al.*, 1990). This can result in pelvic muscle weakness.

Furthermore, obstetric trauma causing perineal pain may inhibit normal pelvic floor action in the first few weeks post partum, resulting in reduced cortical awareness and the inability to perform an active pelvic floor contraction. This deficit may persist in later life if women are not made aware of the pelvic floor muscles and their role in maintaining pelvic support.

Pelvic floor re-education
The healthy pelvic floor muscles serve to provide urethral occlusion and bladder and urethral support. Also, resting tone and active contraction are said to inhibit the overactive bladder. Damage and disuse can lead to atrophy, further reducing muscle awareness. This may predispose a woman to a weak pelvic floor and incontinence.

Aim of pelvic floor re-education
Pelvic floor re-education is aimed at re-establishing cortical awareness, preventing further weakness and strengthening existing muscle fibres.

Method of re-education
Pelvic floor awareness should be established ideally in the antenatal period (or even better, before this). Women need to be taught to identify the muscle. This is best achieved in a crook lying position with the knees slightly apart. It is vital that abdominal, adductor and gluteal contractions are eliminated during this exercise, since this will detract from pelvic floor awareness.

The woman is instructed to imagine that she is controlling the flow of urine (as in a midstream specimen), while simultaneously tightening around the anus as when controlling flatus or 'wind'. As the muscle contracts, an upward and tightening sensation should be felt around the perineum. The contraction should be held for 4 seconds and followed by a short rest. The exercise should be repeated 4–5 times on a regular basis throughout the day. Additionally, incorporated into this programme may be a series of fast maximal contractions which can follow the slow holds. If the pelvic floor is very weak, it is advisable to use a lying or sitting position until the muscles are able to lift the pelvic contents against gravity.

A new mother can be encouraged to exercise at regular intervals during the day, e.g:

- when baby is fed
- after passing urine
- when turning on a tap.

NB. Breast-feeding mothers must be particularly vigilant with pelvic floor re-education because of lowered oestrogen levels.

The physiotherapist must emphasize to the new mother the importance of drawing attention to any abnormal bladder function, i.e. frequency or urgency; any pelvic floor dysfunction, i.e. inability to contract the pelvic floor or any disturbances in bowel control at their postnatal examination.

Assessing progress in pelvic floor re-education
Assessment of progress can be made by the patient when passing urine (Gosling, 1979). The ability to stop midstream in a controlled manner signifies improvement. However, this stop–start regimen on a regular basis should be discouraged since it can be habit forming (Chiarelli, 1991). This should also be avoided if the patient has problems of urine retention, severe pain, bruising and swelling post partum. Additionally, advice should be given to avoid stretching the pelvic floor when passing faeces as this can exacerbate damage and increase pain.

Posture and back care advice

Hormonal change during and after pregnancy causes the ligaments to become lax, making the mother susceptible to low back pain (Calguneri *et al.*, 1982). Laxity is present during the whole of the breast-feeding period and the mother should be made aware of the need to adopt safe posture and lifting practices and to avoid strenuous sporting activities which may put the back at risk.

Relaxation techniques (Chapter 13)
Resuming sports activity

The need to resume sporting activity is important to some mothers and as a general guide swimming can be started as soon as the mother feels confident. Moderately strenuous activities such as yoga and keep fit can usually begin 6 weeks after delivery. If bottle feeding, it is usual to advise the mother to avoid high-impact exercise such as squash, badminton and aerobics for the first 3 months after delivery. These would certainly not appeal to the breast-feeding mother!

DESIGN OF THE HOME EXERCISE PROGRAMME

Those who have been unfortunate to have required rehabilitative exercise will at some time be familiar with the self-discipline and motivation required to persist with the exercise programme prescribed. It is often less painful and certainly much quicker to 'nearly do' the 10 press-ups or sit-ups which feature in the programme. This scenario will no doubt be familiar and it will also be very familiar to many patients who are sent home with a list of exercises commonly referred to as the 'home exercise programme' (HEP).

The HEP is an integral part of the treatment regimen and must be carefully designed with the final goal of rehabilitation in mind. It should:

- be a programme that the patient will be motivated to carry out regularly
- tax patients sufficiently but should not leave them feeling exhausted
- avoid the need for complex apparatus
- be reviewed regularly and progressed gradually to allow the patient to reach his/her full rehabilitation potential.

It should not:

- contain exercises which place the patient at risk from injury
- contain an endless list of exercises (maximum of five different exercises).

Planning the home exercise programme

The HEP should contain a variety of exercises which re-educate balance, co-ordination, strength and flexibility. The physiotherapist should teach all of the exercises to the patient and provide a written *aide mémoire* of how to do each exercise, number of repetitions and the times per day (e.g. two or three) that the programme should be carried out.

Home exercise programme example

The aim of this programme is to regain flexibility and increase the strength and co-ordination of the shoulder girdle of a middle-aged housewife who is recovering from a frozen right shoulder.

1. General mobility to warm-up: 'soup stirring' (Fig. 8.1a).
2. Specific range of motion: 'wall walking' for flexion and abduction

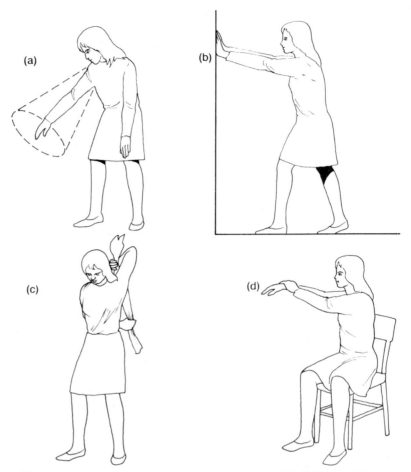

Fig. 8.1 Home exercise programme example to regain flexibility and increase the strength and co-ordination of the shoulder girdle: (a) 'soup stirring'; (b) 'wall walking'; (c) medial rotation using a towel; (d) gaining flexion using other arm

(Fig. 8.1b). Incorporate strength by telling the patient to attempt to hold position when moving a few centimetres away from the wall.

3. Auto-assisted range of motion to gain medial rotation using a towel (Fig. 8.1c).
4. Auto-assisted range of motion to gain flexion using the other arm (Fig. 8.1d). Attempt to 'hold' by reducing amount of support from non-injured limb.

Instructions to patient

1. Spend 10 minutes twice per day carrying out the exercise programme.
2. Try to aim higher up the wall when doing the wall walking exercise.
3. You should not experience pain when doing the exercises but may

feel a strain/pull around the shoulder, i.e. you should feel 'strain but not pain'.

4. If you experience any problems, contact the physiotherapy department.

Maximize patient compliance by

- asking the patient to keep a diary of the number of repetitions (in the case of the above programme the distance reached on the wall), together with how he/she felt when doing the exercises
- providing the patient with follow up appointments initially to correct technique and latterly, prior to discharge, to check for progress and functional independence
- explaining the rationale for the exercise programme and ensuring that the patient understands the value of doing the exercises
- providing an exercise sheet illustrating the exercise programme.

CIRCUIT TRAINING

Circuits can be designed for patients of any age and level of disability although most people tend to visualize fit young people exercising maximally perhaps with the aid of loud music!

During a circuit training programme, patients move around a number of exercise stations. A variable rest period is incorporated after each exercise has been completed. The amount of work done at each station will differ from patient to patient depending on factors such as level of fitness and motivation.

Before entering a circuit, medical clearance is essential and a fitness test desirable, so that an appropriate exercise level can be specified for the patient. For example, the resistance used in a typical training circuit should be between 40% and 60% of the patient's one repetition maximum (1 RM) (Kraemer and Koziris, 1992) (Chapters 6 and 10).

Exercises which may be included in circuit training classes are bench press, dumbbell activities, step-up and trampette work. The design of the circuit will, however, be determined by the type of patient attending the session, e.g. elderly, wheelchair bound, final stage of lower limb rehabilitation, etc and by the available equipment.

To monitor individual progress, patients should be encouraged to record their pulse rate. This should return to near normal limits after 2 minutes following cessation of each exercise (Chapter 7).

SUMMARY

- Group exercise provides a method of treating several patients with similar conditions at the same stage in the rehabilitation programme together in a controlled but relaxed environment.

- The class should always include a warm-up and cool-down period and consist of a series of exercises designed to address patient needs.
- Safety is of prime consideration. Numbers in the class should be limited according to the amount of space available; patients should all be at a similar level of rehabilitation and the physiotherapist should maintain control at all times.
- Several classes have been described: Lower limb, back and Colles' fracture, postoperative, antenatal and postnatal.

REFERENCES

Allen, R.E., Hosker, G.L., Smith, A.R.B. and Worrell, D.W. (1990) Pelvic floor damage in childbirth: a neurophysiological study. *British Journal of Obstetrics and Gynaecology*, **97**, 770–779

Calguneri, M., Bird, H.A. and Wright, V. (1982) Changes in joint laxity occurring during pregnancy. *Annals of the Rheumatic Diseases*, **41**, 126–128

Chiarelli, P.E. (1991) *Women's Waterworks: Curing Incontinence*, Gore and Osment, Australia (distributed in the UK by the Incontinence Advisory Service at the Disabled Living Foundation, London)

de C. Williams, A.C., Nicholas, M.K., Richardson, P.H. *et al.* (1993) INPUT: Evaluation of a cognitive behavioural programme for rehabilitating patients with chronic pain. *British Journal of General Practice*, **43**, 513–518

Frost, H. and Klaber Moffett, J. (1992) Physiotherapy management of chronic low back pain. *Physiotherapy*, **78**(10), 751–754

Frost, H. and Klaber Moffett, J. (1995) *British Medical Journal*, **310**, 151–154

Gosling, J. (1979) The structure of the bladder and urethra in relation to function. *Urological Clinics of North America*, **6**(1), 31–38

Hayne, C.R. (1984) Back schools and total back care programmes: a review. *Physiotherapy*, **70**(1), 14–17

Kiff, E.S. and Swash, M.M. (1984) Slowed conduction in the pudendal nerve in idiopathic (urogenic) faecal incontinence. *British Journal of Surgery*, **71**, 614–616

Klaber Moffett, J.A., Chase, S.M., Portek, I. and Ennis, J.R. (1986) A controlled prospective study to evaluate the effectiveness of a back school in the relief of chronic low back pain. *Spine*, **11**, 120–122

Kraemer, W.J. and Koziris, L.P. (1992) Muscle strength training: techniques and considerations. *Physical Therapy Practice*, **2**(1), 54–68

Laycock, J. (1987) Graded exercises for the pelvic floor muscles in the treatment of urinary incontinence. *Physiotherapy*, **73**(7), 371–373

McKenzie, R.A. (1981) *The Lumbar Spine: Mechanical Diagnosis and Therapy*, Spinal Publications, New Zealand

Norris, C.M. (1993) Abdominal muscle training in sport. *British Journal of Sports Medicine*, **27**(1), 19–27

Polden, M. and Mantle, J. (1990) *Physiotherapy in Obstetrics and Gynaecology*, Butterworth-Heinemann, Oxford

Snooks, S.J., Barnes, P.R.H. and Swash, M.M. (1984a) Damage to the innervation of the voluntary anal and periurethral sphincter musculature in incontinence: an electrophysiological study. *Journal of Neurology, Neurosurgery and Psychiatry*, **47**, 1269–1273

Snooks, S.J., Swash, M., Setchell, M. and Henry, M.M. (1984b) Injury to innervation of the pelvic floor musculature in childbirth. *Lancet*, **2**, 8 Sept., 546–550

White, A.H. (1983) *Low Back Schools and Other Conservative Approaches to Low Back Pain*, C.V. Mosby, St. Louis

9.   *Balance and proprioception*

DEFINITION OF BALANCE

A body is in a state of balance or equilibrium when the projection of its centre of gravity falls within its base of support and when the resultant of all of the forces acting on it is zero.

Balance control in the normal adult usually takes place at subconscious level. For example, when attempting to cross a river by stepping stones above water level one does not have to think, 'I need to put my arms out to the side to help me balance'. Thankfully this is done automatically. Although very much an automatic reaction for the adult with a mature nervous system, this action will have, at some time, been learned, practised and finely tuned in many different situations until it can be executed without thought. This is the same for all the balance mechanisms which develop from a very crude start in the new born, mature and continue to be refined throughout adulthood (see Chapter 4).

This chapter will outline:

- classification of balance
- factors which affect balance
- standing balance
- testing balance in the clinical situation
- methods of retraining balance
- proprioception in rehabilitation
- testing for a proprioceptive deficit
- re-educating proprioception
- proprioceptive neuromuscular facilitation.

CLASSIFICATION OF BALANCE

For convenience, balance is often classified as being either *static* or *dynamic*.

Static balance is said to exist when an adopted position is maintained for a period of time, as in the obvious example of the childhood game of musical statues. Alternatively, *dynamic balance* is the maintenance of balance when on the move, e.g. when getting up from a chair, walking or moving from sitting to lying. The optimum in dynamic balance is portrayed by a world class gymnast who brings to life the saying 'poetry in motion'.

This classification of balance can however be misleading, because rarely if ever does the human body remain perfectly still. Realistically, it continues to make virtually imperceptible postural adjustments which ensure the optimum position for energy conservation. Even during quiet standing, small adjustments are made, particularly around the ankle joint to ensure maintenance of the upright posture. These minute adjustments can be measured and observed using a force platform and are referred to as postural sway (Hellebrandt, 1939; Hasan *et al.*, 1990). Likewise, when working at a computer terminal which, according to the above description, is technically a static posture, the trunk continually makes small adjustments to stabilize the upper limb and thus enable the fingers to find the correct letters on the keyboard.

These postural adjustments which maintain balance are known as *equilibrium reactions*. If the projection of the centre of gravity (CoG) comes to lie outside the base of support, the body is then unstable and it will, in order to prevent injury, need to take action. Usually executed automatically, this action is known as a *saving reaction*. This may take the form of putting both hands out in front to protect the face, as in a fall forwards, or stepping forwards if pushed from behind. In neurological disease, equilibrium and saving reactions are often lost or slowed and as a consequence their re-education by facilitation techniques feature prominently in the rehabilitation process.

FACTORS WHICH AFFECT BALANCE

The ability to maintain balance depends on the integrative functioning of many factors (Figure 9.1). Sensory information is received from visual, vestibular, proprioceptive, exteroceptive and tactile sources, and provided that this information can be decoded and used by the intact nervous system, balance becomes a function which is assumed and taken for granted. It is only when the balance mechanism fails that its importance is realized.

STANDING BALANCE

Although time is devoted to the re-education of sitting balance in the physiotherapy department, particularly in patients with neurological disorders, it is probable that more time is spent teaching patients to achieve standing balance in preparation for walking.

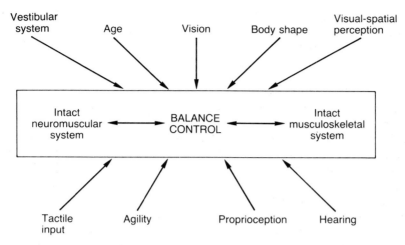

Fig. 9.1 Factors which affect balance

The characteristics of standing balance in the normal subject are referred to as postural sway and have been studied extensively. A brief overview is given below.

Historical aspects of measuring standing balance

From as early as 1853 the characteristics of upright bilateral stance have been reported in the literature, with particular attention being paid to the measurement of postural sway (Romberg, 1853). Early studies tended to concentrate on body segment displacement, with a number of researchers directing their attentions to movements of the head during upright stance. Hinsdale (1887) graphically recorded sway by attaching a smoking paper to the top of the subject's head, whereas other researchers used accelerometers, light beams (Orma, 1957; Kapteyn, 1972), strings and mechanical structures (Nashner, 1971; Gantchev *et al.*, 1972). Later, attention was directed at the trunk, with several studies reporting the use of accelerometers (Thomas and Whitney, 1959), light sources (Hirasawa, 1976), sound sources (Stevens and Tomlinson, 1971), magnetic coils and video systems (Kapteyn, 1973).

The control of upright stance

The maintenance of upright stance, i.e. standing balance is an automatic, active sensorimotor process which maintains the body's CoG over its base of support. Three components of the sensory system provide integrated information which ensures the maintenance of upright stance:

- visual input
- vestibular input
- proprioceptive kinaesthetic reflex activity.

It is generally assumed that this triad form a multi-loop control of postural stabilization (Walsh, 1973; Dichgans *et al.*, 1976; Dornan *et al.*, 1978); however, it remains unclear whether the selection of these sensory inputs occurs in a hierarchical way, i.e. is one of these postural control mechanisms of more importance than another? It has been suggested that the vestibular system is the most dominant of the three in the normal subject and that the visual system is responsible only for the fine tuning of posture as a result of vestibular feedback (Lee and Lishman, 1975; Nashner *et al.*, 1982).

TESTING BALANCE IN THE CLINICAL SITUATION

There are a number of methods which are available for assessing balance:

- platform stabilometry
- the Romberg Test
- performance of functional activity
- kinematic methods (Chapter 15).

Platform stabilometry

At present, a popular method of studying sway characteristics during quiet standing is to use a force platform. The technique known as platform stabilometry was first introduced by Hellebrandt (1939).

Force platforms consist of a rigid rectangular plate (typically 0.6×0.4 m) mounted on force transducers which are positioned under each corner. Each transducer measures forces along the three orthogonal axes (i.e. axes arranged at 90° to each other). The signals from the transducers are combined to give the vertical, anteroposterior (AP) and mediolateral (ML) forces and the point of application of the force on the platform together with any torque applied at this point about the vertical axis.

The point of application of the force on the platform is known as the centre of pressure (CoP) (Hasan *et al.*, 1990). Excursions of the CoP can provide important information on the sway characteristics of upright stance such as sway path, sway area and sway frequency.

If no sway was occurring, which incidentally is never the case, the line of gravity would project vertically through the CoG to the CoP.

Although during quiet standing the CoP and the projection of the CoG do show similarities, their recorded curves are never identical (Murray *et al.*, 1975). These differences are explained by the smaller amplitude and frequency of the real movements of the CoG compared with those of the CoP (Thomas and Whitney, 1959; Murray *et al.*, 1975; Spaepen *et al.*, 1977).

Platform stabilometry unfortunately will not readily be available in most physiotherapy departments, since its cost is prohibitive. It also takes a considerable time to set up and requires the presence of an experienced operator. However, from the research point of view its inclusion in balance investigations should always be a prime consideration because it is capable of providing information on:

1. Forces across the plate in the AP and ML direction. These are shear forces and are often referred to in the literature as the Fx and Fy components. However, lack of standardization in naming, e.g. AP as Fy and ML as Fx, can lead to confusion and care should therefore be exercised when interpreting force platform literature.
2. Vertical force application. This can be known as the Fz or the Fy component.
3. The resultant of all the above forces which is termed the centre of pressure.
4. The moment (torque) about an axis which is perpendicular to the surface of the force platform.

The Romberg Test

In 1853, a German neuropathologist by the name of Romberg was the first to observe that in some neurological diseases such as ataxia and tabes dorsalis (posterior column disorder) the patient's body presented relatively pronounced, visually detectable oscillations in different directions. Romberg went on to introduce a test widely used in clinical neurology today, known as the Romberg Test.

The test provides qualitative information by evaluating quiet standing with the eyes open and with the eyes closed. It can also be performed on a force platform to provide quantitative data on aspects of postural sway, as described above. The test results are interpreted in Figure 9.2.

Method

(a) The patient is instructed to stand with the heels together and with the eyes open.
(b) Postural sway behaviour is observed.
(c) Instruction is given to close the eyes.
(d) Postural sway behaviour is observed.

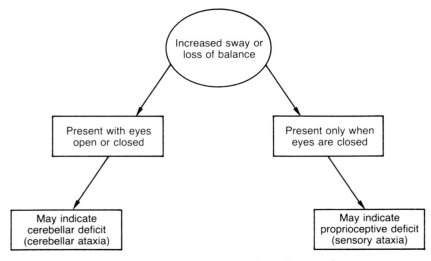

Fig. 9.2 Interpretation of Romberg Test results

Performance of functional activity

A balance assessment is performed to provide information on a patient's ability to balance in a given position. Some of the reasons why balance is assessed in the clinical situation are provided below. The list is far from comprehensive and there are many other clinical situations where balance assessment will be important:

• patients who are getting out of bed for the first time after prolonged bedrest – postural hypotension may cause dizziness
• patients who are getting up on crutches for the first time
• elderly patients who require a walking aid
• patients with severe head injuries who must learn to regain balance
• elderly patients who need to get up from sitting to standing without overbalancing – postural hypotension should also be considered in this situation.

A balance assessment should always be tailored to the patient's existing functional level. The assessment is often carried out with the patient either sitting or standing (in parallel bars).

Guidelines for assessment

The test does not need to take a lot of time – a couple of minutes will usually suffice. Initially, the therapist will observe the patient to ascertain whether postural alignment is normal and how much support is required by the patient to enable maintenance of the assessment position.

Once these criteria have been established, the general guidelines listed below can be followed:

- With the patient in the chosen assessment position, apply gentle but firm pressure in one direction only at a point on the body near to the CoG and give the patient the instruction 'don't let me move you'. The patient will, if able, respond by resisting the applied pressure.
- Apply similar pressure in a known sequence in other directions, e.g. forwards, sideways.
- Progress by applying the same amount of pressure but in random order, so that the patient is not able to predict the direction of resistance to be applied.
- Progress by increasing the length of the lever arm, i.e. applying pressure further away from the CoG.
- Alter the position of the patient's line of gravity in relation to the base of support by, for example, instructing placement of both hands in front of the body, to the side or above the head.
- Instructing the patient to perform the test with eyes closed will provide an additional source of information (see Romberg Test above).
- Incorporate a functional activity at the end of the assessment, e.g. observing the patient tying shoe laces or getting dressed while sitting in the chair. This will provide information on how he/she balances while not concentrating. It also gives information to the therapist about whether associated abnormal reactions are present. For example when attempting to pick up a cup, it is not normal to adduct the upper arm and flex the elbow and fingers of the other upper limb. Neither would it be normal to observe full extension of the left lower limb when reaching a short distance to pick up an object from a nearby table. These are examples of *associated reactions*. They are abnormal stereotyped patterns which present in patients with hypertonicity. Associated movements are also seen in normal people during strenuous activity.

Kinematic methods (Chapter 15)

METHODS FOR RETRAINING BALANCE

The following aspects should be considered when designing a progressive balance re-education programme:

- proceed from bilateral to unilateral activities
- perform activities with eyes open and then eyes closed
- progress from a stable to an unstable surface/base.

During the initial stages, the patient should be encouraged to concentrate on a sense of body position, muscle control and joint movement to enhance sensory information (Irrgang *et al.*, 1994). Latterly, activities such as catching or throwing a ball while on a wobble board should be incorporated into the balance regimen, as these distract the patient from concentrating on balance.

Early stage balance re-education

Stationary balance activities

'Static' balance activities can be used in preparation for dynamic activities which will be incorporated later in the rehabilitation programme. Education of static balance can be started early in a treatment regimen of musculoskeletal and neurologically impaired patients. The footballer who has recently had a below-knee plaster of Paris applied following an ankle injury will need to be mobilized on crutches, non-weight-bearing. The patient who has suffered a stroke will have to regain standing balance before learning to walk.

Programmes for the re-education of balance should be tailored to individual patient needs. The requirements of the county gymnast will be different from those of the elderly patient. The former will need to be able to perform a number of 'static' balance routines when on the beam, mat or parallel bars, whereas the latter may need to be able to walk and turn around safely with or without a walking frame.

Stationary balance re-education is necessary in the following examples:

- the patient who has to be mobilized on crutches without weight-bearing
- the stroke patient who needs to regain sitting balance
- the elderly patient who needs to get up from sitting to standing safely
- the patient with a damaged spinal cord who is learning to do transfers from bed to chair, etc.

Suggested balance activities

Starting position

This should be functional and tailored to patient needs. A wide base of support with a low centre of gravity is preferable initially, to allow the patient to gain confidence. Progress can then be achieved by reducing the size of the base of support and raising the height of the centre of gravity.

Clinical example

The re-education of sitting balance in the spinal cord injured patient who is paralysed from the waist downwards. Starting position: long sitting (i.e. legs straight out in front) on the plinth or floor.

These patients have no sensation below the level of their injury and initially would not be able to sit on a plinth with their legs over the edge. This position would be very frightening for them. Therefore balance re-education should be commenced on a large plinth (or the floor) with the patient in long sitting and therapist supporting from behind.

Method
Gentle pressure should be applied in a known direction to the middle trunk. The patient is told to remain still. This procedure is repeated. A mirror positioned in front of the patient may be used to provide visual feedback, although some patients might find this confusing.

Progression is indicated when the patient is able to sustain balance in the anterior, posterior, medial and lateral direction in response to gentle displacements (perturbations) from the therapist. This takes the form of random applications of pressure to the mid-trunk with the patient responding accordingly.

Additionally, the perturbations can be applied distally at the shoulders or to the outstretched arms.

Progression to sitting on the edge of the plinth is considered to be appropriate when the patient has good balance control in long sitting. Perturbations are then applied as before, ensuring that the patient is safely positioned and supported. This method of balance retraining is important for the patient's functional activities, e.g. dressing, transfers and reaching forwards.

It is possible to use the proprioceptive neuromuscular technique, known as rhythmic stabilizations, in the example above. This involves placing one hand on the anterior surface of one side of the trunk and the other hand on the posterior surface of the opposite side and carrying out the technique as described later in this chapter (Voss *et al.*, 1985).

Later stage balance re-education

Dynamic balance
Dynamic activities are included in the later stages of rehabilitation when range of motion, endurance and strength have been improved. The activities should be functionally designed and orientated to patient requirements. They should allow the patient repetitively to lose and regain balance without falling. Additionally they should be designed to enable recruitment of the injured muscle groups with proper timing and in the correct sequence to facilitate smooth and co-ordinated movement. Repetition is needed to develop skill and to engage the appropriate motor pathways. Initially, this will require conscious effort, but later the patient should be able to perform the activities automatically.

PROPRIOCEPTION IN REHABILITATION

Proprioception is the body's awareness of posture, movement and changes in equilibrium. The normal subject with intact proprioceptive pathways is able to gauge position, weight and resistance of objects in relation to the body.

A patient who has a proprioceptive deficit will lack these abilities which will be manifest as co-ordination and balance problems. A proprioceptive loss or impairment can be localized to one limb, as occurs following an ankle sprain, or it may involve the whole body such as in the neurological condition of sensory ataxia.

Proprioceptive information is usually transmitted via a three-order neuron system from the muscles, the muscle spindles, the Golgi tendon organs and the joint capsules. Very little proprioceptive information reaches conscious level. The majority is concerned with reflex activity and is mediated via the spinal cord and cerebellum with the prime function of controlling posture and movement (Chapter 4).

Proprioceptive retraining has until recently been overlooked in rehabilitation. However, its importance, particularly with reference to the anterior cruciate ligament (Beard *et al.*, 1994) and functional instability of the ankle joint (Freeman *et al.*, 1965), has now been recognized.

Borsa *et al.* (1994) describe the function of proprioceptive retraining in two ways, as enhancing (a) the cognitive appreciation of the respective joint relative to position and movement, and (b) the muscular stabilization of the joint in the absence of structural restraints.

A repeat of injury and falls may be prevented by re-educating proprioception.

TESTING FOR A PROPRIOCEPTIVE DEFICIT

Testing proprioception in the neurologically impaired patient

The standard neurological test used to determine proprioceptive deficit is described below for the joints of the hand. The same procedure can be repeated for all upper and lower limb joints. If proprioception is found to be impaired distally, then it is necessary to move more proximally and to repeat the test. The patient is positioned in supine lying with head supported on a pillow.

(a) The sides of the patient's thumb or fingers are held by the operator who then demonstrates flexion and extension movements passively. Care should be taken to avoid holding the leading surfaces during the test since this can influence the patient's decision because they may feel the pressure exerted in the direction of movement.

(b) The patient is then told to close the eyes and the process is repeated. The patient is instructed to indicate the direction of movement by stating whether the digit is being moved 'up' or 'down'. Failure to provide the correct answer consistently may indicate a proprioceptive deficit because the patient is unable to recognize the direction of joint movement. A positive response to this test would be recorded as loss of *joint position sense*.

Patients with neurological deficit are likely to have loss of joint position, whereas patients with non-neurological injuries such as muscle injuries do not usually demonstrate this type of proprioceptive loss. The reason is that proprioceptive deficit in muscle injuries is localized to the damaged area and does not involve disruption of any of the CNS control mechanisms.

Testing proprioception in the non-neurologically impaired patient

Practical example

To understand the concept of proprioceptive loss in a non-neurological condition, ask a colleague who had an ankle sprain or a lower limb muscular injury within the last few years, for which medical advice was sought, to adopt a unilateral stance, i.e. to stand on one leg. When they are comfortably balanced, instruct them to close the eyes and then repeat the process with the other leg. Failure to maintain stability on the injured side indicates proprioceptive loss. The subject will 'wobble' more when standing on the injured side and may even have to execute a saving reaction by toe touching with the unaffected side.

RE-EDUCATING PROPRIOCEPTION

Wobble board

The wobble board (Burton, 1986) is the most common form of apparatus used for proprioceptive re-education. Several commercial devices are available. The most popular board is the multidirectional board (Figure 9.3), but there are also boards which are limited to a unidirectional tilt in either an AP or ML direction and these are often preferable during early stage rehabilitation.

Wobble board construction for home use

Commercial boards can be expensive for patients who require to use them at home or for students who intend to research into their effectiveness as a proprioceptive re-education tool. It is possible to construct a wobble board relatively easily (Figure 9.4).

The board surface can be made from plywood, while the balancing apparatus used underneath the board can be constructed using golf

Human Movement Explained

Fig. 9.3 A multidirectional wobble board

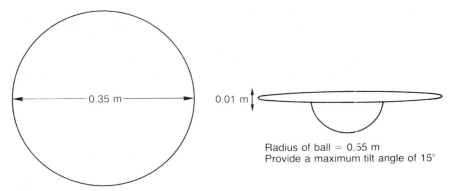

0.35 m

0.01 m

Radius of ball = 0.55 m
Provide a maximum tilt angle of 15°

Fig. 9.4 Suggested dimensions for construction of a wobble board

balls or solid rubber balls. The surface of all boards should be covered with non-slip material such as vinyl rubber or varnished and then sprinkled with sand.

Using the wobble board effectively

The wobble board is a versatile proprioceptive re-education medium. Many progressive programmes like the one detailed below can easily be incorporated into a lower limb rehabilitation programme. The time spent on the board will depend on patient motivation and concentration span. It is suggested that frequent short periods of training are preferable to longer periods once or twice a week.

Progressive wobble board re-education regimen for the lower limb

Initially, the patient is instructed to stand bilaterally on the wobble board but may progress later to unilateral stance.

1. Stand on a unidirectional board placed between parallel bars and move the board using ankle movements. This technique is used to mobilize the ankle joint. The patient is corrected if wobble board movement is produced by trunk oscillations rather than ankle movement.
2. Bilateral stance on a unidirectional board keeping the wobble board level, i.e. parallel to the ground. Here the board is being used to re-educate balance.
3. Bilateral stance on a multidirectional board for mobilizing and balance re-education. This can be attempted between parallel bars initially, but it is more effective if attempted without the bars for support. Care should be taken to place the board on a non-slip surface. Proprioceptive awareness can be facilitated quickly and without conscious thought by throwing and catching a ball while standing on the balanced board.
4. Bilateral stance on a multidirectional board with the eyes closed. The patient is told to keep the board parallel to the ground when attempting this exercise.
5. Unilateral stance on a multidirectional board with the eyes open – progressing to an eyes closed stance is difficult to achieve even for the normal subject.

Other methods for retraining proprioception

Proprioceptive retraining does not necessarily require specialized equipment. The aim of the retraining programme is to facilitate improved co-ordination and stability in the affected body part (usually non-neurological) or in the total body (usually neurological).

A lower limb proprioceptive environment can be created easily using a little imagination. Rough terrain can be imitated by placing a few objects such as balls, hoops or walking sticks underneath a gymnastic mat. Instructing the patient to go for a walk on uneven ground such as a sand-dune or grass bank will have a similar effect, as will instructing them to practise unilateral stance regularly.

Additionally, a simple exercise known as 'braiding' can provide proprioceptive input. This is achieved by instructing the patient to stand at ease with hands on hips. Instruction is then given to cross the right foot in front of the left. The left is then moved laterally and the right foot is then crossed behind the left. The left is moved laterally once more and the whole process is repeated until the patient has 'braided' across the room. Returning to the starting position provides an opportunity to repeat the process with the left leg leading.

Walking along a narrow line, imaginary or real, on the floor of the gymnasium is another example of a simple but effective proprioceptive exercise, as is unilateral or bilateral stance on a trampette.

A circuit consisting of a series of lower limb exercises for proprioception can be devised. Many of the exercises above would be suitable for inclusion in such a programme.

PROPRIOCEPTIVE NEUROMUSCULAR FACILITATION

It must be stressed that the information given in this chapter serves only as a very brief introduction to the concept of proprioceptive neuromuscular facilitation (PNF) and should in no way be used as a substitute for the classic text of Voss *et al.* (1985).

Definition

PNF is a method used by physiotherapists 'to promote or hasten(ing) the response of the neuromuscular mechanism through stimulation of the proprioceptors' (Voss *et al.*, 1985).

Developed by Dr Herman Kabat and his Chief Physical Therapist, Margaret Knott, during the late 1940s, it consists of a number of techniques which can be used to improve:

- strength
- range of movement
- endurance
- co-ordination.

Each technique has a name, e.g. hold-relax, slow-reversals, repeated contractions and rhythmic stabilizations, and is performed manually by the physiotherapist with the patient's co-operation.

To appreciate the underlying philosophy which has led to the development of PNF techniques, it is necessary to revise the anatomical concepts of planes and axes of movement described in Chapter 2.

Anatomy texts describe human movement with reference to the three planes – the sagittal (AP), the coronal (ML) and the transverse planes. Movement in these planes is said to occur about coronal, sagittal and vertical axes, respectively. The human body would be somewhat robotic if all movements were performed solely in these planes. Thankfully this is seldom the case.

For example, when reaching forward to pick up an object, this rarely if ever takes place purely in the sagittal plane. Instead, the upper limb movement with respect to the trunk involves a somewhat complex combination of flexion, abduction or adduction, with an element of rotation, to accomplish the desired action. Likewise, a footballer about to shoot at goal does not just move his lower limb from a position of extension to flexion in the sagittal plane.

As experience is gained, one observes that the body's physiological movements are mainly *rotational* and *oblique* and that they tend to take place in a *diagonal plane*. This concept can be illustrated easily by performing a series of tasks:

- *Task 1*. Observe a colleague walking away from you and then back towards you. Pay specific attention to the movements of the shoulder and pelvic girdles.

Rotation of the shoulder girdle combined with the opposite rotation of the pelvic girdle during all functional movements provides the important background for facilitation of limb movement in the diagonal.

- *Task 2*. To appreciate the concept of diagonal limb patterns of movement, perform the following exercise.

Stand with the feet slightly apart and swing the arms backwards and forwards as if walking quickly. Observe the arm swing and notice that it occurs most comfortably in a plane which is neither sagittal nor coronal. During the anterior swing there is likely to be a tendency for the arms to adduct across the trunk. Conversely, on the back, swing arm movement is away from the trunk (abduction).

- *Task 3*. Begin the arm swing as before, but this time limit the movement to the sagittal plane only. 'Uncomfortable', 'abnormal', 'hardwork', 'robotic' and 'mechanical' are a few words which probably come to mind.

Based on similar but highly skilled observations of normal human function, Kabat and Knott concluded that all *functional body movements* were based on and could be derived from a series of diagonal patterns of movement.

The anatomical attachment of muscle has also somewhat dictated these patterns, since virtually all muscles tend to work most efficiently along the line of direction of their fibres. Very few muscles have attachments which are aligned perpendicularly – most have a proximal attachment which is offset from the distal attachment.

Kabat and Knott went on to identify the diagonal patterns for all major body segments – the head, neck, trunk and limbs, all of which are described fully in Voss *et al.* (1985). The diagonal patterns of the upper and lower limbs are illustrated in Figures 9.5 and 9.6, respectively.

The techniques of PNF are all performed in pattern in these diagonal functional planes. Successful implementation relies heavily on the correct sensory input provided by the physiotherapist.

For example, by placing the hands in certain predetermined positions on the limb or trunk, the therapist provides guided resistance and thus facilitates the desired movement in the required pattern (see Tables 9.1 and 9.2). Additionally, by use of the voice and instructing the patient to watch the limb, the movement performed is reinforced.

Proprioceptive stimuli provide another source of sensory facilitation. A quick, but controlled, stretch effected via the muscle spindle can be

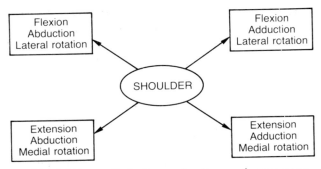

Fig. 9.5 Upper limb diagonal patterns of movement

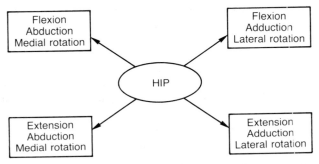

Fig. 9.6 Lower limb diagonal patterns of movement

applied by the therapist to initiate most movement patterns. Compression through the long axis of the limb is thought to facilitate the joint proprioceptors (mechanoreceptors) and it is used to initiate the extensor patterns of the lower limb. Traction is used to facilitate the flexor patterns.

Proprioceptive neuromuscular techniques in the re-education of balance

PNF techniques which facilitate co-ordination can be employed to treat the patient with diminished balance. Among these are rhythmic stabilizations and slow reversals.

Rhythmic stabilizations

This technique is used to achieve isometric co-contractions of all muscle groups around a joint complex to ensure joint stability. Rhythmic stabilizations are particularly useful in the treatment of the neurologically impaired patient (see 'Early stage balance re-education' for the spinal cord injured patient, p. 183) and are frequently employed to gain stability of the trunk where ataxia is present. Additionally, the technique can be used immediately after regaining joint range of movement to enhance joint stability in the new range and to treat pain in musculoskeletal conditions.

Method

The limb is positioned at the point of instability within the functional diagonal. Gradually an isometric 'build' of the agonists is facilitated until a maximum contraction is achieved without pain. Slowly, the manual contacts are then changed to resist and facilitate an antagonist 'build'. Manual contacts are changed slowly from agonist to antagonist as required and within levels of patient tolerance. Rapid change or abrupt release of contact will result in loss of co-contraction and may cause pain or result in loss of balance.

Some clinical applications of rhythmic stabilizations

1. To strengthen in a specific part of the range following a soft-tissue injury around the shoulder joint, e.g. damage to the rotator cuff muscles.
2. To gain proximal stability, i.e. in the trunk of the ataxic patient or during recovery from a polyneuropathy such as Guillain–Barré syndrome.
3. To strengthen the shoulder girdle of the paraplegic patient in preparation for transfer practice.

Slow reversals

The success of all normal functional movements relies on an intricate balance between agonist and antagonist. Slow reversal is a technique which can be used to facilitate contraction of a weak pattern to increase strength, endurance and to improve co-ordination.

A through-range isotonic contraction of the antagonists is followed by a through-range isotonic contraction of the agonists. The movement is performed in one of the functional diagonal planes. A 'hold' can be incorporated in the pattern at a point of weakness. Unlike repeated contractions which involve a passive return to the starting position, slow reversals involve the application of resistance in both directions.

Tables 9.1 and 9.2 give a set of tabled instructions which summarize the procedure for carrying out each upper limb slow reversal technique. These tables can also be used for quick reference when carrying out any of the other techniques and similar ones can be constructed for the lower limb. It is suggested that they are used in conjunction with Voss *et al.* (1985).

Human Movement Explained

Table 9.1 Working on the *right* limb – pattern 1

Name:	Extension/adduction/medial rotation	Flexion/abduction/lateral rotation*
Starting position	Flexion/abduction/lateral rotation	Extension/adduction/medial rotation
Initiation of pattern	Traction into F/ABD/LR to provide stretch response	Traction into E/ADD/MR to provide stretch response
Proximal hand	(L) hand: Anteromedial aspect of upper arm	(R) hand: Posterolateral aspect of upper arm
Distal hand	(R) hand: Palm to palm with patient	(L) hand: Dorsum of hand
Commands	Grip my hand and pull down and across	Fingers back, wrist back and point your thumb to the floor
Hold (required to enable hand change to take place without loss of contraction)	Distal hand maintains resistance	Distal hand maintains resistance

*This pattern is likened to the act of flinging the arms upwards and away from the body, as in proclaiming 'Oh what a lovely morning'!

Table 9.2 Working on the *right* limb – pattern 2

Name:	Extension/abduction/medial rotation	Flexion/adduction/lateral rotation
Starting position	Flexion/adduction/lateral rotation	Extension/abduction/medial rotation
Initiation of pattern	Traction into F/ADD/LR to provide stretch response	Traction into E/ABD/MR to provide stretch response
Proximal hand	(L) hand: Posterolateral aspect of upper arm	(R) hand: Anteromedial aspect of upper arm
Distal hand	(R) hand: Dorsum of hand	(L) hand: Palm to palm with patient
Commands	Fingers back, wrist back and chop down	Grip my hand and pull up and across
Hold (required to enable hand change to take place without loss of contraction)	Distal hand maintains resistance	Distal hand maintains resistance

Points to remember
- always name the pattern according to the muscles being used (i.e. in the direction in which the limb is moving)
- the therapist must stand in the same diagonal as the pattern in which they are working
- when performing the techniques, the patient should be positioned close to the operator to avoid undue stress on the spine
- manual contact should be placed on the leading surface (i.e. the

surface of the working muscle) to provide resistance and facilitation to the movement

- apply appropriate resistance according to whether isotonic or isometric movement is required
- initiate each pattern with either traction (upper limb and lower limb flexion patterns) or compression (lower limb extension patterns) and maintain throughout
- the operator should move (flow) with the patient and use the whole body to control the pattern and not just use upper limb strength, to ensure that the diagonal is maintained; resistance should be altered appropriately through range
- the therapist should facilitate movement by correct use of the voice, e.g. 'now pull', verbalized at the same time as the stretch is applied to facilitate the movement
- all patterns are initiated distally to proximally. This is an important contrast with other facilitation techniques such as Bobath (1990) which aims to achieve proximal stability first
- all patterns and techniques can be performed in a variety of starting positions such as lying, sitting and standing.

Table 9.3 summarizes the application of a selection of techniques in the clinical setting. It is stressed that this is only a guide and that there are a number of other techniques available which may be considered more appropriate to use following patient assessment.

Table 9.3 Some PNF techniques and their clinical uses

Technique	Strength	Endurance	Range of movement	Co-ordination
Hold-relax	*	*	***	*
Rhythmic stabilizations	***	***	**	***
Repeated contractions	***	***	***	*
Slow reversals	***	***	*	***

Key: * Least useful
*** Most useful

SUMMARY

- Balance occurs when the centre of gravity of the body falls within its base of support
- Balance may be quasi static or dynamic and is maintained by postural adjustments called equilibrium reactions
- The ability to maintain balance depends on a number of integrated functions of the visual, vestibular and proprioceptive systems
- Balance can be assessed using either platform stabilometry, the Romberg test, functional activity or kinematic analysis

- When retraining balance, starting position, visual input and the stability of the supporting surface may all be varied to progress the exercise programme
- Proprioception is the body's awareness of posture, movement and changes in equilibrium. Proprioceptive retraining is fundamental to many areas of rehabilitation
- Proprioceptive neuromuscular facilitation (PNF) is a treatment method which may be used to re-educate balance and joint position sense.

REFERENCES

Beard, D.J., Dodd, C.A.F., Trundle, H.R. and Simpson, A.H.C.W. (1994) Proprioception enhancement for anterior cruciate ligament deficiency – a prospective randomised trial of two physiotherapy regimes. *Journal of Bone and Joint Surgery*, **76B**, 654–659

Bobath, B. (1990) *Adult Hemiplegia*: *Evaluation and Treatment*, 3rd edn. William Heinemann, London

Borsa, P.A., Lephart, S.M., Kocher, M.S. and Lephart, S.P. (1994) Function assessment and rehabilitation of shoulder proprioception for glenohumeral instability. *Journal of Sport Rehabilitation*, **3**, 84–104

Burton, A.K. (1986) Trunk muscle activity induced by three sizes of wobble (balance) boards. *Journal of Orthopaedic and Sports Physical Therapy*, **8**(1), 27–29

Dichgans, J., Mauritz, K.-H., Allum, J.-H.-J. and Brandt, T.H. (1976) Postural sway in normals and atactic patients: analysis of the stabilising and destabilising effects of vision. *Agressologie*, **17(C)**, 15–24

Dornan, M.B., Fernie, G.R. and Holliday, P.J. (1978) Visual input: its importance in control of postural sway. *Archives of Physical Medicine and Rehabilitation*, **59**, 586–591

Freeman, M.A.R., Dean, M.R.E. and Hanham, I.W.F. (1965) The aetiology and prevention of functional instability of the foot. *Journal of Bone and Joint Surgery*, **47-B**(4), 678–685

Gantchev, G.-N., Draganova, N. and Dunev, S. (1972) The role of visual information and ocular movements for the maintenance of body equilibrium. *Agressologie*, **13(B)**, 55–61

Hasan, H.S., Lichtenstein, M.J. and Shiavi, R.G. (1990) Effect of loss of balance on biomechanics platform measures of sway: influence of stance and a method for adjustment. *Journal of Biomechanics*, **23**(8), 783–789

Hellebrandt, F.A. (1939) The influence of sex and age on the postural sway of man. *American Journal of Physical Anthropology*, **24**(3), 347–360

Hinsdale, G. (1887) Station of man, considered physiologically and clinically. *American Journal of Medical Science*, **93**, 478–485

Hirasawa, Y. (1976) Study of human standing ability multi-point XY-tracker and pedoscope. *Agressologie*, **17(B)**, 21–27

Irrgang, J.J., Whitney, S.L. and Cox, E.D. (1994) Balance and proprioceptive training for rehabilitation of the lower extremity. *Journal of Sport Rehabilitation*, **3**, 68–83

Kapteyn, T.S. (1972) The stabilogram; measurement techniques. *Agressologie*, **13(C)**, 75–78

Kapteyn, T.S. (1973) Afterthought about the physics and mechanics of the postural sway. *Agressologie*, **14(C)**, 27–35

Lee, D.N. and Lishman, J.R. (1975) Visual proprioceptive control of stance. *Journal of Human Movement Studies*, **1**, 87–95

Murray, M.P., Wood, B.S., Seireg, A.A. and Sepic, S.B. (1975) Normal postural stability and steadiness: quantitative assessment. *Journal of Bone and Joint Surgery*, **75-A**(4), 510–516

Nashner, L.M. (1971) A model describing vestibular detection of body sway motion. *Acta Oto-laryngologica*, **72**, 429–436

Nashner, L.M., Black, F.O. and Wall, C. (1982) Adaptation to altered support and visual conditions during stance: patients with vestibular deficits. *Journal of Neuroscience*, **2**(5), 536–544

Orma, E.J. (1957) The effects of cooling the feet and closing the eyes on standing equilibrium – different patterns of standing equilibrium in young adult men and women. *Acta Physiologica Scandinavica*, **38**(3&4), 288–297

Romberg, M.H. (1853) *Manual of Nervous Diseases of Man*, Sydenham Society, pp. 395–401

Spaepen, A.J., Vranken, M. and Williams, E.J. (1977) Comparison of the movements of the centre of gravity and of the centre of pressure in stabilometric studies. *Agressologie*, **18**(2), 109–113

Stevens, D.L. and Tomlinson, G.E. (1971) Measurement of human postural sway. *Proceedings of the Royal Society of Medicine*, **64**, 14–17

Thomas, D.P. and Whitney, R.J. (1959) Postural movements during normal standing in man. *Journal of Anatomy*, **93**, 524–539

Voss, D.E., Ionta, N.K. and Myers, B.J. (1985) *Proprioceptive Neuromuscular Facilitation*, 2nd edn, Harper and Row, Philadelphia

Walsh, E.G. (1973) Standing man, slow, rhythmic tilt, importance of vision. *Agressologie*, **14**(C), 79–85

10.   *Strength*

DEFINITION OF MUSCLE STRENGTH

Strength is the ability of a muscle or group of muscles to produce
tension and a resulting force in one maximal effort, either dynamically
or statically, in relation to the demands placed upon it (Kisner and
Colby, 1990).
  This chapter will outline:

- factors which determine muscle strength
- types of muscle strength
- principles of strength training
- muscle contraction types and exercise
- physiological effects of strength training
- re-educating strength during early stage rehabilitation
- principle of specificity of training
- factors to consider when designing a strength programme
- methods used to train muscle strength
- clinical examples of muscle strengthening exercises
- training eccentric muscle work
- plyometrics
- endurance exercise.

FACTORS WHICH DETERMINE MUSCLE STRENGTH

Muscle strength is a complex concept which depends on a number of
factors (Figure 10.1).
  The cross-sectional area determines the strength of a muscle. The
greater the cross section, the larger is the maximal force generated.
Force of contraction also depends upon the *integrity of the connective
tissue* and tendons which assist in the transfer of muscle tension to
external force. Damage to these tissues will invariably result in loss of
muscle strength.
  Activation of motor units and their *rate of firing* are considered to be
one of the more important factors associated with muscle strength.

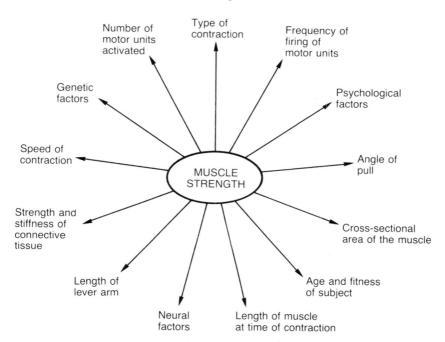

Fig. 10.1 Factors determining muscle strength

Initially as a load is applied, the required number of motor units are recruited and as the load increases more motor units are recruited. When the load becomes heavier still, the rate rather than the number of motor units firing becomes the most important mechanism for the development of muscle force (Astrand and Rodahl, 1986).

The speed of contraction has an important influence on muscle strength. At lower speeds of contraction the opportunity for motor unit recruitment is greater. This is demonstrated by the production of greater torque at lower angular velocities in a concentric contraction (Kisner and Colby, 1990).

Force generation is also dependent upon muscle *length* at the time of contraction (Currier and Kumar, 1982). A muscle is able to generate greater force during an isometric contraction when in a slightly lengthened state (Williams and Stutzman, 1959). Force generation is lower in muscles that are already shortened due to disturbances in sarcomere structure and failure of the activation processes involved in crossbridge formation (Berne and Levy, 1990).

The three *types of muscle contraction* produce variations of force: greatest force is produced during an eccentric (lengthening) contraction, less in an isometric (static) contraction and least in a concentric (shortening) contraction (Singh and Karpovich, 1966).

The role of the *central and peripheral nervous systems* in the determination of muscle strength has long been recognized. Neurophysiological explanations which underlie strength gain have included selective fibre type recruitment, central inhibition on the motor neuron, motor unit synchronization, impulse irradiation and CNS maturation, but currently their individual mechanisms of function are not fully understood (Klausen, 1990).

Level of fitness and *age* of the subject can dictate strength gains. Strength peaks during the early twenties (Galley and Forster, 1987) and gradually declines thereafter. The strength of the female is approximately two-thirds that of the male (Astrand and Rodahl, 1986), with male strength usually attributed to the male sex hormones.

Psychological factors are also associated with gains in muscle strength. A subject must be motivated to produce a maximum contraction, whether this is achieved during an isokinetic evaluation programme in the physiotherapy department or whether it is a case of 'needs must', as demonstrated by 'superhuman' feats like lifting a car. During these situations, it is known that the levels of the hormones epinephrine and norepinephrine (adrenaline and noradrenaline) increase, but the mechanism of their effect still remains elusive (Astrand and Rodhal, 1986).

Biomechanical aspects of muscle strength, such as angle of pull and the principle of leverage, have been detailed in Chapter 1.

TYPES OF MUSCLE STRENGTH

Muscle strength can be defined as either static or dynamic. Static strength is demonstrated in an isometric contraction, whereas dynamic strength is observed during an isotonic concentric or eccentric contraction. Functional activities involve a combination of all contraction types. For example, loss of static strength around the shoulder joint will result in the loss of proximal stability on which to perform dynamic activities such as writing, painting and playing racket sports.

Measuring muscle strength

See Chapter 18.

Muscle power and work

The term 'muscle power' is frequently used when discussing muscle strengthening programmes. Physics textbooks describe power as 'the rate of working' and work as the 'distance moved by a force' and both are defined according to the formulae where:

$$\text{Work} = \text{Force} \times \text{Distance moved by the force}$$

$$\text{Power} = \frac{\text{Force} \times \text{Distance moved by the force}}{\text{Time}}$$

By convention, during a concentric contraction a muscle is doing positive work (i.e. work is being done by the muscle on the object being moved). Conversely, when contracting eccentrically it is doing negative work (i.e. it has work done to it while exerting a force on the object being moved). Power is thus 'produced' during a concentric contraction and 'absorbed' in an eccentric contraction (Enoka, 1988).

PRINCIPLES OF STRENGTH TRAINING

The overload principle forms the basis of strength training. It is based on prescribing exercises which tax the muscle groups towards their maximum capacity and beyond their usual functioning capacity (Enoka, 1988). The model in Figure 10.2 illustrates the overload principle.

Initial strength gains are rapid but if the therapist fails to recognize plateauing of progress (i.e. stage 2), the training programme will then serve only to improve endurance and have little if any effect on strength.

MUSCLE CONTRACTION TYPES AND EXERCISE

A muscle is capable of contracting in three different ways: isotonically, isometrically and isokinetically.

Isotonic contraction

Muscles are contracted against a constant load, with the body segment moving against the load through a range of movement.

Although the 'functional' contraction of the human body, the main disadvantage of using an isotonic contraction to re-educate strength is

Stage 1	Stage 2	Stage 3	Stage 4
Initial rapid increase in strength demonstrated by increasing ability to lift a known training load	Plateauing of improvement with same known training load	Endurance increase only with no further increase in strength if load is not progressed	Training load must be increased for further increase in strength

Fig. 10.2 The overload principle

that it does not impose maximal tension and work demands throughout the entire range of motion.

The resistance offered to the moving body segment throughout the range remains constant (i.e. a 2 kg load), but the resistance offered to the muscle is not constant because of the modifying effects of the lever system through which it must pass.

The resistance has greatest mechanical advantage on the muscle at the extremes of joint range because it is here that the muscle is either maximally lengthened or contracted. Close to mid-range the lever is more efficient, the fixed resistance has least effect and the load on the muscle is proportionately less.

For the purposes of exercise, the largest load to be applied to a muscle is that which the muscle is able to lift at its weakest point. The muscle is normally weakest at its extremes of range.

Advantages of isotonic exercise

- the programme can be easily regulated
- measured increments of progression are readily available
- a variety of equipment can be used to provide resistance
- relatively inexpensive
- can be used as a home exercise
- patients can see strength increases and are motivated to achieve
- involves concentric and eccentric contractions
- allows objective documentation
- closely approximates to functional activity
- can be used to increase muscular endurance
- can be used to reduce swelling and improve circulation.

Disadvantages of isotonic exercise

- maximum loading occurs only at the weakest part of the range of motion
- does not accommodate for pain, fatigue, or the musculoskeletal leverage system
- fatigue causes a decrease, or compromise, in range of motion
- difficult to measure velocity, work and power – problems with reproducibility
- difficult to achieve controlled exercise at faster velocities
- momentum factor cannot be controlled
- eccentric exercise can cause delayed onset of muscle soreness (DOMS) (Chapter 11).

Isometric contraction

An isometric contraction occurs when the resistance acting on the skeletal lever is of sufficient magnitude to prevent motion.

A near maximum voluntary contraction (MVC) maintained for at least 3 seconds is considered to be the most efficient way of training isometric strength. Additionally, the number of contractions per training session should be reasonably high (Klausen, 1990). Currently, there is no consensus in regard to: (a) the number of contractions per session, (b) the number of exercise sessions which should be carried out per week, and (c) the length of the rest periods.

Advantages of isometric exercise

- involves little or no equipment
- easily taught and performed
- can be carried out at home
- assists in the prevention of muscle atrophy during periods of immobilization
- can be used when movement is contraindicated, e.g. pain, recent surgical procedure
- assists in the promotion of circulation
- can be performed at selected points within range.

Disadvantages

- an isometric contraction performed in isolation is not a functionally normal activity
- difficult to quantify objectively
- minimal if any improvement in muscular endurance
- it can increase heart rate and blood pressure
- strength gains are specific to the training angle only (Lindh, 1979)
- patient does not see improvement – motivation to continue may be difficult
- the concept can be difficult for the patient to understand
- the exercise can be boring to perform

Isometric protocol example

The isometric protocol suggested by Davies (1985) is summarized as:

Sets = 10, Repetitions = 10, Contraction Time = 10 seconds, Range of motion = performed throughout range at 10° increments

Isotonic protocols will be dealt with later in the chapter.

Isokinetic contraction (see Chapter 11)

An isokinetic contraction is a dynamic contraction which takes place at a constant angular velocity. A muscle cannot perform an isokinetic contraction without mechanical assistance.

PHYSIOLOGICAL EFFECTS OF STRENGTH TRAINING

Hypertrophy

Hypertrophy which results in an increase in cross-sectional area of the muscle is a usual consequence of strength training; however, it is not always a prerequisite since increased strength has been demonstrated without accompanying hypertrophy of individual muscle fibres (McArdle *et al*., 1994).

The mechanisms of hypertrophy are not fully understood, but several theories feature in the current literature. It has been explained by:

(a) An increase in the size of individual muscle fibres but with no increase in fibre number.

(b) An increase in fibre number.

Hyperplasia is hypertrophy caused by an increase in fibre number. Two explanations of hyperplasia have been put forward: Gonyea (1980) explains hyperplasia as a longitudinal splitting of the muscle fibre; Salleo *et al*. (1980) put forward another theory – the development of new fibres from satellite cells which when activated divide, fuse and form new muscle fibres. Evidence for hyperplasia has been reviewed on the basis of animal studies only.

(c) An increase in protein synthesis (Goldberg *et al*., 1975)

Capillary density and mitochondrial volume

Strength training may also cause a decrease in capillary density and mitochondrial volume density changes which incidentally are detrimental to endurance performance (Sale *et al*., 1990).

With an increase in fibre size there is an increased area which must be supplied by the capillary. This might partly explain why muscles trained purely for strength have a relatively poor aerobic (endurance) capacity.

Conversely, endurance training will increase mitochondrial volume and capillary density. However, some strength training programmes have increased endurance and have produced small increases in VO_2 max and some endurance programmes have resulted in increased strength and muscle fibre size (Sale *et al*., 1990).

RE-EDUCATING STRENGTH DURING EARLY STAGE REHABILITATION

Early stage rehabilitation is defined here where muscles have atrophied to grade 2 and below on the Oxford scale, i.e. from no voluntary contraction to a full range contraction with gravity counterbalanced.

The general principles of strength training in the early stages involve:
- reducing pain and improving circulation
- maintaining range of movement and circulation using passive movements (Chapters 5 and 13)
- progressing to auto-assisted and manually assisted movement
- ensuring correct postural alignment and support to maximize proprioceptive input and to prevent further damage

To facilitate movement:
- counterbalance gravitational resistance by appropriate positioning of the patient, e.g. perform abduction and adduction in supine lying
- minimize frictional resistance by using a sliding board with/without talcum powder
- provide external assistance if necessary using, for example, suspension therapy, hydrotherapy
- use the stretch reflex to initiate movement, i.e. a controlled short sharp stretch at the start of the movement may enable the patient to initiate a muscle contraction and then be able to continue with the movement through the desired range
- use 'quick ice', brushing and/or mechanical vibration (Goff, 1969, 1972)
- use electrotherapy modalities such as neurotrophic stimulation (Kidd *et al.*, 1988) and interferential therapy (Low and Reed, 1990).

PRINCIPLE OF SPECIFICITY OF TRAINING

Specificity of training means that exercise adaptations are directly related to the nature of the exercise stimulus (Kraemer and Koziris, 1992). For example, if strength is trained isometrically and measured isotonically and isometrically, the latter form of measurement will show the greatest gains.

During any strength programme there is never 100% carry-over to a sporting or functional activity. Some programmes will provide more carry-over than others, since their patterns of neuromuscular recruitment are similar to those of the required activity (Kraemer and Koziris, 1992). The only way of achieving 100% carry-over is to train for a sport or functional activity by actually using that specific activity.

This could imply that all that is necessary in rehabilitation is to teach the patient to practise the sport or functional task in order to regain optimum performance capabilities, assuming of course that the injury permits training at functional level. One could not help but question the need for exercise prescription and exercise in therapy if this was the case. It would be very convenient to go and play football all day long with the player who needs to be ready for the start of the season. It would not be unreasonable to walk up and down a ward with a patient who needs to be able to walk to the bottom of the garden and back!

There are many reasons why training using the functional activity is not the ultimate solution to successful rehabilitation. The main ones are outlined here:

1. The patient would only gain sufficient strength to carry out the activity for which training had been given. There would be nothing in reserve to cope with additional demands on strength, co-ordination and flexibility, etc.
2. It is difficult to apply the principle of overload – the main principle of strength training – in the context of the motor patterns of movement as used in the functional or sporting activity.
3. Practising the activity may involve a higher risk of injury than a controlled exercise regimen which has been designed to work the muscles in the same way, i.e. same range, same type of contraction, same speed of contraction, etc., as in the functional activity.

Clinical example to illustrate the above points: stair climbing
1. The patients develop sufficient strength, musculoskeletal and cardio-vascular endurance to allow them to climb the department stairs. On discharge they arrive home to find that the lift is out of action and they live on the third floor. Do they sit and rest when they arrive at the first floor? Yes, if they have to, but with a well-planned rehabilitation programme they may now have the necessary reserve to get home.
2. Asking an elderly patient – who may have balance problems – to walk up and down the stairs in the department wearing weighted vests or ankle bands may not be the ideal way to approach rehabilitation. Even if initially acceptable, progression using the overload principle could lead to many problems!
3. Can the therapist be certain at the start of the rehabilitation programme that quadriceps femoris working eccentrically will be able to control the descent or could this result in injury?

FACTORS TO CONSIDER WHEN DESIGNING A STRENGTH PROGRAMME

Kraemer and Koziris (1992) suggest five factors which contribute to the design of successful resistance training programme. These are summarized in Figure 10.3.

When planning a strength training programme, it is highly unlikely for a patient to require strength in isolation and factors such as endurance, flexibility and balance should also be considered.

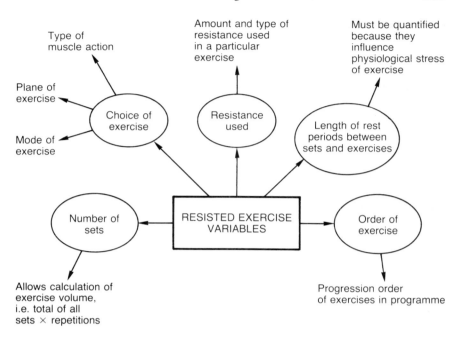

Fig. 10.3 Factors to consider when designing a strength programme

Choice of exercise

This depends on the nature of the injury and on the muscle groups involved. Knowledge of normal biomechanical function of the damaged area, together with an appreciation of the rehabilitation needs of the patient, will assist in the design of the optimal exercise.

The type of muscle action, the range through which it must function and the plane of movement in which the exercise should be performed are the main considerations in the choice of exercise.

The type of exercise chosen will probably dictate the method of application of the resistance. For example, a patient who has recently sustained a fractured humerus will need to work on shoulder stability. This can be achieved using isometric contractions of all shoulder muscles, preferably performed in all planes within the existing range of movement. Manual resistance ensures that the therapist is able to ensure adequate control in this situation.

Order of exercise

Historically, and as a general rule when training for strength, large muscle groups are exercised before the smaller groups. Complex exercises also tend to precede the simpler ones and push and pull movements are often paired.

Exercise intensity

The intensity of the exercise is determined by the mass to be lifted. The yardstick by which this is measured is known as the one repetition maximum (1RM) – see later in chapter.

Length of rest periods

Rest periods of longer than 2 minutes between sets are used for the optimum development of strength and power. A set consists of the number of repetitions performed without a rest, e.g. one set may consists of six repetitions. The longer rest period allows for greater recovery and the ability to lift heavier loads with each set of exercise. If hypertrophy is the main aim, then a high volume of exercise (sets × repetitions) and short rest periods of 1 minute is performed.

Number of sets

Multiple sets of between three and eight, rather than single sets, have been used. The optimal number of exercise sets depend on (a) the type of muscle group, (b) the training level of the patient, and (c) the training aim for the muscle group.

Table 10.1 gives an example of how to break down and plan an exercise programme. This example relates to the analysis of quadriceps femoris during star climbing. Similar analyses can be applied to any functional activity.

Table 10.1 Exercise analysis

Basic analysis	Exercise prescription	Exercise example
Concentric quadriceps to extend the knee in space to allow the foot to strike the step	*Open kinetic chain concentric quadriceps	Inner range quadriceps using a small wedge
Minimal controlled knee flexion as foot makes contact to absorb shock	*Closed kinetic chain eccentric quadriceps	Weight-bearing 'collapse and reform' knee dips
Knee extension as whole body arrives on step	*Closed kinetic chain concentric quadriceps	Cycling or weight-bearing 'collapse and reform' knee dips
Combine all of the above	Functional activities	Practise stair climbing

* See Chapter 6.

METHODS USED TO TRAIN MUSCLE STRENGTH

Subject's own body weight

Auto-resistance (using the patient's own body weight) is one of the most convenient methods of applying static, concentric and eccentric resistance and yet there is a tendency to overlook its use in favour of the more complex and glamorous equipment.

Activities such as press-ups, chin-ups, knee dips and ski sits are examples of body weight resistance. The effect of these exercises can be enhanced by wearing weighted vests (Bosco *et al.*, 1984), by holding dumbbells or by using a medicine ball.

For a somewhat different reason, weighted body belts have also been used to treat the neurologically impaired patient suffering from ataxia, the aim being in this case not to strengthen but to facilitate an increase in proprioceptive input (Hewer *et al.*, 1972).

Auto-resistance can also be provided with latex rubber strips which are frequently used in aerobics classes! Theraband, available in colour coded rolls of varying resistances, is an example of this type of resistance (Medical Equipment Services Ltd, Nottingham).

Free weights

Free weights, such as sandbags, dumbbells, weighted belts and boots, feature prominently in many physiotherapy departments. They are easy to use and are of a known fixed weight. Progression is possible by adding additional weights to the weighted belt or boot or by changing the dumbbell for the next size. The main disadvantage of this type of apparatus is that it is only able to apply maximal resistance at the weakest point in the muscle range. However, by altering patient position it is possible to apply maximal resistance at a different point in the range.

For example, by exercising the knee flexors in standing with a weighted ankle cuff, zero torque will be produced when the knee is fully extended. As the patient begins to bend the knee, maximum torque is produced when the knee is flexed to 90° and this then decreases as knee flexion increases. In prone, maximum torque is produced at full knee extension. This then decreases as the knee is flexed towards 90°. Knee flexion past 90° is no longer controlled by the hamstrings but by eccentric activity of the quadriceps. With a little thought, the same principle can be applied to most muscle groups.

Progressive resistance exercise (PRE)

The term 'progressive resisted exercise' is synonymous with the terms 'heavy resisted exercise' and 'progressive resistance training'. Devised by De Lorme in 1946 the PRE method involves using the 10RM to carry out a set programme of isotonic exercise:

- the 1RM is the maximum load which can be lifted once through a prescribed range
- the 2RM would be the maximal load which can be lifted twice and the 3RM three times and so on up to 10 and beyond.

The 1RM is used as a measure of the maximal concentric force developed in a muscle group. Its assessment involves a series of single repetitions with progressively heavier weights until the 1RM is achieved.

The 1RM test for quadriceps femoris

The 1RM test can be performed in an open or closed environment, i.e. as an open or closed kinetic chain exercise.

Test warm-up
An adequate warm-up should precede the 1RM test. It is suggested that the warm-up should consist of a series of quadriceps and hamstring stretches through full range, a 5-minute moderately hard session at 60 rpm on a cycle-ergometer and several rehearsals of the lift without the weight.

Apparatus
The leg press training apparatus (closed kinetic chain) can be used to test the 1RM. Alternatively, but possibly at greater risk for injury, an open chain kinetic method can be employed using the De Lorme boot. First the boot is attached to the patient's shoe and then weights are applied to it, while taking care to support fully the thigh and to place both feet on a stool.

Method
- Leg press. A series of single repetitions of knee extension with progressively heavier weights are performed until the 1RM is achieved.
- De Lorme boot. The upper body is supported and the patient is instructed to extend the knee fully ('lock out') and to hold until the assessor is satisfied that the attempt has been successful.

Using the 1RM to train strength
The starting point for the rehabilitation of strength will depend on factors such as existing level of fitness, nature of injury and functional requirements for the patient.

Repetition maxima of up to 10RM are used to train strength. Repetition maxima of greater than 12 are used to train muscular endurance. Kraemer and Koziris (1992) classify RMs according to Table 10.2.

It has been shown that by varying RM loads during a training programme, greater strength gains occur than if a constant repetition/load protocol is used. Three popular protocols, the MacQueen, De Lorme and Watkins and Zinovieff, are outlined below. The MacQueen (1954, 1956) power and hypertrophy protocols provide examples of varying and constant repetition/load protocols, respectively (Table 10.3). Table 10.4 gives the De Lorme and Watkins (1951) protocol. The Oxford protocol, designed by Zinovieff (1951), involves a progressively decreasing regimen of 10 lifts of 10RM, followed by a rest, and then 10 lifts of 10RM minus 1 lb, the same rest period, and then 10RM minus 2 lb etc. The total number of lifts per session performed is 100 and this is carried out daily for 5 days. Performing the Oxford technique today would probably require a quick conversion calculation of pounds (lb) to kilograms (kg), where 2.2 lbs = 1 kg.

Table 10.2 Classification of repetition maxima

3–5RM	Heavy load	Strength
8–10RM	Moderate load	Strength
12–15RM	Light load	Endurance

Table 10.3 MacQueen protocols

Power protocol *(varying)*	*Hypertrophy protocol* *(constant rep./load)*
10 lifts with 10RM	10 lifts with 10RM
8 lifts with 8RM	10 lifts with 10RM
6 lifts with 6RM	10 lifts with 10RM
4 lifts with 4RM	10 lifts with 10RM
2 lifts with 2RM	

Table 10.4 De Lorme and Watkins protocol

De Lorme and Watkins programme	*Progression*
10 lifts with $\frac{1}{2} \times$ 10RM	Exercise performed 5 times per week
Rest	1RM and 10RM retested each week
10 lifts with $\frac{3}{4} \times$ 10RM	
Rest	
10 lifts with 10RM	

Isokinetic exercise regimens

For a detailed description of isokinetic exercise the reader is referred to Chapter 11.

Manual resistance

A well-known manual method for gaining strength are the techniques of proprioceptive neuromuscular facilitation (PNF). For a detailed description of all of the techniques the reader is referred to Voss *et al.* (1985).

One of the techniques, repeated contractions, is summarized below. Two others, rhythmic stabilizations and slow reversals, together with a brief summary of the philosophy of PNF, are described in Chapter 9.

Repeated contractions

This technique is used to promote motor learning and to develop strength and endurance into the shortened range. There are three methods of repeated contractions which are used clinically.

Method 1
The primitive repeated contraction involves an initial stretch with an isotonic contraction through range, followed by a passive return to the starting position.

Method 2
The first advanced method of repeated contractions also involves an isotonic contraction through range, but this time the movement is interspersed with a number of isometric 'holds'.

The pattern is initiated with a quick stretch to stimulate the stretch reflex and once started the therapist instructs the patient to 'hold' at a certain point within the range for a few seconds.

The movement in the same direction is then continued with the instruction to 'now pull' (if moving in the direction of flexion) or 'now push' (if moving in the direction of extension). A second or even a third 'hold' can be incorporated into the pattern, depending on the available range and on patient tolerance. Movement after each hold is initiated by a facilitatory stretch.

Method 3
The second advanced method involves an initial stretch and then an isotonic movement through range until a point is reached within range where muscle strength is weak. At this point the therapist facilitates an isometric contraction by instructing the patient to 'hold'. Thereafter a series of isotonic movements preceded by a quick stretch into range is

encouraged using the command 'now pull', 'now pull', 'now pull' or 'now push', 'now push', 'now push'.

Unlike method 2, where there were a series of isometric contractions through range, there is only one in this method.

Mechanical resistance

Pulleys

Pulley systems such as the Westminster pulley (Duffield Medical Ltd, Derbyshire) (Figure 10.4) provide a useful method of strengthening upper and lower limbs. They can be free standing, but are often wall mounted where they do not need as much room.

The patient should never be told to 'go on the pulleys for ten minutes' without supervision following thorough assessment and evaluation of the current level of muscle function. When used correctly, pulleys enable the patient to exercise concentrically, eccentrically or isometrically whether on bedrest, confined to a wheelchair or fully mobile.

Failure to recognize that the functional diagonal patterns of PNF (Voss *et al.*, 1985) can be used with pulley systems is common among therapists. Although it is difficult to achieve controlled limb rotation, it is possible to set up and incorporate most of the limb diagonals safely into the exercise regimen.

Springs

Springs can be used in suspension therapy and incorporated into pulley circuits. They can also be used in isolation to provide mechanical resistance. Their importance in rehabilitation, however, is tending to be superseded by multi-gym equipment which is readily available and easy to set up. The clinical use of springs has been extensively covered, along with that of pulleys and suspension, in Hollis (1981).

Suspension therapy

This method of treatment was very popular, as evidenced by the number of metal frames (known as 'Guthrie Smith' frames) suspended from the ceilings of physiotherapy departments. Within the past 10 years, however, popularity has decreased and it is now only used selectively.

Uses
During late stage rehabilitation it is of minimal importance, but it can be used successfully in the early stages when lengthening and strengthening soft tissue in preparation for surgery or to gain range in

Fig. 10.4 The Westminster pulley

a controlled environment following a joint replacement. In these examples it functions to eliminate frictional resistance from the moving body part and can promote a feeling of weightlessness such as that experienced by buoyancy in the hydrotherapy pool. It is also beneficial when used to provide support to enable exercise of a painful joint to take place when patients are unable to provide their own support.

Method
Suspension therapy involves 'suspending' a limb above a supporting surface using slings and ropes. To set up a patient for suspension therapy, the physiotherapist should take into account all biomechanical and anatomical aspects which may affect the suspended limb including, e.g.:

• localization of hip abduction in supine by placing the non-exercising limb over the edge of the plinth
• prevention of lumbar spine involvement during hip flexion and extension by supporting the pelvis manually during exercise
• determination of the plane of upper limb exercise, i.e. in the plane of the scapula or in the anatomical plane
• determination of patient comfort and whether the movement 'looks right'.

Types of suspension
Axial fixation. The fixation point for all ropes attached to the slings which support the limb to be moved is located immediately above the centre of the moving joint (Figure 10.5). For example, when mobilizing the hip joint and/or strengthening the hip abductors, the point of suspension is positioned vertically above the anterior marking of the centre of the hip joint which is located 1.2 cm below the middle third of the inguinal ligament (Warwick and Williams, 1973).

Axial fixation allows maximum movement of the hip joint to occur. During the exercise, when passing through the middle of the arc of movement, the lower limb movement is parallel to the plinth.

Vertical fixation. The fixation points for limb suspension are placed immediately above the centre of gravity of the limb segment, i.e. at approximately the junction of the upper and middle third. The arc of movement transcribed by the limb will be orientated vertically to the plinth rather than horizontally as was the case with axial fixation.

Vertical suspension is used primarily for support and it is often combined with axial fixation to achieve the required outcome such as when using suspension therapy to re-educate range of movement and strength at the knee joint. In this example, the patient is positioned in side lying and the thigh is supported using vertical suspension. The fixation point of the thigh lies above its centre of gravity, while the rope supporting the foot is suspended axially above the centre of the knee joint (Figure 10.6).

Re-educating strength with suspension therapy
1. Apply manual resistance to the suspended limb. Use the principle of leverage to progress the exercise (Chapter 1).

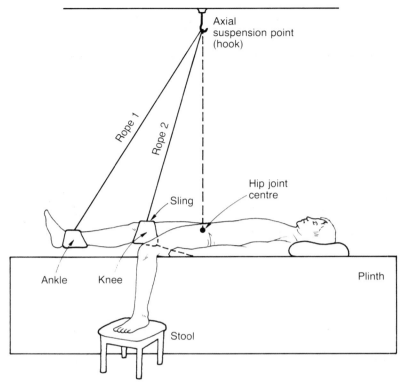

Fig. 10.5 Axial suspension for hip abduction/adduction

2. Displace the fixation point to elongate (move towards outer range) or to shorten (move towards inner range) the exercising muscle.
3. Incorporate spring resistance into the suspension regimen.
4. Use voice, e.g. 'swing and swing and swing and hold'.

When using suspension therapy, all apparatus and accessories must be tested before use and its adjustment when the patient has been positioned for treatment should be kept to a minimum.

Care must be exercised when suspending the patient. Hyperextension of the knee and elbow joints must not be allowed when raising or lowering the apparatus. To avoid this, body segments must be raised and lowered in the correct order. For example, when suspending the lower limb with the patient in supine, the proximal sling should be raised first, followed by the distal sling.

Localization – manually or by positioning – of the joint being treated is essential to avoid unnecessary strain on joint structures. For example, if mobilizing abduction and adduction of the hip joint, the other lower limb should be abducted over the edge of the plinth and the foot supported on a stool. This position will in effect 'fix' the pelvis and assist localization of the movement.

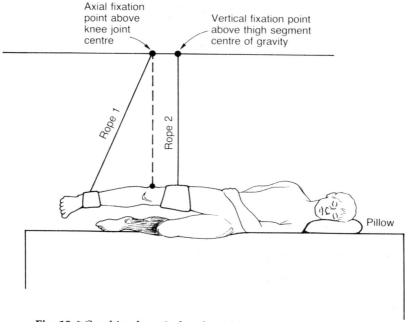

Fig. 10.6 Combined vertical and axial suspension for the knee

CLINICAL EXAMPLES OF MUSCLE STRENGTHENING EXERCISES

Lower limb

1. Controlled side lunges are preferable to deep knee squats. The foot on the side of the lunge should be pointing in the direction of the knee flexion.
2. Pseudo closed kinetic chain dips from a step using body weight as a resistance.
3. 'Collapse and reform' partial knee squats or dips using a weight to provide additional resistance.
4. Open chain quadriceps exercises in a semi-reclined position using a wedge.
5. Plyometric type zigzag bounce over bench (see later).
6. Leg press using mechanical gymnastic equipment.
7. Isokinetic dynamometry (Chapter 11).
8. Closed kinetic chain knee extensions using a rowing machine.
9. Open chain pulley circuit knee extension resistance.

Upper limb

1. Standing – medicine ball thrust forwards, upwards, to the left and to the right.

2. Press-ups – half or three-quarter if unable to do a full press-up.
3. Mechanical gymnastic equipment adductor/abductor press.
4. Westminster pulley using the proprioceptive neuromuscular diagonals of the upper limb and the anatomical planes.
5. Free weight dumbbells lifted anteriorly (overhead press) and laterally (flys).
6. Isokinetic dynamometer limb diagonal patterns.
7. Rowing machine.
8. Arm curls using dumbbells or sandbags.

ECCENTRIC TRAINING

Interest in eccentric training is relatively recent, having been inspired with the introduction of isokinetic dynamometers which enable its quantification.

Several aspects of eccentric contraction are important for rehabilitation and these are outlined:

1. The majority of eccentric muscle work of the lower limb takes place in the closed kinetic chain situation. For example, during gait at heel strike, tibialis anterior contracts eccentrically by lowering the foot to the ground to prevent 'foot slap' and quadriceps femoris acts as a shock absorber by allowing the knee to flex slightly again at heel strike (Chapter 15).
2. Eccentric mechanisms feature prominently in the biomechanics of muscle injury (Duncan *et al.*, 1989). Additionally, delayed onset muscle soreness (DOMS) is a manifestation of eccentric exercise. The signs of damage begin at the end of the exercise period and continue for a period of days following the cessation of all activity (Jones *et al.*, 1986).
3. Greatest force is developed in an eccentric contraction, less in an isometric and least in a concentric contraction (Singh and Karpovich, 1966).

Albert (1991) describes the physiological, biomechanical and clinical characteristics of eccentric muscle loading. These are summarized briefly in Table 10.5.

Using eccentric exercise in the clinical setting

Eccentric exercise training should not be used exclusively but be incorporated into a strengthening programme. A variety of apparatus and techniques are available for strengthening eccentrically, including free weights, isokinetic dynamometry, elastic bands and proprioceptive neuromuscular facilitation.

Table 10.5 The characteristics of eccentric loading

Electromyogram	EMG activity lower than for concentric contraction but similar to that of an isometric contraction
Oxygen utilization	Uses less oxygen than a concentric contraction at same work intensity levels
Blood pressure	A rise occurs, but this tends to depend on effort rather than on muscular activity
Body temperature	Skin and muscle temperature are slightly higher (3 °C, 1.2 °C, respectively) during eccentric rather than concentric work
Psychological parameters	Loads do not 'feel' as heavy, but the subject has feelings of weakness and instability after exercise Pure eccentric training regimens feel unnatural
Endurance factors	Slow-velocity contractions very fatiguing High-velocity contractions more fatigue resistant
Force/velocity relationship	Controversial at present A plateauing or increase in force generation with an increased speed of movement is suggested
Neuromuscular control	Poorly understood ? peripheral or central Known to provide protective mechanism during peak loads which prevents complete muscle rupture
Energy utilization	Lower ATP-CP utilization in an anaerobically contracting muscle

Clinical example

Manual facilitation of the eccentric contraction can be particularly effective in the early stages following lower limb surgery when the patient is required to achieve a straight leg raise before being allowed up. They often respond well to an initial 'assisted' eccentric quadriceps contraction (i.e. lowering the leg slowly back down onto the bed guided and supported by the therapist). The concentric contraction which is attempted immediately afterwards is often then easier to achieve.

When to start eccentric contractions following soft-tissue injury

It has been suggested that eccentric contractions should not be performed within the first 30 hours following injury and then they should only be performed when a pain-free concentric contraction has been achieved (Walmsley *et al.*, 1986).

To re-educate eccentric contractions, begin with submaximal slow-velocity contractions and gradually progress according to patient tolerance. The question 'is it pain or is it strain?' provides a guide. If the contraction is painful, then it is likely that damage is occurring. Strain is acceptable and a healthy manifestation of rehabilitation.

Examples of eccentric exercises for the lower limb

- Closed kinetic chain: rowing machine, cycle-ergometer, ski-sit against a wall.
- Open kinetic chain: controlled knee extension to knee flexion using free weight, functional activities wearing ankle weights, e.g. walking, spring- and pulley-resisted exercise.

PLYOMETRICS

The principle of plyometric training is to achieve a vigorous lengthening (eccentric contraction) of a muscle group and immediately follow this with a maximal concentric contraction. Strength training using plyometrics is virtually exclusive to the lower limb muscles and it is primarily used in the sporting world to gain explosive power for, e.g., sprinting and long and high jump ability (Blattner and Noble, 1979).

To train plyometrically, the subject is required to drop from a height into a full squat and then to jump as high as possible back into a standing position onto another support placed about 1–2 m away (Figure 10.7).

Plyometric training is not without risk and subjects are prone to muscle injury, particularly at touch-down and take-off from the floor. As a rehabilitation tool it should be used with caution, but its use will

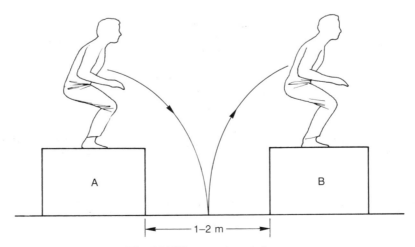

Fig. 10.7 Plyometric training

obviously depend on patient needs and only be used if a high level of rehabilitation is required for sporting activity.

ENDURANCE EXERCISE

There are two types of endurance: muscular and cardiovascular.

Muscular endurance is the ability of a muscle to contract repeatedly or to generate tension and sustain that tension over a prolonged period of time (Kisner and Colby, 1990). Local endurance can be built up in isolated muscle groups by specific repetitive low-intensity exercise. If many muscles are involved in an exercise, then cardiovascular endurance becomes an important factor.

All rehabilitation programmes must contain an element of endurance exercise because endurance is necessary to enable patients to function in day-to-day activities. Cardiovascular and muscular endurance can diminish very quickly in patients who have spent time on bedrest.

Figure 10.8 summarizes the short-term and long-term effects of endurance training.

Exercise prescription for endurance

Endurance programmes consist of submaximal, repetitive exercise, i.e. high-repetition, low-resistance exercise. Their design, as with any exercise programme, will depend on individual requirements.

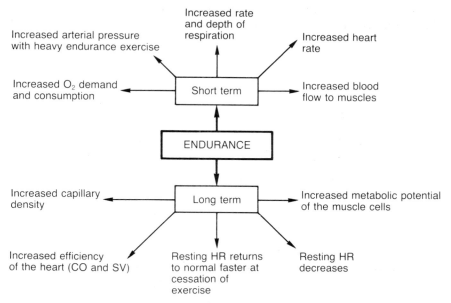

Fig. 10.8

Examples of endurance exercise

Cycle-ergometry
The cycle-ergometer is often used to promote lower limb and cardio-vascular endurance. It is advantageous since compression weight-bearing forces are eliminated, and so it can be suitable for patients who are unable as yet to weight-bear following lower limb injury.

Exercise prescription should preferably be based on the results of a fitness test (Chapter 7) and generally involves the patient cycling at about 70–75% of their maximal heart rate. Initially the exercise time should be short – 5–10 minutes – but progressively increased according to patient tolerance.

Swimming
Swimming is another activity which can promote muscular endurance. As in cycle-ergometry, patients who are unable to weight-bear on dry land will benefit from hydrotherapy.

Treadmill
Walking and running on a treadmill are performed to train endurance. Care should be taken when teaching patients to use a treadmill, as some patients will be unsuited to this form of training such as elderly patients with balance problems and young children.

Stepping exercises
Endurance can be progressed and assessed by performing the Harvard Step Test described in Chapter 7.

Isokinetic exercise (Chapter 11)
Isotonic open or closed kinetic chain exercise using in excess of 12RM
The 1RM should be reassessed each week and the 12RM progressed accordingly.

Functional activities
Functional activity can also be used to train endurance, but this should be supplemented with more specific exercise.

Endurance and changes in muscle fibre type

For a detailed description of the types of muscle fibre refer to Chapter 3. There are three types of muscle fibre (Ganong, 1991):

- the red, type I, slow oxidative (SO) postural muscles
- the white type IIB, fast glycolytic (FG) skill muscles
- the red type IA, fast oxidative muscles (rare in humans).

Slow oxidative fibres in the sedentary adult male and female make up about 45–55% of the total muscle composition. This figure is very different in the trained athlete, however, with power athletes demonstrating considerably less, and endurance athletes notably greater, percentages.

A constant source of interest in muscle physiology is whether specific training changes fibre types from (SO) to (FG) and vice versa. The consensus at present supports the transformation of muscle fibre in both directions with prolonged specific training (McArdle *et al.*, 1994). Additionally, these authors suggest that some of the type IIA fibres might actually be 'transition fibres' which are capable of taking on the characteristics and functions of SO fibres with training.

SAFETY ASPECTS OF THE STRENGTH PROGRAMME

Dangers and precautions associated with exercise planning are described in Chapter 6. Several aspects, outlined below, are considered to be specifically relevant to strength training:

1. A thorough warm-up and cool-down is essential.
2. Ensure that the exercise is biomechanically safe to perform.
3. Avoid exercise regimens which fail to promote muscle balance, e.g. postural imbalance can result from a concentrated unilateral strengthening regimen in racket players.
4. Consider the principle of specificity of training and provide strengthening exercises which mimic muscle function during a particular activity, e.g. a painter will spend a lot of time using deltoid in inner range if he paints many ceilings! Exercises performed in the department should be designed to strengthen in this range and enable him to use the same muscle work that he would use when painting. For example, open kinetic, free weight dumbbell exercise used to work deltoid in mid to inner ranges of abduction and flexion would be suitable, because painting is an open kinetic chain activity.
5. Stretch muscles to be strengthened before and after the exercise programme.
6. Progress slowly and do not allow the patient to attempt to overcome resistances which may cause damage to the exercised muscles.
7. Demonstrate and teach a new exercise. Initially, closely supervise and correct as necessary.

SUMMARY

- Strength is the ability of a muscle to produce tension and force in relation to the demand placed on it.
- The overload principle is the basis of strength training. It involves taxing a muscle group towards its maximal capacity and beyond its usual functional capacity.

- A muscle is capable of contracting in 3 ways: isotonically, isometrically and in certain circumstances, isokinetically.
- The principle of specificity of training means that exercise is adapted in direct relation to the exercise stimulus.
- Methods available to train muscle strength include: use of body weight, free weights, progressive resisted exercise, manual and mechanical resistance.

REFERENCES

Albert, M. (1991) *Eccentric Muscle Training in Sports and Orthopaedics*, Churchill Livingstone, Edinburgh

Astrand, P.-O. and Rodahl, K. (1986) *Textbook of Work Physiology: Physiological Bases of Exercise*, 3rd edn, McGraw-Hill, New York

Berne, R.M. and Levy, M.N. (eds) (1990) *Principles of Physiology*, Wolfe, London

Blattner, S.E. and Noble, L. (1979) Relative effects of isokinetic and plyometric training on vertical jumping performance. *Research Quarterly*, **50**, 583–588

Bosco, C., Zanon, S., Rusko, H. *et al*. (1984) The influence of extra load on the mechanical behaviour of skeletal muscle. *European Journal of Applied Physiology*, **53**, 149–154

Currier, D.P. and Kumar, S. (1982) Knee force during isometric contraction at different hip angles. *Physiotherapy Canada*, **34**(4), 198–202

Davies, G.J. (1985) *A Compendium of Isokinetics in Clinical Usage and Rehabilitation Techniques*, 2nd edn, S. & S. Publishing, La Crosse, WI

De Lorme, T.L. (1946) Heavy resistance exercise. *Archives of Physical Medicine*, **27**, 607–625

De Lorme, T.L. and Watkins, A. (1951) *Progressive Resistance Exercise*, Appleton-Century, New York

Duncan, P.W., Chandler, J.M., Cavanaugh, D.K., Johnson, K.R. and Buehler, A.G. (1989) Mode and speed specificity of eccentric and concentric exercise training. *Journal of Orthopaedic and Sports Physical Therapy*, **11**(2), 70–75

Enoka, R.N. (1988) *Neuromechanical Basis of Kinesiology*, Human Kinetics Publishers, Illinois

Galley, P.M. and Forster, A.L. (1987) *Human Movement: An Introductory Text for Physiotherapy Students*, 2nd edn, Churchill Livingstone, Edinburgh

Ganong, W.F. (1991) *Review of Medical Physiology*, 15th edn, Prentice-Hall, London

Goff, B. (1972) The application of recent advances in neurophysiology to Miss Rood's concept of neuromuscular facilitation. *Physiotherapy*, **58**(12), 409–415

Goff, B. (1969) Excitatory cold. *Physiotherapy*, **55**(11), 467–468

Goldberg, A.L., Etlinger, J.D., Goldspink, D.F. *et al*. (1975) Mechanism of work induced hypertrophy of skeletal muscle. *Medicine and Science in Sports and Exercise*, **7**, 185–198

Gonyea, W.J. (1980) Role of exercise in inducing increases in skeletal muscle fibre number. *Journal of Applied Physiology*, **48**(3), 421–426

Hewer, R.L., Cooper, R. and Morgan, M.H. (1972) An investigation into the value of treating intention tremor by weighting the affected limb. *Brain*, **95**, 579–590

Hollis, M. (1981) *Practical Exercise Therapy*, 3rd edn, Blackwell Scientific Publications, Oxford

Jones, D.A., Newham, D.J., Round, J.M. and Tolfree, S.E.J. (1986) Experimental muscle damage: morphological changes in relation to other indices of damage. *Journal of Physiology*, **375**, 435–448

Kidd, G.L., Oldham, J.A. and Stanley, J.K. (1988) Eutrophic therapy and atrophied muscle: a pilot clinical trial. *Clinical Rehabilitation*, **2**, 219–230

Kisner, C. and Colby, L.A. (1990) *Therapeutic Exercise: Foundations and Techniques*, 2nd edn, F.A. Davis, Philadelphia

Klausen, K. (1990) Strength and weight training. In *Physiology of Sports* (eds T. Reilly, N. Secher, P. Snell and C. Williams), Chapman and Hall, London

Kraemer, W.J. and Koziris, L.P. (1992) Muscle strength training: techniques and considerations. *Physical Therapy Practice*, **2**(1), 54–68

Low, J. and Reed, A. (1990) *Electrotherapy Explained: Principles and Practice*, Butterworth–Heinemann, Oxford

Lindh, M. (1979) Increase of muscle strength from isometric quadriceps exercise at different knee angles. *Scandinavian Journal of Medicine*, **11**, 33

MacQueen, I.J. (1954) Recent advances in the technique of progressive resistance exercise. *British Medical Journal*, **2**, 1193–1198

MacQueen, I.J. (1956) The application of progressive resistance exercise in physiotherapy. *Physiotherapy*, **40**, 83–93

McArdle, W.D., Katch, F.I. and Katch, V.L. (1994) *Essentials of Exercise Physiology*, Lea and Febiger, Philadelphia

Sale, D.G., Jacobs, I., MacDougall, J.D. and Garner, S. (1990) Comparison of two regimens of concurrent strength and endurance training. *Medicine and Science in Sports and Exercise*, **22**(3), 348–356

Salleo, A., Anastasi, G., La Spada, G., Falzea, G. and Denaro, M.G. (1980) New muscle fibre production during compensatory hypertrophy. *Medicine and Science in Sports and Exercise*, **12**, 268–273

Singh, M. and Karpovich, P.V. (1966) Isotonic and isometric forces of forearm flexors and extensors. *Journal of Applied Physiology*, **21**, 1435–1437

Voss, D.E., Ionta, N.K. and Myers, B.J. (1985) *Proprioceptive Neuromuscular Facilitation*, 2nd edn, Harper and Row, Philadelphia

Walmsley, R.P., Pearson, N. and Stymiest, P. (1986) Eccentric wrist extensor contractions and the force velocity relationship in muscle. *Journal of Orthopaedic Sports Physical Therapy*, **8**, 292

Warwick, R. and Williams, P.L. (1973) *Gray's Anatomy*, 35th edn, Longman, Edinburgh

Williams and Stutzman (1959) Strength variation through the ROM. *Physical Therapy Review*, **39**(6), 145–152

Zinovieff, A.N. (1951) Heavy resistance exercise: the Oxford technique. *British Journal of Physical Medicine*, **14**, 129

11. *Isokinetic exercise*

DEFINITION OF ISOKINETIC EXERCISE

Isokinetic exercise, from the Greek derivation *iso* = equal, and kinetics from *kine* = move, is defined as a method that relies on the use of a machine to control the speed of movement over a range of motion (Atha, 1981). It enables the evaluation and training of muscular strength under conditions of constant angular velocity (Dvir, 1991).

This chapter will outline the:

- development of isokinetics
- concept of isokinetic exercise
- importance of gravity correction
- method of performing an isokinetic evaluation
- definitions of evaluation parameters
- interpretation of an evaluation torque tracing
- disadvantages of isokinetic exercise
- advantages of isokinetic exercise
- reliability of isokinetic devices
- role of visual feedback in isokinetic exercise
- isokinetic exercise using diagonal functional movement patterns
- role of the isokinetic dynamometer in the rehabilitation environment
- aspects of isokinetic exercise in the eccentric mode.

DEVELOPMENT OF ISOKINETICS

The isokinetic concept was originally patented in 1962 by Cybex (Ronkonkoma, New York, USA). The patent expired in 1982 which paved the way for other manufacturers to experiment in the field of isokinetic exercise. In 1982 the first Kin-Com (Kinetic Communicator) machine was introduced, to be followed by the Lido in 1984 and the Biodex isokinetic dynamometer in 1986. Currently there are more than eight different companies which manufacture isokinetic dynamometers.

The isokinetic dynamometer was first introduced by Hislop and Perrine (1967). Since then it has been used in rehabilitation, research, orthopaedic and exercise physiology centres throughout the world to provide objective and quantifiable information on dynamic musculo-skeletal performance.

Before its invention, cable tensiometry was used to test muscle strength but this was only capable of measuring isometric (static) strength (Cabri, 1991). The manual muscle grading scale – the Oxford scale (Chapter 18) – is used clinically but it is somewhat subjective and incapable of providing interval or ratio strength data.

The isokinetic system

All isokinetic dynamometers are based on the same basic design principle which consists of a fixed axis with rotating mechanical lever driven either electrically or hydraulically. The lever arm accommodates the movement generated by the contracting muscles in such a way that the distal segment moves along a sector of the joint range at a constant angular velocity. The machine does not provide resistance or measure torque until the limb segment attempts to exceed the preset speed (Rothstein *et al.*, 1987).

All systems require a large amount of floor space and their weight may necessitate floor reinforcement prior to installation. Recent systems consist of a single chair with a tilting back-rest and a dynamometer. The chair and dynamometer can be aligned easily, since they are both capable of rotating 360°.

The modern systems are user friendly and one incorporates a touch

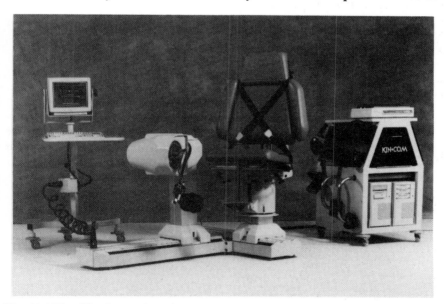

Fig. 11.1 The Kin-Com AP isokinetic dynamometer (courtesy of Chattanooga UK, Oxford)

screen facility which avoids the need for the operator to be computer literate. All exercise and evaluation positions can be set up easily on this machine, with practice, by following the simple instructions on the screen (Kin-Com, Chattanooga Group Inc., Hixon, TN, USA).

Currently, the muscle groups which enjoy most attention from the isokinetic dynamometer are those of the lower limb and particularly those which act on the knee. The musculature of the trunk has also featured in the literature since the advent of specialized isokinetic back machines (Tis *et al.*, 1991).

Passive and active isokinetic systems

Systems are described as being either passive or active. Passive systems were the original dynamometers and were capable only of measuring the moment generated during concentric (shortening) and isometric (static) contractions.

Active systems are the more recent and can still operate passively, but their main asset is that they are capable of driving the distal limb segment at a preset speed in the presence of an eccentric muscle contraction.

Data handling
Data generated by either type of dynamometer are relayed via an analogue to digital converter to a computer which is able to generate the load calculations and display them graphically on the screen. Data can also be printed out as a hard copy and some systems are capable of downloading in ASCII format to allow import into a spread sheet for research purposes.

CONCEPT OF ISOKINETIC EXERCISE

All isokinetic dynamometers allow the therapist to select the appropriate velocity, range of movement, damp setting, moment/force threshold values, number of repetitions and the contraction mode for the patient.

Velocity

Measured in degrees per second (°/sec) current dynamometer velocities range from 1 to 500(°/sec) which incidentally are a lot lower than the velocities generated in many sporting events. For example, velocities of 6180°/sec have been recorded in top-flight baseball pitchers (Perrin, 1993). However, isokinetic velocities are not functional because no muscle contracts through range at a constant velocity.

Angular velocities at which an exercise or evaluation can be performed on current machines are classified as slow (1°–60°/sec), inter-mediate (60°–240°/sec) and fast (over 240°/sec). The most usual clinical testing and training velocities range from 30 to 240°/sec.

The ability of a muscle to generate force varies at different testing velocities. As a general rule, greatest force is generated at the slowest angular velocities and least force at the fastest velocities during a concentric contraction.

Eccentric force generation in the lower limb can decrease (Chandler and Dunan, 1987), but it usually increases (Komi, 1973) or remains constant (Chandler and Duncan, 1987) when the velocity is increased. Eccentric force velocity relationships in the upper extremity tend to demonstrate greater force with an increasing velocity (Walmsley *et al.*, 1986).

Muscle fibre type and velocity of contraction

An important consideration in the selection of a training velocity is the relative contribution made by the different muscle fibre types (Chapter 3). Slow twitch – slow oxidative (SO) – fibres have motor units which respond at lower thresholds; they have low conduction velocities and long twitch contraction times. Fast twitch fibres – fast glycolytic (FT) – have motor units which respond at higher thresholds, have higher conduction velocities and short twitch contraction times.

From this description it would be expected that at slow angular velocities SO fibres would be recruited and, at fast velocities, FT fibres would function. However, current theories favour the recruitment of both fibre types irrespective of exercise velocity (Rothstein *et al.*, 1987). Additionally, while there is a distinct order of recruitment in the concentric contraction, i.e. slow to fast twitch, recruitment in an eccentric contraction appears to be more selective favouring the type IIB (FG) fibres (Korvanen *et al.*, 1984).

Range of movement

The range of movement through which a joint can be exercised is controlled by the isokinetic system and can be preset by the therapist. Mechanical stops positioned at approximately 5° beyond the computer-set start and stop angles are available on some machines and all incorporate a patient-controlled cutout switch. These safety mechan-isms are required in the unlikely event of software control mechanism failure.

Damping and preload setting

All isokinetic systems demonstrate a phenomenon called torque over-shoot. This is a combination of excessive acceleration of the limb segment above the terminal isokinetic velocity and the system's attempt to reduce this by the introduction of a deceleration force. The result is a sudden peak in the torque readings in the first few hundredths of a second at the onset of the movement. To deal with this, the manufacturers have introduced two systems to help minimize the production and collection of artefact data of this type: damping and preload.

There are two forms of damping – damping of the signal and damping of the resistance. The former is simply the omission of the first few hundredths of a second of the torque data. The latter is the gradual introduction of a breaking force up to the isokinetic velocity. The term 'ramping', which is often confused with damping, is the time allotted for this breaking force to occur. In general, the larger the amount of force involved the greater the damp setting required.

Disadvantages of damping

1. Measurement of angle specific torque is confounded by the damp setting – the amount of accuracy is dependent upon the amount of the damp setting.
2. Damping may eliminate some aspects of the torque signal itself.
3. The changes in the analogue scale due to damping are not known and therefore unquantifiable.

In contrast to damping, preload is introduced before movement of the limb segment occurs (Jensen *et al.*, 1991). A preload allows the limb to be pre-accelerated to its terminal isokinetic velocity.

Damping introduces shear forces during the movement, whereas preload allows the shear to take place before the limb segment is released. From a clinical standpoint, the effects of damping and preload on joint impact have yet to be evaluated.

Moment/force threshold values

Dynamometers have torque limits, i.e. the maximum amount of resistance that they can provide. This varies between machines and, when exceeded, an error message is relayed via the screen.

Number of repetitions

The number of sets and the number of repetitions in each set can be programmed by the therapist. It is possible to design an individual protocol for a patient and then to save it on disk until the patient's next attendance. Protocol modification can be carried out easily when required.

Contraction mode

Modern machines provide a wide range of exercise modes – passive, isometric, isotonic and isokinetic – and all can be combined if required in a tailor-made exercise protocol.

Passive mode

The velocity remains constant and no voluntary force is required by the patient to initiate movement. Progress to active assisted movement can be made when in this mode.

Isometric mode

The therapist is able to preset a series of isometric hold angles within the patient's available range of motion.

For example, to train quadriceps at specific angles in range at 30°, at 45°, at 60° and at 75°, the therapist programmes these angles before starting the exercise. The machine passively moves the patient's limb to the first preset angle (30°) and instructs the patient via a robotic voice or via the screen to contract isometrically for a predetermined time period (e.g. 5 seconds). The machine instructs the patient to relax and the limb is then allowed to reposition or is moved passively to the next hold angle and the process repeated.

Isotonic mode

The term 'isotonic' is best regarded only as a 'working definition' because muscle tension never remains constant throughout range as implied by its name (*iso* = same, *tonic* = tension) (Chapter 3).

When using the dynamometer in isotonic mode, the exercise velocity is controlled by the patient and the muscle tension varies throughout the available joint range. Maximum effect of the resistance will be confined to the weakest point in range.

Isokinetic mode

Isokinetic differs from isotonic exercise in that the resistance is variable or accommodated to the force-producing capacity of the muscle. The external load against the moving limb segment always remains in tune with the maximal capacity of the muscle whether the contraction is concentric or eccentric. Additionally, the angular velocity remains constant, unlike during isotonic exercise where it is variable and controlled by the patient.

IMPORTANCE OF GRAVITY CORRECTION

The effect of gravity on limb movement should be taken into account when performing an isokinetic evaluation. Criticism has been levelled at isokinetic studies which have failed to do this (Winter *et al.*, 1981).

For movements in the horizontal plane, as gravity is counterbalanced, there will be no gravitational errors. In vertical movements, the limb and dynamometer are either resisted or assisted by the force of gravity. Torque measurements tend to overestimate the strength of muscles assisted by gravity and to underestimate the strength of muscles opposed by gravity (Perrin *et al.*, 1992).

Modern dynamometers provide the option of being able to correct for the effects of gravity.

Example of the gravity correction procedure

1. Set up the patient for the evaluation as described below.
2. Select the gravity correction option from the menu screen.
3. Follow the instructions to:
   (a) move the limb to an angle within the range to be tested
   (b) record this angle relative to the horizontal plane (conventionally, angles above the horizontal plane are recorded as positive, those below as negative)
   (c) instruct the patient to relax so that the limb rests on the load cell of the dynamometer
   (d) the force of the limb due to gravity is then measured and recorded.

Gravity corrections are calculated by the software using the following relatively simple mathematical formula:

Gravity correction = Force sampled − [(Weight of limb) × cosine θ]

where θ = the angle between the lever arm and the horizontal; force = the force sampled at 1/100th of a second; and limb weight = force in Newtons.

METHOD OF PERFORMING AN ISOKINETIC EVALUATION

A number of factors influence the selection of the test protocol, the most important being:

• diagnosis, e.g. post-surgery, cardiovascular status
• patient age and physique
• stage of rehabilitation
• muscle group to be tested.

Basic rules to be followed when carrying out an evaluation

1. Ensure that the system has been calibrated. Recent machines have inbuilt calibration facilities.
2. Alignment of the axes of rotation of the limb and dynamometer are crucial for reliable and repeatable evaluation outcomes.
3. Always standardize the patient testing position.

Evaluation protocol – an example of an isokinetic lower limb evaluation

1. Assess the patient – subjective and objective assessment.
2. Familiarize the patient with the isokinetic dynamometer.
3. Explain the test aims.
4. Ensure that the patient warms up without dynamometer, e.g. stretches, cycle-ergometer.
5. Position and stabilize the patient accurately on the dynamometer.
6. Test the contralateral limb first.
7. Align the joint and dynamometer axes of rotation as closely as possible.
8. Use gravity correction if testing in a gravity-dependent position.
9. Select the test type, e.g. concentric/eccentric for knee extensors.
10. Select the test velocity, e.g. 30°/sec.
11. Warm up on the dynamometer using the 'warm-up' mode.
12. Perform the maximal test at the chosen velocity, e.g. perform 3 concentric/eccentric repetitions with overlay facility with a 30 sec or 1 min rest between repetitions.
13. Record test details to ensure replication on retest.
14. Retest at the same time of day as the original evaluation was performed.

DEFINITIONS OF EVALUATION PARAMETERS

Enoka (1988) describes human movement as a rotation of body segments about their joint axes and states that the movement is caused by interaction of the forces which are associated with muscle contraction and external loading. He states that it is 'an imbalance between the components of these forces that produce rotation'.

Torque

Torque is the capability of these forces to produce rotation. It is calculated according to the formula:

Torque = Force measured at the load cell × Moment arm

Isokinetic dynamometers can measure either force or torque. If the load cell is located on the lever arm, the dynamometer will directly measure force. If it is positioned at the axis of rotation, it will measure torque. If the distance of the moment (lever) arm from the axis of rotation is known, it is possible to convert force to torque using the above formula.

Torque can be evaluated as: (a) a peak value, (b) as an average peak value, and (c) as an average value. Peak torque is the highest point on the isokinetic curve. The average peak is perhaps a more representative indication of patient performance, since it is the average of a number of peak values taken over several consecutive torque curves. The third value, the average torque, is measured using the entire tracing of one or several consecutive isokinetic curves.

To obtain comparative average values during consecutive evaluations, it is important to ensure that the range of motion through which the limb is tested remains constant throughout. The advantage of obtaining averaged values is that artefacts due to deceleration of the limb and lever arm might have an overall less effect than they would on the peak values (Perrin, 1993).

Torque ratios

Torque values are either often expressed in terms of ratios between the agonist and antagonist or for comparisons of contraction types within the same muscle (Albert, 1991). The use of ratios is somewhat controversial due to lack of standardization of (a) the gravity correction procedure, and (b) the ratio calculation. Normative ratio values are so far unavailable for any muscle group, although torque ratios for the hamstrings and quadriceps are the most frequently reported, with ratios of between 0.43 and 0.90 cited for these muscles (Albert, 1991).

Modern dynamometers are also capable of measuring work and power.

Work

Some dynamometers provide computer software which enables the calculation of the area under a single isokinetic curve. This area is representative of the work done by the muscle group and is reported in the SI unit of measurement known as the joule (J).

Power

Power is determined by measuring the length of time taken to perform either one or a series of contractions. Measured in watts (W) its calculation also relies on the appropriate software.

All of the parameters described above can be evaluated at slow, intermediate or fast angular velocities.

INTERPRETATION OF AN EVALUATION TORQUE TRACING

A typical averaged torque tracing for a normal young adult female is given in Figure 11.2. The test was carried out using the Kin-Com AP (Chattanooga Group, Inc., Hixon, TN, USA).

Before analysing the curves, the printout should be checked to ensure that the basic information is correct, e.g. patient name, date, test details, etc., and then:

1. Examine for matched phase curves, i.e. concentric with concentric and eccentric with eccentric.

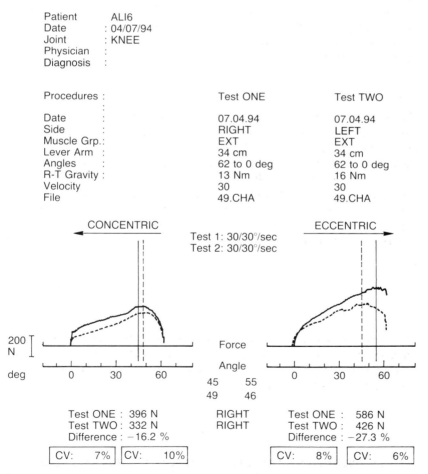

Patient	ALI6
Date	: 04/07/94
Joint	: KNEE
Physician	:
Diagnosis	:

Procedures :	Test ONE	Test TWO
Date :	07.04.94	07.04.94
Side :	RIGHT	LEFT
Muscle Grp.:	EXT	EXT
Lever Arm :	34 cm	34 cm
Angles :	62 to 0 deg	62 to 0 deg
R-T Gravity :	13 Nm	16 Nm
Velocity	30	30
File	49.CHA	49.CHA

CONCENTRIC

Test 1: 30/30°/sec
Test 2: 30/30°/sec

ECCENTRIC

200 N

deg   0   30   60

Force

Angle
45   55
49   46

0   30   60

Test ONE : 396 N   RIGHT   Test ONE : 586 N
Test TWO : 332 N   RIGHT   Test TWO : 426 N
Difference : −16.2 %   Difference : −27.3 %

CV: 7% CV: 10%   CV: 8% CV: 6%

Fig. 11.2 Kin-Com test result – version 4.02H53 (courtesy of Chattanooga UK, Oxford)

2. Check whether the data are raw or averaged, i.e. one curve for one test or the averaged curve from a series of tests.
3. The peak torque should be coincident with the maximal biomechanical advantage for the joint in the normal curve. A peak torque which occurs slightly before the maximal biomechanical advantage is possibly explained by a highly motivated and aggressive evaluation attempt by the patient. If slightly behind, this may indicate reduced motivation or that the subject is a little wary of giving a maximal effort.
4. The amplitude of the curve gives an indication of the force velocity generating capacity of the tested muscle group. Comparisons between ipsilateral and contralateral limbs will demonstrate differences in the abnormal evaluation.
5. The coefficient of variation gives an indication of the reproducibility in overlay test mode, i.e. if three maximal eccentric contractions are performed and averaged, the coefficient of variation indicates how closely all have approximated to a hypothetical optimum maximum voluntary contraction (MVC). Tests with coefficient of variations of below 10% are generally acceptable (Sapega, 1990).
6. The peak eccentric force (tension) for any given muscle will be 1.8 times the isometric force of the same muscle (Abbott *et al.*, 1951).
7. Levels of absolute force generation in the eccentric are greater than those of the concentric contraction. Consequently, 'problems', e.g. pain inhibition, are usually identified as breaks in the eccentric curve first (Dvir, 1991).
8. Currently definitive diagnosis with curve analysis is not considered to be accurate (Dvir, 1991). However, with the relevant research this method may gain acceptance in the future.

DISADVANTAGES OF ISOKINETIC EXERCISE

1. It is initially time consuming to learn how to use the dynamometer.
2. To obtain maximum benefit from the machine a member of staff should preferably be assigned to use it regularly so that expertise is developed. Very few departments can afford this luxury.
3. Isokinetic movement is an artificial constraint. Normal functional movement does not occur at a fixed velocity (Dvir, 1991).
4. Strength evaluation is the prime function of the isokinetic dynamometer, but strength is only one facet of normal movement.
5. Malalignment of the axes of rotation of joint and dynamometer will not provide a true reflection of muscle performance (Rothstein *et al.*, 1987). Alignment can be difficult when complex joints are involved.
6. Isokinetic systems are relatively expensive.
7. Eccentric testing predisposes to the phenomenon of delayed onset of muscle soreness (DOMS) (Newham, 1983).
8. Testing an MVC in continuous data collection mode can result in

artificially high concentric curves immediately following an eccentric contraction. This occurs because of the storage and recovery of elastic energy which is coupled with reflex potentiation via muscle spindle discharge (Komi and Bosco, 1978; Bosco *et al.*, 1982), i.e. as if providing a facilitatory prestretch to the muscle.
9. If pain is present then it is not possible to collect MVC data successfully.

ADVANTAGES OF ISOKINETIC EXERCISE

1. Isokinetic dynamometry provides an objective and quantifiable measure of concentric and eccentric dynamic muscle strength. It may provide a more meaningful index of function than that provided in the isometric test.
2. Isokinetic dynamometers are safe to use because during concentric exercise they apply no extra load to the limb and any resistance encountered by the musculature is a function of the force applied by the patient to the apparatus (Newton's Third Law).
3. They provide optimal and efficient loading of muscles and joints through range, thereby minimalizing potential for injury.
4. Strength evaluation is not limited to the weakest point in the range, since the resistance is accommodating.
5. Facilitates standard communication of muscle strength data between professionals.
6. Isokinetic unlike isotonic exercise does not require the same degree of skill and co-ordination to stabilize a free weight. Therefore it may reduce the risk of muscle and joint strain which can result from efforts to control a weight in an 'open', i.e. unconstrained, environment (Osternig, 1986).
7. An isokinetic evaluation can identify muscle weakness at a certain point in range. Specific targeting of this range when designing an exercise protocol may reduce treatment time.
8. Provides the patient and therapist with multiple training capabilities: isokinetic, passive, isometric and isotonic modes.
9. May indicate capability and compliance of the evaluated joint and its supporting structures.

RELIABILITY OF ISOKINETIC DEVICES

The following factors have all been shown to afffect reliability in isokinetic dynamometry (Albert, 1991):

- machine calibration
- joint axis alignment with that of the dynamometer
- gravity correction (Winter *et al.*, 1981)
- stabilization of joint to be tested

- patient position and level of stabilization
- design of protocol
- therapist commands
- visual feedback from screen
- subject co-operation
- patient fatigue
- psychological state of the patient.

Reliability is generally good, especially following proper patient instruction, familiarization with the equipment and testing procedure (Walmsley and Pentland, 1993). Intra-reliability is higher than inter-reliability (i.e. when the same subject is tested on different days on the same machine than when the same subject is tested on two different machines). Some of the reasons for this have been attributed to:

- differences in positioning
- differences in stabilization of patients
- the physical differences between the isokinetic machines
- differences in the way the computer manipulates raw data.

Many of the reliability studies have been confined to the lower limb and so currently there is a need for similar studies on other joints (Walmsley and Pentland, 1993).

Validity
Does the dynamometer have the ability to measure what it is supposed to measure? Studies concerning validity are not prevalent in the literature, but a major criticism was addressed by Winter *et al*. (1981). They expressed concern that the effect of gravity was not taken into account during isokinetic testing procedures (see above). A gravitational correction factor has now been incorporated into the recent dynamometers.

ROLE OF VISUAL FEEDBACK IN ISOKINETIC EXERCISE

The advent of computer technology has enabled the incorporation of visual feedback into isokinetic systems. Although much research has been carried out in isokinetics, little is available which recognizes the effect of visual feedback on patient performance during evaluation and rehabilitation.

Evaluation

Carlson *et al*. (1992) compared the effect of visual feedback on concentric and eccentric motor behaviour in the quadriceps muscle. They conducted their tests at slow (30°/sec) and moderate (120°/sec) speeds and found that subjects produced consistently higher forces when they were able to see the monitor. These results suggest a need for consistency in either excluding or including visual feedback during testing. The performance-enhancing effect of visual feedback used inconsistently could lead to inaccuracy in determining an MVC.

Rehabilitation

Exercise on the isokinetic dynamometer can also be affected by visual feedback (Brown *et al.*, 1984). The aim of an exercise programme is to gain strength as quickly and as safely as possible. There is a strong correlation between increased strength and intensity of exercise and it is probable that the incorporation of visual feedback may decrease rehabilitation time. However, it should be remembered that not all pathologies subjected to isokinetic regimens require or are capable of attaining high force output.

ISOKINETIC EXERCISE USING THE DIAGONAL FUNCTIONAL MOVEMENT PATTERNS

A major criticism of isokinetic dynamometry until recently has been that it was impossible to set up a patient to exercise in one of the more functional diagonal body planes (Voss *et al.*, 1985). All available exercise positions were defined according to the sagittal, coronal or transverse planes of movement, i.e. the anatomical planes. Gradually manufacturers have begun to accommodate the needs of patient and clinician and to design isokinetic machines with dynamometers which are now capable of spinning 360° in the transverse and vertical planes of movement.

ROLE OF THE ISOKINETIC DYNAMOMETER IN THE REHABILITATION ENVIRONMENT

Isokinetic dynamometers potentially have three uses in rehabilitation. They can be used as a treatment tool, as a diagnostic aid and for research purposes.

Treatment

The versatility of the isokinetic dynamometer in 'experienced hands' allows the design of a rehabilitation programme which can approximate functional activity.

Guidelines for treatment progression

Progression largely depends on patient assessment and capability, but a basic guide follows, for the inexperienced operator:

- acclimatize the patient to the dynamometer using the passive mode – slower velocities allow motor learning and are therefore preferable before progressing to the faster speeds (Griffin, 1987)
- incorporate active assisted exercise into the passive mode
- add a number of isometric holds at different points within the range – the isometric exercise mode is used if pain is present during voluntary movement
- isokinetic exercise allows dynamic contractions to take place in a finely controlled loaded situation
- isotonic exercise is added because this approximates to normal function.

With regard to this last point, isotonic contractions performed on a dynamometer most closely approximate but do not replicate the demands experienced through normal function. Instead, they occur in a controlled and protected environment. This environment possibly facilitates a standardized replication of the motor programmes which may approach those of the unconstrained functional activity.

Traditionally, isokinetic systems have been used for testing and rehabilitation, but their ability to be used in the early stages of a treatment programme has not yet been exploited. The reasons for this are suggested in Figure 11.3.

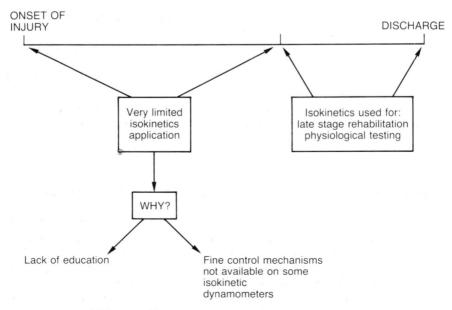

Fig. 11.3 The treatment spectrum of isokinetics

Diagnosis

An isokinetic evaluation is a valuable adjunct to diagnosis because it:

- provides a baseline measurement from which to progress and evaluate therapy
- provides information about muscle imbalance
- provides information on the precise point in range where pain or weakness is present
- can be used to localize treatment to a specific point in range
- allows maximal strength training effects to occur because of the accommodating resistance
- may produce different recruitment patterns for different velocities.

Research

The isokinetic dynamometer has provided a relatively simple method for allowing the quantification of the characteristics of dynamic muscle contractions. The scope for using isokinetics to research the behaviour and effects of the dynamic muscle contraction is vast and the current research status is summarized as follows:

- studies have focused primarily on the reliability and validity of the isokinetic dynamometer to quantify dynamic muscle strength in the lower limb
- very few studies have examined isokinetic dynamometry in the upper limb
- there is limited information on age-related changes and muscle strength using isokinetic dynamometry
- there is minimal information on using isokinetic dynamometry in disability
- the characteristics of the eccentric contraction can now be evaluated using the active dynamometer
- muscle groups may produce their own typical curves on evaluation
- recruitment patterns may be specific for different velocities.

ASPECTS OF ISOKINETIC EXERCISE IN THE ECCENTRIC MODE

1. Soft-tissue non-contact injuries tend to occur during the eccentric phase of a muscle contraction (Garrett, 1986). Unlike a concentric contraction where muscle fibre recruitment is sequenced from slow oxidative (type I) to fast glycolytic (type IIB) muscle fibres, eccentric contractions are known to recruit fibres selectively and those which are recruited first are the type IIB fibres. It is these fibres which show highest damage in a non-contact soft-tissue injury and therefore it is

important to tension these fibres appropriately to ensure maximal rehabilitation (Albert, 1991).

2. Joints which have poor eccentric capacity (shock absorption) tend to be predisposed to osteoarthritic joint damage (Marks, 1993). Therefore, patients with osteoarthritis could benefit from rehabilitation-specific eccentric contractions (Duncan *et al.*, 1989).

Delayed onset of muscle soreness

Delayed onset of muscle soreness (DOMS) is associated with eccentric exercise. It is characterized by pain, muscle swelling, loss of active range of movement, decreased peak torque and elevated EMG activity.

Several theories have been put forward to explain this phenomenon, which occurs as a consequence of eccentric loading 24–48 hours following exercise (Newham *et al.*, 1983). The theories are summarized by Albert (1991) as:

- lactic acid build-up
- selective damage of the Z bands which are weakest in type IIB fibres
- occurrence of tonic muscle spasms
- connective tissue damage in type IIB fibres
- an increase in tissue fluid which does not occur in concentrically exercised muscles.

The theories which are currently favoured are those of connective tissue damage and an increase in tissue fluid (Albert, 1991).

SUMMARY

- Isokinetic exercise enables the evaluation and training of muscular strength under conditions of constant angular velocity.
- The therapist may use the isokinetic dynamometer to vary the velocity, range of movement, moment and force threshold values, repetitions and contraction mode for each patient.
- Isokinetic dynamometry provides an objective and quantifiable measure of static and dynamic muscle strength.
- This chapter summarizes the advantages and disadvantages of isokinetic exercise.
- The isokinetic dynamometer has a role in clinical rehabilitation, diagnosis and evaluation for research purposes.

REFERENCES

Abbott, B.C., Aubert, X.M. and Hill, A.V. (1951) The absorption of work by a muscle stretched during a single twitch or a short tetanus. *Proc. R. Soc. B.*,

139, 86. In Albert, M. (1991) *Eccentric Muscle Training in Sports and Ortho-paedics*, Churchill Livingstone, Endinburgh

Albert, M. (1991) *Eccentric Muscle Training in Sports and Orthopaedics*, Churchill Livingstone, Edinburgh

Atha, J. (1981) Strengthening muscle. In *Exercise and Sports Science Reviews* (ed. R. Terjung), Collamore Press, Lexington, 1–73

Bosco, C., Viitasalo, J.T., Komi, P.V. and Luhtanen, P. (1982) Combined effect of elastic energy and myoelectrical potentiation during stretch shortening cycle exercise. *Acta Physiologica Scandinavica*, **114**, 557–565

Brown, B.S., Daniel, M. and Gorman, D.R. (1984) Visual feedback and strength improvement. *National Strength and Conditioning Association Journal*, Feb.–Mar., 22–24

Cabri, J.M.H. (1991) Isokinetic strength aspects in human joints and muscles. *Applied Ergonomics*, **22**(5), 299–302

Carlson, A.J., Bennett, G. and Metcalf, J. (1992) The effect of visual feedback in isokinetic testing. *Isokinetics and Exercise Science*, **2**(2), 60–64

Chandler, J.M. and Duncan, P.W. (1987) Eccentric vs. concentric force–velocity relationships of the quadriceps femoris muscle. Unpublished manuscript. Duke University, Durham, NC. In Cress, N.M., Peters, K.S. and Chandler, J.M. (1992) Eccentric and concentric force–velocity relationships of the quadriceps femoris muscle. *Journal of Orthopaedics, Sports and Physical Therapy*, **16**(2), 82–86

Duncan, P.W., Chandler, J.M., Cavanaugh, D.K., Johnson, K.R., Albert, M.S. and Buehler, M.A. (1989) Mode and speed specificity of eccentric and concentric exercise training. *Journal of Sports Physical Therapists*, **11**(2), 70–75

Dvir, Z. (1991) Clinical applicability of isokinetics: a review. *Clinical Biomechanics*, **6**, 133–144

Enoka, R.N. (1988) *Neuromechanical Basis of Kinesiology*, Human Kinetics Publishers, Illinois

Garrett, W.E. (1986) Basic science of musculotendinous injuries. In *The Lower Extremity and Spine in Sports Medicine* (eds J.A. Nicholson and E.B. Hershman), C.V. Mosby, St. Louis

Griffin, J.W. (1987) Differences in elbow flexion torque measured concentrically, eccentrically and isometrically. *Physical Therapy*, **67**, 1205–1209

Hislop, H. and Perrine, J.J. (1967) The isokinetic concept of exercise. *Physical Therapy*, **47**, 114–117

Jensen, R.C., Warren, B., Laursen, C. and Morrissey, M.C. (1991) Static pre-load effect on knee extensor isokinetic concentric and eccentric performance. *Medicine and Science in Sports and Exercise*, **23**, 10–14

Komi, P.V. (1973) Relationship between muscle tension, EMG and velocity of contraction under concentric and eccentric work. In *New Developments in Electromyography and Clinical Neurophysiology* (ed. J.E. Desmedt), Basel

Komi, P.V. and Bosco, C. (1978) Elastic energy in leg extensors by men and women. *Medicine and Science in Sports and Exercise*, **10**, 261–265

Korvanen, V., Suominen, H. and Heiddinen, E. (1984) Mechanical properties of fast and slow skeletal muscle with special reference to collagen and endurance training. *Journal of Biomechanics*, **17**, 725

Marks, R. (1993) Muscles as a pathogenic factor in osteoarthritis. *Physiotherapy Canada*, **45**(4), 251–259

Newham, D.J., Mills, K.R., Quigley, B.M. and Edwards, R.H.T. (1983) Pain and fatigue after concentric and eccentric muscle contractions. *Clinical Science*, **64**, 55–62

Osternig, L.R. (1986) Isokinetic dynamometry: implications for muscle testing and rehabilitation. *Exercise and Sports Science Reviews*, **14**, 45–80

Perrin, D.H. (1993) *Isokinetic Exercise and Assessment*, Human Kinetics Publishers, Leeds

Perrin, D.H., Hellwig, E.V., Tis, L.L. and Shenk, B.S. (1992) Effect of gravity correction on shoulder rotation isokinetic average muscle force and reciprocal muscle group ratios. *Isokinetics and Exercise Science*, **2**(1), 30–33

Rothstein, J.M., Lamb, R.L. and Mayhew, T.P. (1987) Clinical uses of isokinetic measurement. *Physical Therapy*, **67**, 1840–1844

Sapega, A.A. (1990) Current concepts review. Muscle performance evaluation in orthopaedic practice. *Journal of Bone and Joint Surgery*, **72-A**, 1562–1574

Tis, L.L., Perrin, D.H., Snead, D.B. and Weltman, A. (1991) Isokinetic strength of the trunk and hip in female runners. *Isokinetic and Exercise Science*, **1**(1), 22–25

Vos, D.E., Ionta, N.K. and Myers, B.J. (1985) *Proprioceptive Neuromuscular Facilitation*, 2nd edn, Harper and Row, Philadelphia

Walmsley, R.P., Pearson, N. and Stymeist, P. (1986) Eccentric wrist extensor contractions and the force velocity relationship in muscle. *Journal of Orthopaedic Sports Physical Therapy*, **8**(6), 288–293

Walmsley, R.P. and Pentland, W. (1993) An overview of isokinetic dynamometry with specific reference to the upper limb. *Clinical Rehabilitation*, **7**, 239–247

Winter, D.S., Wells, R.P. and Orr, G.W. (1981) Errors in the use of isokinetic dynamometers. *European Journal of Applied Physiology*, **46**, 397–408

12.   *Flexibility*

DEFINITION OF FLEXIBILITY

Flexibility is characterized by the ability of a muscle to relax and yield to a stretch force (Basmajian and Wolf, 1990) and also by the range of motion available in a joint or group of joints (Holland, 1968).

This chapter will outline:

- types of flexibility
- benefits of flexibility training
- factors affecting flexibility
- the collagen time scale
- methods used to increase flexibility: exercise, passive stretching, serial casting, proprioceptive neuromuscular facilitation, ice and heat
- the modified Sit and Reach Test
- examples of stretching exercises.

Rationale for stretching

Stretching prepares a muscle for activity. It enables the muscle to 'rehearse' its role in readiness for the 'performance'.

Flexibility allows the tissues to accommodate more easily to stress, to dissipate shock from impact and to improve the efficiency and effectiveness of movement, thus minimizing or preventing injury (Zachazewski, 1989). It also maintains range (e.g. passive movement) and can be used to increase joint range of movement (e.g. passive stretch).

Failure to stretch adequately can prove a costly experience for the patient who is recovering from injury and, at the other end of the spectrum, for the top class athlete who is attempting to win a gold medal at the Olympic Games.

TYPES OF FLEXIBILITY

There are two types of flexibility: *static flexibility* which relates to range of motion about a joint with no emphasis on speed, and *dynamic flexibility* where the joint is required to take part in the performance of physical activity at normal or rapid speed (Alter, 1988).

Both types are required for normal function and should be included in any rehabilitation programme. It is usual to re-educate static flexibility initially using selective stretching exercises and then to progress to dynamic flexibility activities. For example, following a calf injury the patient is initially taught to stretch the calf muscle in a controlled environment, by isolating stretches for soleus and gastrocnemius, before progressing to dynamic, functional flexibility activities such as running, hopping and step-up exercises.

The amount of flexibility available is unique to the individual and is specific to the joint concerned. Some people will always demonstrate more flexibility than others and some joints will have greater range of movement than others. Although the state of generalized hyper-mobility is more usual, hyper-mobility of one joint does not necessarily signify that the other joints will be as flexible. For example, it is not uncommon for a subject to demonstrate a degree of hyper-mobility at the elbow or knee joint and to have barely full range of movement at other joints.

The available range is often a result of habitual activity. The javelin thrower will probably have a greater range of movement around the shoulder than will the sedentary worker, but each is a normal range for the subject.

The need to gain flexibility is an essential part of the rehabilitation programme. The physiotherapist should pay particular attention to the cause of the loss of flexibility and ascertain whether it is due primarily to loss of joint range or to loss of muscle extensibility.

Flexibility must never be gained at the expense of joint stability. An unstable joint can create major problems for the patient and may, in extreme cases, lead to the need for permanent orthoses. Caution should also be observed in the acute stages of inflammation, infection and trauma.

BENEFITS OF FLEXIBILITY TRAINING

There are many advantages in gaining flexibility and Figure 12.1 shows these benefits.

FACTORS AFFECTING FLEXIBILITY

The factors which may affect flexibility are outlined below. Virtually every patient treated who has a musculoskeletal and/or neuromuscular disorder will require advice on how to maintain or regain flexibility.

Flexibility reduces with increasing age and, by 70 years of age, sedentary individuals will have lost 20–30% of their flexibility (Adrian,

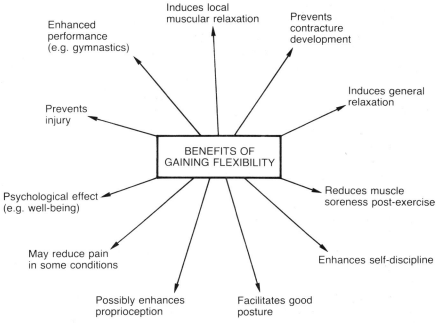

Fig. 12.1 Benefits of gaining flexibility

1981). It continues to decline, with the result that the less active elderly person soon begins to present with mobility and self-care problems.

All of the following factors will affect flexibility in both function and dysfunction of the body:

1. Type of body tissue:
   (a) bone (skeleton): trauma, disease, loss of alignment
   (b) connective tissue (ligament, joint capsule, tendon): collagen time scale (see below), adhesion formation, immobilization, trauma, disease
   (c) muscle: imbalance, immobilization, inadequate motor control, trauma, disease.
2. Nervous system: normal functioning and disease of the nervous system, e.g. states of altered tone.
3. Psychology: motivation, mental state.
4. Physical characteristics: sex, body shape, level of fitness.
5. Other: pain, oedema, orthoses, prosthetics, age.

THE COLLAGEN TIME SCALE

Immediately following soft-tissue injury, many changes occur at cellular level. Healing cannot be accelerated by any form of therapeutic intervention. However, physiotherapy can be used to encourage optimum conditions and reduce negative influences such as swelling and pain.

The mechanisms which lead to healing are categorized by Evans (1980) as occurring in three stages: (a) at the time of injury, (b) during the period of inflammation, and (c) at the repair stage. Details of these stages can be obtained from Evans (1980), but a brief summary is provided below.

Within a short period of trauma, i.e. 10–15 minutes, the injury site contains damage extracellular tissue and blood. The area becomes inflamed – the normal reaction of the body to this damage. Inflammation is characterized by four components: heat, redness, swelling and pain. Fibrous repair of the damaged structures follows which involves the laying down of scar tissue (Table 12.1). The most important and yet easily forgotten point is that new scar tissue (collagen) will always contract if it is not stretched.

It begins to form as a result of fibroblast activity in the fifth day following injury and continues to be laid down until the sixth month, long after the patient has left the physiotherapy department. As it is laid down in a haphazard manner, it is not as strong as it could be. To ensure that maximal strength is gained, gentle stress should be placed on it by regular stretching (Mason and Allen, 1941). This encourages organization of the tissue, with collagen fibres aligning along the line of applied stretch.

Stretching should continue for at least 6 months after injury, with patients advised to emphasize their stretching routines, especially prior to any sporting activity.

METHODS USED TO INCREASE FLEXIBILITY

Currently, there are a number of methods available which can be used to improve flexibility. Those described in this chapter are:

- exercise
- passive stretching
- serial casting
- proprioceptive neuromuscular facilitation.

Two other modalities which promote relaxation in preparation for flexibility training are the therapeutic application of ice and heat and their effects are also briefly discussed. Additional techniques, namely, manual therapy (e.g. Maitland, 1991), hydrotherapy (Davis and Harrison, 1988) and massage (Hollis, 1987) are beyond the scope of this chapter.

Table 12.1 Summary of the collagen time scale

	Characteristics	*Onset*	*Duration*
Stage 1: Injury	Damaged tissue contains extra-cellular tissue and blood	10–15 minutes after injury	Acute signs fade during the inflam-mation period
Stage 2: Inflammation	Heat, redness, swelling and pain	2 hours after injury	5 days or more post-injury if a lot of damage
Stage 3: Repair	Fibroblasts pro-duce fibrils which polymerize to form bundles of collagen	5 days post-injury col-lagen begins to be laid down	Fully formed col-lagen occurs from 3rd week to the 6th month and continues to con-tract if not stretched

Exercise

Stretching exercises can be used to regain muscle and joint flexibility. Exercise prescription will depend on patient needs and will vary from the restoration of upper limb function in the elderly patient following a cerebrovascular accident (CVA) to lower limb late stage rehabilitation of the competitive sports man and woman.

Method of stretching

Stretches should be performed slowly and the patient should experi-ence no pain. The patient must never be instructed to push through pain. For the stretch to be effective, a pulling, mildly uncomfortable sensation should be experienced. The duration of stretch should be approximately 10–20 seconds.

The answer to the following question will usually indicate whether a stretch is being performed properly: 'Is it pain or is it strain?' 'Strain' is an acceptable answer, but 'pain' is not.

Patient advice

The information below can be used as a guide for advising a patient how to perform a stretch. Obviously this will vary according to patient requirements and therapist experience.

- warm up with light exercise, before embarking on a stretch routine
- begin with gentle stretches and progress, using the question 'Is it pain or is it strain?': never push through pain
- stretch before and after exercise

- emphasize muscle groups which feature specifically in the activity or exercise programme
- use static stretches during the warm-up programme
- ballistic stretches (i.e. bouncing) can push further into the range of joint movement, but they raise the excitability of the stretch reflex which can result in injury (Chapter 4); generally this type of stretch should be avoided (see below).
- hold each stretch for between 10 and 20 seconds
- perform the stretch routine in a warm and relaxing environment
- passive stretching by a responsible partner may be effective for obtaining maximal stretch
- stretch on a mat rather than the floor.

Passive stretching

The term 'passive stretching' should not be confused with that of 'passive movement' (Chapter 6) which is only performed within the existing available range of joint motion. During passive stretching of structures, the full active range of motion of the structure may be increased.

Method

To perform a passive stretch the muscle is gently, firmly and slowly taken to the end of its existing range. At the end of the range, a long, slow stretch is applied. This position is maintained for a count of approximately 10–20 seconds or long enough to ensure that full stretch has been obtained.

Precautions

Anatomical points to be taken into account by the therapist include: whether a muscle crosses two joints, its attachments and the positions which constitute middle, full outer and full inner range of muscle contraction and joint movement.

Additionally, knowledge of open (loose) and close packed joint positions will assist the therapist in determining the degree of passive stretch that can safely be implemented (Chapter 2). An over-zealous stretch can not only be painful but also may result in detrimental symptoms such as those of myositis ossificans (Chapter 6).

Care should also be taken to ensure that the passive stretch is performed slowly. This avoids increasing muscle spindle excitability in the muscle being stretched which would have the opposite effect – that of generating an increase rather than a decrease in muscle tension. Neurophysiologically, the relaxation effect on the stretched muscle is achieved through inhibition via the type Ib fibres from the Golgi tendon organ (GTO) (Chapter 4).

Clinical application

1. The passive slow stretch technique is used to treat a number of conditions, the prime example being that of the paralysed patient on long-term bedrest. On spinal injury units, passive stretches may be incorporated into the passive movement routine, particularly if high tone is a problem. They are used to ensure that joints and muscles retain their full range. Particular attention is paid to hamstring and Achilles tendon length. The former is essential to allow the patient to regain sitting balance, while the maintenance of Achilles tendon length allows the paralysed foot and ankle to be placed in the plantigrade position, which is important for the maintenance of upright stance in a standing frame, on a tilt table or when wearing an orthosis.

   NB. An exception is the tenodesis action of the wrist which allows a tetraplegic patient to achieve a functional but relatively weak grip. In this instance it is detrimental to stretch the wrist flexors and extensors which are used in the tenodesis action which involves a combination of wrist flexion with finger extension followed by a wrist extension and finger flexion.

2. A long slow passive stretch is also used to reduce hypertonicity in the patient with other neurological disorders such as multiple sclerosis and CVA. The relaxation effect on the spastic muscle is achieved via GTO inhibition described above. Unfortunately, it is not possible to attain a lasting reduction of spasticity with this technique, but by stretching on a regular basis and timing with nursing procedures, patient hygiene can be maintained and contracture formation prevented or at least minimized.

3. The ballistic stretch has until recently been used during warm-up routines in preparation for sport activity. It is a high-intensity, short-duration 'bouncing' stretch which is potentially very dangerous. Neurophysiologically, the bounce at the end of range heightens the excitability of the muscle spindle which causes an increase rather than a decrease in tension of the stretched muscle (Sady *et al.*, 1982). Its use is not recommended.

Serial casting

Also known as serial splinting, this technique is primarily used in conjunction with manual passive stretching to prevent or control the formation of contractures. A contracture is the term used to describe a shortening of muscle or other tissues that cross a joint which results in loss of joint motion (Cherry, 1980). Several causes of contracture are recognized, the main ones identified by Cherry (1980) being:

- intrinsic adaptive change in response to prolonged positioning which often occurs after orthopaedic immobilization
- poor positioning during prolonged bedrest
- dynamic muscle imbalance, as in neurological conditions when a spastic agonist tends to be unopposed by its weaker antagonist
- habitual poor posture

Serial cast

Clinically, there are three main instances when the application of a serial cast should be considered:

(a) to regain range when contractures have become the main source of mobility loss
(b) to manage spasticity of the lower limb following head injury (Zacha-zewski *et al.*, 1982; Booth *et al.*, 1983)
(c) to reduce flexion contractures of joints for patients with rheumatoid arthritis (Ward and Tidswell, 1992).

Method of application
Serial casts are particularly useful in the treatment of lower limb conditions to prevent/control knee flexion and ankle plantar flexion deformities:

1. Initially, bony points are protected and cotton padding or 'stock-inette' applied to the whole limb.
2. The limb is positioned and supported at the end of its available range of movement.
3. Plaster of Paris (POP) is applied as a full-length cylinder for treat-ment of knee contractures, or as a below-knee cast for controlling ankle position, and allowed to dry for 2 days before weight-bearing is allowed.
4. The cast is changed at approximately 7–10 days and a new one applied if necessary, which takes account of any range gained (this time period may vary depending on clinician preference and patient diagnosis).

The drop-out (cut-out) cast

Drop-out casts are modifications of cylinder casts and are considered to be a progression from the cylinder cast. Their use has tended to focus on the management of neurological dysfunction (Cherry 1980; Booth *et al.*, 1983).

The knee drop-out cast is an adaptation of the full length cylinder. To convert the cylinder to a drop-out cast, the lower anterior portion of the cast is removed from a point just superior to the patella and from a point slightly anterior to the knee joint (Figure 12.2).

The advantage of this cast is that when the patient is lying prone, the knee will be able to extend and possibly gain range in response to the effect of gravity, and when in supine, the foot will be supported fully in the cast and the existing range will be maintained.

Another advantage of the drop-out cast is that it is possible to observe the leg for complications such as swelling, colour alteration and signs of pressure. This can be a problem with the application of a cylinder. Patients should always be monitored for complications for the entire period that the cast is worn, but particularly during the first 24 hours or until the cast has dried, since slight shrinkage can occur during drying.

Resting splints

Resting splints are often used by patients who have rheumatoid arthritis. They are worn at night to ensure that joints are maintained in a functional position.

Stages 1–3 are performed as for the lower limb serial cast, but the difference is that the POP is bivalved (slit longitudinally) immediately when dry and converted into a resting splint.

The posterior section in which the limb lies is neatened by removing the rough edges and made comfortable with a small amount of additional padding if necessary. This section can be bandaged in place when the patient is not receiving exercise or worn for support during weight-bearing activity if considered necessary.

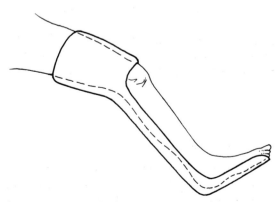

Fig. 12.2 Knee drop-out cast with the anterior portion of cast removed below the knee

Upper limb resting splints for the wrist and hand can be made as a back slab initially, rather than as a cylinder, and this avoids the need to bivalve.

Clinical applications of casting

Casts provide a prolonged static stretch to tight structures. They are useful when used to increase range of motion which is limited by hypertonicity or pain. They provide effective inhibition of tone in the neurologically impaired patient, although the mechanism by which this is achieved remains elusive. One explanation could be that weight-bearing through a lower limb in normal postural alignment may produce a normal proprioceptive input, with the subsequent output of a normal postural response.

Proprioceptive neuromuscular facilitation

Hold–relax

Hold–relax is a proprioceptive neuromuscular facilitation (PNF) technique which is frequently used to gain range of motion where loss is attributed to a tight antagonist muscle. The aim is to achieve a maximal contraction followed by a maximal relaxation of the antagonist and then to facilitate either a passive, active or resisted movement of the agonist into the new range.

Method

1. Select the most appropriate pattern and diagonal. Muscle groups tend to work more functionally in one particular diagonal. For example, biceps brachii is a shoulder flexor, an elbow flexor and a forearm supinator. The pattern which incorporates this activity is flexion, adduction and lateral rotation of the upper limb (Chapter 9) (Voss *et al.*, 1985).
2. *Place* the limb in pattern so that the antagonist is in a maximally extended position.
3. Manually resist to obtain a maximal isometric contraction of the antagonist using the command 'hold'.
4. Instruct the patient to relax.
5. Move the limb either passively, actively or by applying a guided resistance into the new range.
6. If moving actively or resisting the movement, it will be necessary to gently change manual contacts to the opposite surfaces to enable this to take place.
7. Repeat the procedure if range continues to be gained and if the patient is not experiencing pain.

Some clinical applications of Hold–Relax

1. To gain range of knee flexion using hold–relax on quadriceps femoris.
2. To gain range of elbow extension by applying hold–relax to biceps brachii.
3. To perform a manually assisted or auto-assisted stretch to hamstrings in preparation for sports activity.

Ice

Ice or cryotherapy is used to reduce muscle spasm which is associated with muscle strain and post-exercise muscle soreness. Its application immediately following injury to the elevated limb tends to reduce local pain and swelling (Low and Reed, 1990).

As soon as gentle stretching commences at about 48 hours after injury, its continued application appears to facilitate a reduction in pain during the stretching phase. However, its application should be carried out with caution to avoid further trauma to the injury site, since it has a numbing effect on the area.

Ice can also be used to relieve temporarily the spasticity which results from an upper motor neuron lesion (Newton and Lehmkuhl, 1965). It is applied for prolonged periods (e.g. greater than 15 minutes) to large areas of the body in an ice bath or by ice towels laid along the length of the muscle belly. It is particularly effective for the relief of spasticity when combined with weight-bearing through the long axis of the limb, but the mechanism of its effect is not known.

Heat

When the acute stage of a soft-tissue injury has subsided, the application of heat is favoured, since it can be effective in inducing local muscle relaxation and providing pain relief (Low and Reed, 1990). Auto-assisted and manual passive stretching techniques, together with electrotherapy modalities, can then be used to promote maximal extensibility following soft-tissue injury.

THE MODIFIED SIT AND REACH TEST

The American College of Sports Medicine (1992) describe a modified Sit and Reach Test which is used in the sports environment to assess non-specific flexibility (Figure 12.3). The equipment required is a yardstick and adhesive tape.

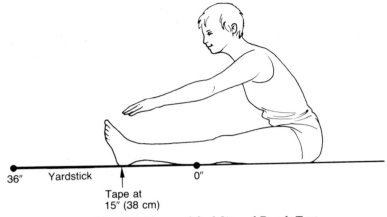

36″ Yardstick

0″

Tape at
15″ (38 cm)

Fig. 12.3 The modified Sit and Reach Test

Method

1. Sit on the floor with legs extended and slightly abducted.
2. Place the yardstick on the floor between the legs so that the zero mark of the yardstick is closest to you.
3. Tape the yardstick to the floor at the 15-in (38-cm) mark.
4. Line your feet up so that they are just in contact with the tape.
5. Without flexing the knees, slowly reach as far forward with both hands and slide the fingertips along the yardstick. The hands should be placed on top of one another so that the middle fingers are accurately aligned and provide the point of contact with the yardstick.
6. Do this 3 times and compare your best score with the tabled normal values (Table 12.2).

Table 12.2 Standards for the modified Sit and Reach Test (from American College of Sports Medicine, 1992, by permission)

	*Score at age**				
	20–29	*30–39*	*40–49*	*50–59*	*60+*
Men					
High	⩾ 19	⩾ 18	⩾ 17	⩾ 16	⩾ 15
Average	13–18	12–17	11–16	10–15	9–14
Below average	10–12	9–11	8–10	7–9	6–8
Low	< 9	< 8	< 7	< 6	< 5
Women					
High	⩾ 22	⩾ 21	⩾ 20	⩾ 19	⩾ 18
Average	16–21	15–20	14–19	13–18	12–17
Below average	13–15	12–14	11–13	10–12	9–11
Low	< 12	< 11	< 10	< 9	< 8

*The scored values are in inches and not centimetres. Conversions can easily be made using the conversion where 1 cm = 0.3937 in.

EXAMPLES OF STRETCHING EXERCISES

For detailed basic lower limb stretches which can be used to increase flexibility (Figure 12.4), the reader is referred to Alter (1988) and Anderson (1980).

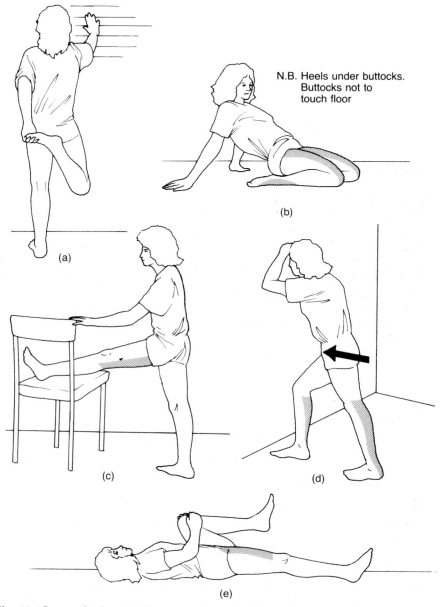

N.B. Heels under buttocks. Buttocks not to touch floor

Fig. 12.4 Lower limb stretching exercises: (a, b) quadriceps stretch; (c) hamstring stretch; (d) gastrocnemius left leg, soleus right leg stretch; (e) stretch for hip flexors

SUMMARY

- Flexibility is the ability of a muscle to relax and yield to a stretch force.
- There are two types of flexibility – static, which relates to range of motion about a joint with no emphasis on speed, and dynamic, where the joint moves at normal or rapid speed.
- Flexibility will vary with many factors including: body type, state of nervous tissue, individual physical characteristics and mental state.
- The Collagen Timescale relates to the changes occurring in soft tissue following injury. Collagen is fully formed between 3 weeks and 6 months post injury and will continue to contract if it is not stretched.
- Flexibility may be increased by active and passive stretching, casting and proprioceptive neuromuscular facilitation techniques

REFERENCES

Adrian, M.J. (1981) Flexibility in the ageing adult, In *Exercise and Ageing: The Scientific Basis* (eds E.L. Smith and R.C. Serfass), Enslow Publishing, New Jersey

Alter, M. (1988) *Science of Stretching*, Human Kinetics Publications, Leeds

American College of Sports Medicine (1992) *ACSM Fitness Book*, Leisure Press, Champaign, Illinois

Anderson B. (1980) *Stretching: Exercises for Everyday Fitness and for Twenty-five Individual Sports*, Pelham Books, London

Basmajian, J.V. and Wolf, S.L. (1990) *Therapeutic Exercise*, 5th edn, Williams and Wilkins, Baltimore

Booth, B.J., Doyle, M. and Montgomery, J. (1983) Serial casting for the management of spasticity in the head injured adult. *Physical Therapy*, **63**(12), 1960–1966

Cherry, D. (1980) Review of physical therapy alternatives for reducing muscle contracture. *Physical Therapy*, **60**(7), 877–881

Davis, B.C. and Harrison, R.A. (1988) *Hydrotherapy in Practice*, Churchill Livingstone, Edinburgh

Evans, P. (1980) Healing process at cellular level: a review. *Physiotherapy*, **66**(8), 256–259

Holland, G.J. (1968) The physiology of flexibility: a review of the literature. *Kinesiology Review*, **1**, 49–62

Hollis, M. (1987) *Massage for Therapists*, Blackwell, Oxford

Low, J. and Reed, A. (1990) *Electrotherapy Explained: Principles and Practice*, Butterworth–Heinemann, Oxford

Maitland, G.D. (1991) *Peripheral Manipulation*, 3rd edn, Butterworth–Heinemann, Oxford

Mason, M.L. and Allen, H.S. (1941) The rate of healing of tendons: an experimental study of tensile strength. *Annals of Surgery*, **113**, 424–456

Newton, M.J. and Lehmkuhl, D. (1965) Muscle spindle response to body heating and localised muscle cooling: implications for relief of spasticity. *Physical Therapy*, **45**, 91–105

Sady, S.P., Wortman, M. and Blanke, D. (1982) Flexibility training: ballistic, static or proprioceptive neuromuscular facilitation. *Archives of Physical Medicine and Rehabilitation*, **63**, 261

Voss, D.E., Ionta, N.K. and Myers, B.J. (1985) *Proprioceptive Neuromuscular Facilitation*, 3rd edn, Harper and Row, Philadelphia

Ward, D.J. and Tidswell, M.E. (1992) Osteoarthritis. In *Cash's Textbook of Orthopaedics and Rheumatology for Physiotherapists*, 2nd edn (ed. M.E. Tidswell), Mosby Year Book, London

Zachazewski, J.E. (1989) Improving flexibility. In *Physical Therapy* (eds R.M. Scully and M.R. Barnes), J.B. Lippincott, Pennsylvania

Zachazewski, J.E., Eberle, E.D. and Jefferies, M. (1982) Effect of tone inhibiting casts and orthoses on gait. *Physical Therapy*, **62**(4), 453–455

Part III
Locomotion and ergonomics

13.   *Relaxation*

INTRODUCTION

This chapter introduces the reader to the idea of a relaxation response as the opposite state to the physiological conditions produced by stress. The physiology of the relaxation response is described. There are a number of techniques that may be used to try to induce a relaxed state. The principal methods that are used by physiotherapists are described, together with the rationale behind them. The evidence for the clinical efficacy of relaxation as a therapeutic intervention is discussed.

This chapter will outline:

- the relaxation response
- the different methods of relaxation and the rationale for them
- clinical applications of relaxation therapy

DEFINITION

Relaxation may be described as a state in which there is total physical immobility and relaxation of the skeletal muscles with a regulatory effect on the sympathetic nervous system. This is accompanied by a feeling of diminished consciousness of the external world, drowsiness, passivity and focusing of the attention on feelings of internal well-being (Rosa, 1976).

RESPONSE OF THE BODY TO STRESS

Stress is a physiological response to a stimulus. Some stress is necessary for an individual to function normally, a mild level of emotional arousal produces alertness and the nervous system needs a certain amount of stimulation to function properly. To a large extent stress is an appropriate and necessary response that enables individuals to cope with life's challenges.

Schmidt (1988) states that arousal may be considered as a neutral state in that it represents the amount of effort that an individual applies to an action. Stress and motivation represent different types of arousal and are considered to have directional components – stress usually being considered as negative and motivation as positive.

The inverted-U hypothesis

The inverted-U hypothesis of arousal and performance states that increasing arousal will increase performance up to a given point. Beyond this point, further increases in the arousing stimulus impair performance (Figure 13.1).

Humans react to stress by triggering the 'fight or flight' response. If this is repeatedly elicited it will produce somatic responses that may be damaging, such as elevated blood pressure and increased respiratory rate. Thus stress that is too prolonged may be damaging physiologically and psychologically. Relaxation is a state diametrically opposite to the fight or flight response. Benson (1976) states that relaxation is an innate integrated set of physiological changes in opposition to those of the fight or flight response.

Muscle tension
Under normal circumstances, muscles are always under some tension, as even when a person is in a fully relaxed state their muscles maintain a state of readiness for action via the muscle spindle to provide muscle tone (Chapter 4).

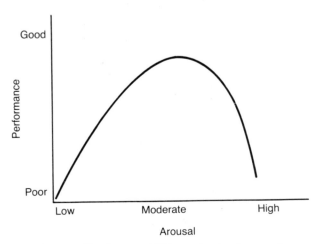

Fig. 13.1 The inverted-U hypothesis of arousal

Postural tone

The muscles concerned with countering gravity are responsible for maintaining postural tone by a stretch reflex. Stretching of a muscle by a force, e.g. gravity, stimulates the sensory afferent fibres to transmit impulses which produce a sustained contraction of the muscle spindle via gamma efferent fibres sufficient to heighten the level of contraction of the extrafusal muscles. Relaxation training is directed at decreasing the level of neuromuscular excitation in the muscle spindle (Chapter 4).

The position of stress

Laura Mitchell (1987) describes the commonly observed postural pattern of someone who is under excessive stress (Figure 13.2). The head

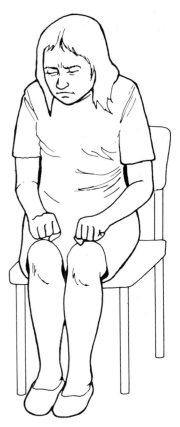

Fig. 13.2 The position of stress

is either carried forward with the chin tucked in or the head moves forward on the neck with the chin jutting out. The shoulder girdle is elevated, with the upper limbs held rigidly close to the chest wall and the elbows flexed, while the trunk is held stiffly and inclined forwards. The hands are held clenched or perform repetitive actions such as jingling keys in a pocket. In sitting, the legs are crossed with repeated dorsiflexion and plantar flexion of the ankle. The face is tense, with furrowed forehead, eyes held tightly closed or wide open, lips are closed and the tongue is held on the roof of the mouth. This position of stress will be easily recognized, either in its entirety or in part, and relaxation programmes are directed at trying to alter these characteristic patterns.

THE RELAXATION RESPONSE

A generalized relaxation response can be shown which is associated with decreased sympathetic nervous system activity and increased parasympathetic nervous system activity. The integrated actions of the automatic nervous system, the endocrine system and the higher centres of the brain result in a generalized trophotropic or relaxation response. In such a trophotropic state there is a quiescent state of parasympathetic functioning, and the following physiological, cognitive and behavioural manifestations, which have been described by Dimotto (1984) and Benson (1976) among others, may be observed.

Physiological manifestations of relaxation
 1. Decreased heart rate.
 2. Decreased blood pressure.
 3. Decreased respiratory rate.
 4. Decreased oxygen consumption.
 5. Decreased carbon dioxide production and excretion.
 6. Decreased blood cholesterol and lactate.
 7. Decreased muscle tension.
 8. Decreased metabolic rate.
 9. Peripheral vasodilatation.
10. Increased peripheral temperature.
11. Pupil constriction.
12. Increased saliva output.
13. Increased digestive activity.
14. Increased urine output.

Cognitive manifestations of relaxation
1. Altered state of consciousness with increased alpha and theta brain waves.

2. Heightened concentration on a single mental image.
3. Increased suggestibility to ideas.

Behavioural manifestations of relaxation
1. Lack of attention and concern for environmental stimuli.
2. No verbal interaction.
3. No voluntary change of position.
4. Passive movement easy.

As automatic reactions vary greatly from one individual to another, just as in stress responses a wide range of clinical features may be observed, so individual responses to relaxation will vary.

Mental relaxation

The cognitive changes that occur with the relaxation response have been described as a state of passive concentration and awareness and observation of bodily and mental sensations. They are not simply a cessation of mental effort.

METHODS OF RELAXATION TRAINING

There are many different techniques available to produce a relaxation response. However, the question of how effective relaxation training is and how the techniques compare with each other is difficult to answer. Hillenberg and Collins (1982) and Keable (1985) produced extensive literature reviews of over 80 studies, but failed to answer the question of how effective the different methodologies had been. This was partly because of the variation in the clinical populations that they studied.

Preparation for relaxation training

Relaxation training should take place with the patient in a comfortable, supportive and restful position. Support will be maximized if the subject lies supine on a firm surface that will mould itself to the subject's body. Lying on the floor or a treatment plinth is unlikely to prove conducive to relaxation. The head should be supported by a small pillow and a pillow may also be placed under the knees to decrease tension in the hamstrings and facilitate a straight lumbar spine.

Alternatively, the subject may lie half supported in half lying, either in an armchair with a footstool or on a bed with the backrest pulled out.

The subject should be as comfortable as possible, tight clothing removed or loosened and the room well ventilated, but warm and free from draughts. The atmosphere should be as restful as possible; while complete silence is not necessary, there should not be a lot of background noise. The session should be free from interruptions by other people or bleeps and telephone calls, thus a cubicle in a busy out-patient department is unlikely to be suitable and a more private location should be sought. The physiotherapist should start the session by explaining what she intends to do to inspire confidence in her from the patient and to try to eliminate tension. The manner of conversation should be calm and clear as the training in whichever technique the therapist chooses is introduced.

The most commonly used relaxation techniques in physiotherapy clinical practice are:

- progressive relaxation (Jacobsen)
- accelerated progressive relaxation (contract–relax)
- reciprocal physiological relaxation (Mitchell)
- meditative relaxation response (Benson)
- autogenic training (Schultz and Luthe)
- biofeedback (Basmajian).

Progressive relaxation

This technique was developed by Jacobsen in the 1920s. It emphasizes the learning of a relaxed state as a muscular skill. With progressive relaxation, the subject initially sequentially tenses and releases the major skeletal muscles, concentrating on the contrasted sensations, and attempts to achieve a more complete muscular relaxation. The training is directed at the recognition of small amounts of muscle tension rather than the exaggerated difference between full muscle contractions and their subsequent release.

In its original form, progressive relaxation involved relaxing two or three muscle groups per session until some 50 groups, covering the whole body, had been relaxed. Typically it took between 3 and 6 months to achieve full mastery of the technique. Hourly practice on one movement such as bending the elbow occurs, progressively increasing and decreasing the tension levels, until a relaxed state can be achieved (Jacobsen, 1938).

Rationale

Jacobsen held that there exists a state of reciprocal influence between the brain and peripheral structures of the body. The level of activity in either of these areas is controlled in part by stimulation from the other, thus the brain is quietened by decreased sensory input and so influenced it will tranquillize the autonomic nervous system. Thus, Jacobsen uses muscle control in his progressive relaxation training to reduce anxiety and tension.

Technique

The patient sits in a well-supported position or lies with eyes closed. The therapist introduces the patient to a relaxed state by leading them through the movements:

I am going to ask you to keep your eyes closed and breathe slowly in and out. . . .
I will ask you to tense different muscles of the body; when I do, tense them as much as you can until I say relax. . . .
Tense the muscles of your right hand by clenching your fist, keep it tight, feel the strain and the muscles working really hard and relax. . . .
Relax completely, let the muscles flop and compare in your mind the feeling of tension you felt in your hand a few moments ago and the relaxed feeling you have in it now. . . .

As the sessions progress, the amount of direction by the therapist decreases and the patient is left to tense the muscle groups on their own, and achieve control over skeletal muscle until they can induce very low levels of tension in groups of muscle.

Accelerated progressive relaxation

This is sometimes also referred to as the 'contract–relax method'. The technique is essentially the same as Jacobsen's method, but it has been systematically abbreviated and revised. Current practice successively relaxes about 15 muscle groups in a half-hour session by the contract–hold–relax technique (King, 1980; Lichstein, 1988). The rationale and basic technique are the same as for Jacobsen's progressive relaxation. However, many researchers suggest that Jacobsen's original method is more likely to be effective at reducing stress symptomology than the modified versions. Accelerated progressive relaxation programmes tend to emphasize large contrasts between the states of tension and relaxation in the muscle group, as opposed to Jacobsen's recognition of the minute amounts of muscle tension that could be detected and lowered (Lehrer, 1982). However, in the clinical situation this method is frequently more practicable than the months of work that are required by the original progressive relaxation method.

Reciprocal physiological relaxation

This method has been devised and developed by the physiotherapist Laura Mitchell and is sometimes referred to as 'simple relaxation' (Mitchell, 1987). The main rules of Mitchell's method of physiological relaxation are:

1. The brain must be given a definite order that it recognizes will produce work, i.e. to agonist groups only. Thus the joint is moved out of the position of stress.

2. Only voluntary activity will reproduce the reciprocal relaxation in the opposite group to the working group. That is, instructions such as '*Pull* your shoulders towards your feet', rather than 'Drop your shoulders' are given.
3. When the command 'stop' is given, the patient stops moving the part – the command is not to relax but to stop moving.
4. The feeling of the new joint position and skin pressure is registered mentally by proprioceptive feedback to enhance the recognition of a relaxed position.
5. Training is in joint and skin consciousness, not muscle consciousness.
6. The brain is trained to discriminate between sensations and to recognize them.
7. Breathing is natural but slow in rhythm and diaphragmatic.

Rationale

Mitchell's method of physiological relaxation is based on Sherrington's law of reciprocal inhibition. The technique involves working the opposite group of muscles to those involved in adopting the classic tense posture. The principle of reciprocal inhibition is used to produce relaxation in the antagonist group, so working the opposite muscle group makes the opposing muscle group, which was causing the tense posture, relax. After the movement is finished, the order 'stop' produces relaxation in both agonist and antagonist muscle groups. Mitchell does not believe that muscle tension can be consciously perceived and concentrates instead on training the sensory region of the brain to recognize the relaxed position by joint position and skin pressure proprioception.

Technique

The technique is carried out in a warm room in one of three positions: lying; sitting in forward supported leaning onto pillows or a table; or sitting upright in a well-supporting chair. The subject is given an order:

Pull your shoulders towards your feet . . . pull until you can go no further . . . stop . . . concentrate on feeling the position your body is now in and register it. . . .

Sequentially the arms, elbows, hands, legs, knees, feet, trunk, head and face are worked through, moving into the opposite position to Mitchell's position of tension described earlier. The technique can also be adapted to coping in a variety of situations and settings in daily life, disregarding the conventional starting positions, for example in the workplace, while driving, etc.

Meditative relaxation response

This has been described by Benson (1976). It is based on meditation principles and requires that the subject has a quiet environment, a comfortable position, a passive attitude and a repetitive mental stimulus to concentrate attention.

Rationale

The rationale for Benson's relaxation response is based on stress research and transcendental meditation (Benson, 1976). He found that such practices involved common elements which he incorporated into his programme. The four components described above are all designed to decrease the activity of the sympathetic nervous system.

Technique

The patient is instructed to:

Sit quietly and comfortably. . . .
Close the eyes. . . .
Relax your muscles, working up from your toes to the top of your head and hold. . . .
Breathe slowly in through the nose, focusing your attention on your breathing and maintaining an easy, natural rhythm. . . .
Silently repeat the word 'one' with every exhalation. . . .

Such instructions are given and the procedure is carried out for around 15 minutes twice a day.

Autogenic training

Autogenic training, or autosuggestion, is based on the work into hypnosis by Schultz and Luthe (1969) pioneered in the 1920s. Autogenic training techniques combine pleasant nature scene imagery with focusing on bodily sensations. It is largely an imaginal method of relaxation in which the subject adopts a passive attitude and creates a peaceful environment and comforting bodily sensations in their mind. The standard exercises of autogenic relaxation training are composed of six relaxation themes:

- heaviness
- warmth
- cardiac regulation
- respiration
- abdominal warmth
- cooling of the forehead.

The first two of these, heaviness and warmth, are applied sequentially to seven body parts:

- right arm
- left arm
- both arms
- right leg
- left leg
- both legs
- arms and legs.

The exercises concentrate on producing heaviness (muscle relaxation) and warmth (peripheral vasodilatation).

Rationale

Autogenic relaxation training induces relaxation through hypnotic methods. It is thought to work due to the influence of the proximal brain structures promoting diffuse relaxation and a trophotropic response. Gelhorn (1967) attributes the response of autogenic training to trophotropic tuning – repeated elicitation of the trophotropic response, by stimulating the anterior hypothalamus by relaxation practice, produces a lower elicitation threshold and a strong response to autosuggestion.

Technique

The introduction involves setting the mood by concentrating on feelings of peace.

I am at peace. . . . Imagine you are lying on a warm beach listening to the waves lapping on the sand. . . .
Think about the heavy feeling you have in your right arm. . . .
My right arm is heavy. . . . My right arm is heavy. . . .
My right arm is warm. . . .
Concentrate on inner rhythms, think of your breath calm and even . . . pulse calm and strong. . . .

All seven areas of the body are worked through, followed by the phrase 'My solar plexus is warm' for the abdominal region and finishing with 'My forehead is pleasantly cool' for the final phase. The session is ended by consciously coming back to reality, opening the eyes, taking a deep breath and stretching the body.

Biofeedback

Biofeedback usually utilizes electromyography and gives the subject auditory or visual signals in response to varying levels of tension in specific muscle groups. It has been extensively described by Basmajian and De Luca (1985) who showed that autonomically controlled variables such as heart rate, blood pressure and muscle tension could be altered by biofeedback techniques. Low and Reed (1990) describe the concept by use of a feedback loop, whereby the physiological change is detected, measured by a device and displayed, so that the subject can perceive the information and consciously attempt to alter it (Figure 13.3).

The electromyograph detects the electrical signal generated when a motor unit fires. When a muscle contracts many motor units fire, producing an electrical potential that can be detected by needle or surface electrodes and displayed on a screen, via a paper recorder or by an auditory signal. The display will bear a direct relationship to the magnitude of the muscle activity. Biofeedback is most suitable for use with superficial muscles. For example, tension headache may be relieved by relaxation of the occipitofrontalis and posterior neck muscles using biofeedback. Thus a patient may use electromyographic equipment to receive information about the muscle tension levels in a group and attempt to relax those muscles by reducing the EMG reading.

Rationale

It is proposed that biofeedback enhances the control of the autonomic nervous system by enhancing introspection and body awareness. The biofeedback amplifies subconscious information and transforms it into a medium that is understandable to the patient. However, this only applies to relaxation of a given muscle group or area, not to a state of total body relaxation.

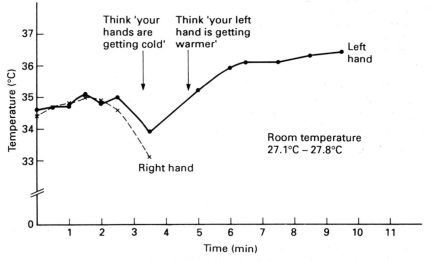

Fig. 13.3 Biofeedback loop

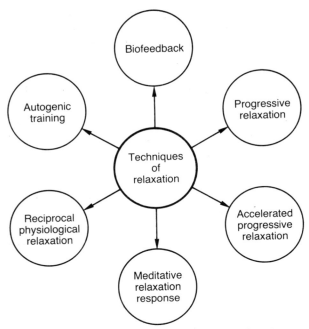

Fig. 13.4 Summary of the techniques of relaxation

Figure 13.4 gives a summary of the techniques of relaxation which were described above.

CLINICAL USE OF RELAXATION THERAPY

In clinical physiotherapy, relaxation therapy may be either a treatment in its own right or a component part of a larger treatment plan. Therapeutically, the use of relaxation techniques is widespread for a variety of conditions including anxiety, asthma, hypertension and pain, in childbirth and muscular dysfunction. Thus it may be an appropriate technique for therapists to incorporate into their treatment programmes for many of these conditions. It is particularly well suited to the class setting and is frequently incorporated into group work such as the back school (Klaber Moffett *et al.*, 1986) and antenatal classes (Lauer, 1992) (Chapter 8).

It may also be a useful adjunct to postoperative prophylactic breathing exercises and mobilization, as Flaherty and Fitzpatrick (1978) found that postoperative patients who used relaxation techniques felt significantly less incisional pain and bodily tension and required less analgesia than those who had not undertaken a relaxation programme.

Jackson (1991) studied the effect of Mitchell's physiological relaxation on patients with rheumatoid arthritis. She found that it was effective at reducing tension and improving body posture. The use of relaxation as a self-help technique for headaches has been documented, by among others, Engel (1992), who found it to be successful at treating tension headaches in both adults and children.

Relaxation training may be particularly appropriate where the condition has a significant chronic nature such as low back pain. Relaxation techniques are used frequently in chronic pain management (Linton, 1986). Strong (1991) and Strong *et al.* (1989) found that relaxation training resulted in a significant decrease in the reported rating of pain in patients with chronic back pain.

Other reports suggest that relaxation may be useful in oncology, psychiatry and certain respiratory conditions. In a meta-analysis on the effects of relaxation training on clinical symptoms covering 48 different studies, Hyman *et al.* (1989) found that all forms of relaxation, except Benson's relaxation response, demonstrated effectiveness, particularly for chronic problems such as pain, headaches and hypertension.

SUMMARY

- Relaxation is a state whereby there is a conscious easing up of effort or attention.
- The body produces certain responses to stress both physiologically and mentally.
- The body also assumes certain postural patterns when under stress.
- A variety of techniques exist to try to produce a relaxed state. These include progressive relaxation, reciprocal physiological relaxation, relaxation response training, autogenic training and biofeedback.
- The clinical benefits of therapeutic interventions of relaxation techniques are well documented across a wide range of conditions. It would appear that chronic conditions respond particularly well to such interventions.

REFERENCES

Basmajian, J.V. and De Luca, C.J. (1985) *Muscles Alive*, 5th edn, Williams and Wilkins, Baltimore

Benson, H. (1976) *The Relaxation Response*, Collins, London

Dimotto, J.W. (1984) Relaxation. *American Journal of Nursing*, **84**(6), 754–758

Duffy, E. (1962) *Activation and Behaviour*, Wiley, New York

Engel, J.M. (1992) Relaxation training: a self-help approach for children with headaches. *American Journal of Occupational Therapy*, **46**(7), 591–596

Flaherty, G.G. and Fitzpatrick, J.J. (1978) Relaxation technique to increase comfort level of post operative patients: a preliminary study. *Nursing Research*, **27**, 352–355

Gelhorn, E. (1967) *Autonomic and Somatic Integrations*, University of Minnesota Press, Minneapolis

Human Movement Explained

Hillenberg, J.B. and Collins, F.L. (1982) A procedural analysis and review of relaxation training research. *Behavioural Research Therapy*, **20**, 251–260

Hyman, R.B., Feldman, H.R., Harris, R.B. *et al.* (1989) The effects of relaxation training on clinical symptoms: a meta-analysis. *Nursing Research,* **38**(4), 216–220

Jackson, T. (1991) An evaluation of the Mitchell method of simple physiological relaxation for women with rheumatoid arthritis. *British Journal of Occupational Therapy*, **54**(3), 105–107

Jacobsen, E. (1938) *Progressive Relaxation*, 2nd edn, University Press, Chicago

Keable, D. (1985) Relaxation training techniques – a review. *British Journal of Occupational Therapy*, **48**(7), 201–204

King, N.J. (1980) The therapeutic utility of abbreviated progressive relaxation: a critical review with implications for clinical practice. In *Progress in Behaviour Modification* (eds M. Hersen, R.M. Eisler and P.M. Miller), Academic Press, New York

Klaber Moffett, J.A., Chase, S.M., Portek, I. and Ennis, J.R. (1986) A controlled, prospective study to evaluate the effectiveness of a back school. *Spine*, **2**, 120–122

Lauer, E.L. (1992) Relaxation: back to basics. *International Journal of Childbirth Education*, **7**(3), 31–32

Lehrer, P.M. (1982) How to relax and how not to relax: a re-evaluation of the work of Edmund Jacobsen. *Behavioural Research Therapy*, **120**, 417–428

Lichstein, K.L. (1988) *Clinical Relaxation Strategies*, Wiley, New York

Linton, S.J. (1986) Behavioural remediation of chronic pain: a status report. *Pain*, **24**, 125–141

Low, J. and Reed, A. (1990) *Electrotherapy Explained*, Butterworth–Heinemann, Oxford

Mitchell, L. (1987) *Simple Relaxation*, John Murray, London

Rosa, K.R. (1976) *Autogenic Training*, Victor Gollancz, London

Schmidt, R.A. (1988) *Motor Control and Learning*, 2nd edn, Human Kinetics Books, Champaign, Illinois

Schultz, J.M. and Luthe, W. (1969) *Autogenic Therapy*, Vol. 1, *Autogenic Methods*, Grune and Stratton, New York

Strong, J. (1991) Relaxation training and chronic pain. *British Journal of Occupational Therapy*, **54**(6), 216–218

Strong, J., Cramond, T. and Maas, F. (1989) The effectiveness of relaxation techniques with patients who have chronic low back pain. *Occupational Therapy Research*, **9**(3), 184–191

14. *Posture*

INTRODUCTION

The evaluation of posture and postural re-education have formed part of physiotherapy education and practice since its earliest days. The importance of normal upright posture has been proposed since the early 1900s. The posture that man adopts should involve little expenditure of energy, minimal stress and strain and be conducive to maximum efficiency of the body. Such a posture would be considered normal, whereas deviations from this ideal may result in postural faults or dysfunctional states. This chapter will outline:

• postural alignment
• changes of posture with standing, sitting and lying
• characteristics of poor posture
• causes of poor posture
• principles of posture re-education
• occupational postural problems
• examination and measurement of posture
• re-education of posture.

DEFINITION

Posture has been described by the American Academy of Orthopaedic Surgeons (1947) as:

. . . the relative arrangement of the parts of the body. Good posture is that state of muscular and skeletal balance which protects the supporting structures of the body against injury or progressive deformity, irrespective of attitude (erect, lying, stooping) in which these structures are working or resting. Under such conditions the muscles will function most efficiently and the optimum conditions are afforded for the thoracic and abdominal organs. Poor posture is a faulty relationship of the various parts of the body which produce increased strain on the supporting structures and in which there is less efficient balance of the body over its base of support.

Posture may be either static or dynamic.

Static posture

For any stationary posture to be maintained, two rules of equilibrium must be satisfied:

- a vertical line, directly through the centre of gravity of the body must fall within the body's base of support
- the net torque (or moment) about each articulation of the body must be zero (Pheasant, 1991).

Thus static posture is that of the body at rest; although obviously the body is never completely still, minute postural adjustments are being made continually.

Dynamic posture

This is the posture adopted while the body is in action, or in the anticipatory phase just prior to an action occurring.

CONCEPTS OF GOOD POSTURE

All postures of the body are maintained by some muscular effort resisting the effects of gravity. Normally this is directly through the line of gravity and the body is balanced around it. However, if one area of the body moves away from this line, it will set off a reaction throughout the rest of the body of a series of small movements until equilibrium has been regained. These movements may be conscious or automatic adjustments.

Movement and posture

Voluntary movement is performed against a near continuous background of postural adjustments. Because the human body is relatively unstable, with its high centre of gravity above a small base of support, it must constantly adapt to counteract the gravitational forces acting on it. Postural adjustments accompany voluntary movements to prevent or minimize the displacement of the centre of gravity and thus allow more efficient use of voluntary muscle, forming a postural control system (Lee, 1989).

For a fuller description of the neurophysiology of movement and postural control mechanisms, the reader is referred to Chapter 4.

Postural alignment

Good posture requires the alignment of the different weight-bearing segments of the body upon each other. Numerous authors have described the ideal position of the body, including Kendall *et al.* (1952),

Joseph (1960) and Woodhull *et al.* (1985) (Figure 14.1). There is general agreement that, in standing, the centre of gravity of the body lies at 55–57% of the height of the person above the ground or at approximately the level of the vertebral body of S_2. The line of weight is perpendicular to the centre of gravity. The importance of this line lies in its relation to the transverse axes of rotation of the joints of the vertebrae and the lower limbs, the body tending to fall anteriorly or posteriorly due to gravitational forces according to whether the line of gravity passes in front of or behind these axes, respectively.

Ankle. The centre of gravity falls in front of the ankle joint approximately midway between the heel and the metatarsal heads, tending to rotate the tibia forwards about the ankle. This is resisted by the plantar flexors, especially soleus. Many standard kinesiology texts continue to depict the centre of gravity as vertically through the ankle, despite evidence that this is incorrect.

Knee. The normal line of gravity runs anterior to the knee joint, keeping it in extension. Stability is provided by the anterior cruciate ligament, and tension in the gastrocnemius and hamstrings. If the knee joint is in full extension, active muscle work is not necessary, but if the knees flex, then quadriceps must work concentrically to maintain an upright stance.

Hip. The line of gravity usually lies approximately 1.8 cm behind the hip joint, but this varies with the body's sway. In this position,

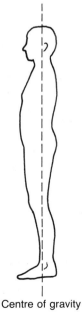

Centre of gravity

Fig. 14.1 Normal postural alignment in standing

posterior rotation of the pelvis is controlled by tension in the hip flexor iliopsoas and ligamentous stability from the iliofemoral ligament. If the line of gravity is directly through the hip joint, there is equilibrium; if it falls anteriorly, it is stabilized by the hip extensors.

Trunk. The line of gravity usually falls through the bodies of the vertebrae of the cervical and lumbar regions and through the physiological curves of the spinal column in such a way that the spine is balanced.

Head. The line of gravity is anterior to the atlanto-occipital joints, tending to rotate the head forwards on the vertebral column. This is balanced by activity of the posterior cervical muscles.

POSTURAL SWAY

In standing, there is continuous movement due to alternating action of antagonistic muscle groups resisting gravitational stresses, endeavouring to keep the total centre of gravity of the body within the body's base of support. This results in a slight anteroposterior swaying of the body of approximately 4 cm excursion (Figure 14.2). In the 1890s Romberg showed that postural sway would increase if the eyes were closed and the base of support narrowed. Force platforms are widely used to evaluate steadiness, showing the deviations from vertical that occur while standing and allowing such movements to be quantified.

Vision is the most important factor in reducing postural sway, fixating on a given point lessening the excursion. Proprioceptive influences from the foot and ankle also contribute to controlling balance.

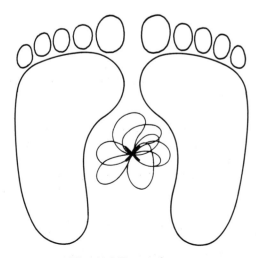

Fig. 14.2 Postural sway

Postural sway varies across the lifespan, being greatest in the young and elderly (Woollacott, 1990) (Chapter 9).

CHANGES OF POSTURE

Standing posture

- Standing upright is physiologically efficient, balance is dynamic and the energy expenditure necessary to maintain such a position is negligible.
- In upright standing the vertebral column shows its characteristic alternating curves shape, with primary kyphotic curves, convex to the rear, in the thoracic and sacral areas, and secondary lordotic curves, concave to the rear in the cervical and lumbar areas.
- In standing the anterior superior iliac spines and pubic symphysis are vertical and the sacrum inclined forward at an angle of approximately 50° to the horizontal.
- There is considerable variation in the sacral angle between individuals and this has an effect on the lumbar spine. If the sacral angle is increased then the lumbar spine will have to assume an increased lordosis to compensate and maintain an erect posture. If the sacral angle is small, the lordosis will be flattened.
- Although there may be much variation in the degree of lumbar lordosis, pelvic tilt and lateral curvature, or scoliosis, all may fall within what would be considered normal upright posture. However, when these become extreme they may lead to characteristic postural dysfunction patterns.

Sitting posture

- The sitting posture is one of the most frequently adopted, particularly in the workplace.
- Sitting necessitates approximately 90° flexion of the knees and of the hips, to bring the trunk forward over the thighs.
- Most people can flex to about 60° at the hip before tension in the hamstrings causes discomfort, so the movement is accompanied by posterior rotation of the pelvis and weight is transferred through the ischial tuberosities.
- To maintain an erect trunk position there must be compensatory flattening of the lumbar spine – equivalent to the posterior rotation of the pelvis (Figure 14.3).
- To maintain an upright sitting posture requires that active muscle effort restores the lumbar lordosis by rocking forward on the ischial tuberosities and that this position is maintained by muscle work, not just ligamentous tension.

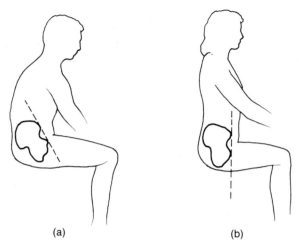

(a)　　　　　　　　　　　　　　　(b)

Fig. 14.3 Position of the pelvis in sitting: (a) relaxed sitting – pelvis tilted backward; (b) upright sitting – trunk and pelvis vertical, lumbar lordosis flattened

- It becomes tiring and uncomfortable to maintain this upright seated position if no support is available from the chair. The individual will assume a slouched position with a kyphotic spine, increasing the stress on the ligaments and the pressure at the intradiscal interface.
- Nachemson and Elfstom (1979) showed that the amount of pressure at the intervertebral disc will vary with posture.

Disc pressures are approximately 40% greater when sitting than when standing, and greater when the lumbar curve is flattened than when the lordosis is maintained. Andersson and Ortengren (1974) carried their work further to show that disc pressures varied with the effect of seat angle and that increasing the seat angle, i.e. the angle between the legs and the trunk, up to 130° decreases the pressure at the disc. Thus they have produced evidence that argues against the maintenance of an upright sitting stance and recommend that for a person leaning back at an angle of 110–120°, with a lumbar support, disc loads will result that are 30–40% lower than that of a standing posture. This position also produced lower muscle stress as measured by electromyography.

In the clinical setting, use of a lumbar support in a chair is recommended as it helps to restore a lumbar lordosis without requiring muscular effort and keeps the back in a supported position.

Clinical note: a rolled-up towel will often provide sufficient support to maintain the lumbar lordosis.

Lying posture

- Lying should represent a position of complete comfort and relaxation.

- The position may be varied, supine, prone or side lying.
- It is the least energy consuming of the positions discussed, as the centre of gravity is low and gravitational forces are resisted by passive mechanisms.
- Lying prone accentuates the lumbar lordosis, while supine lying reduces the lordosis, or may produce a kyphotic lumbar spine, depending on the amount of support given to the spine by the surface that is being laid upon.
- Disc pressure in supine lying is 35% of that of standing, but in side lying it is 75% of that of standing because of the side flexion of the spine that occurs in such a position.
- The spine should remain as neutral as possible, so the head should be supported on pillows to maintain a symmetrical neutral alignment of the head on the spine irrespective of which lying position is adopted.

CHARACTERISTICS OF POOR POSTURE

Poor posture occurs when there is a faulty relationship between differ-ent body parts, resulting in stress and strain on body structures. Kisner and Colby (1990) describe postural faults under the categories dis-cussed below.

Postural pain syndromes

These occur when posture deviates from normal alignment but has no structural limitation. Postural pain results from mechanical stress but will be relieved by activity or a change of position. No abnormalities exist in muscle strength, flexibility or balance, but continued failure to adopt a correct posture may lead to their development, e.g. working on a car engine – leaning under the bonnet.

Postural dysfunctions

There is adaptive shortening of soft tissues and muscle weakness, either due to poor postural habits or trauma to the tissues. Stress on the shortened structures causes pain, and imbalances in muscle strength and flexibility may result in further stresses to the area, e.g. the dominant side of someone who plays a lot of racket sports.

Poor postural habits

Poor postural habits, especially in the developing child, may lead to adaptive changes in muscle and soft tissues and abnormal stresses on growing bone.

CHARACTERISTIC POSTURAL ABNORMALITIES

These are extensively described by Kendall *et al.* (1952), in the classic work *Posture and Pain*, and by Kisner and Colby (1990).

Relaxed or swayback posture (Figure 14.4a)

The pelvis is shifted forwards, resulting in hip extension, and an exaggerated compensatory kyphosis in the thoracic region and flexion of the thorax on the lumbar spine. There is an increased lumbar lordosis and increased thoracic kyphosis. It is usually accompanied by a lateral curve or scoliosis, with weight being taken on one leg or alternating weight distribution from one side to the other. It is caused by failure to use the back musculature to support the spine and hanging on the ligaments. Stability is provided by the passive checks at the end of joint range of the ligaments, bony approximation and capsular tension. The commonest causes are attitude, fatigue or muscular weakness.

Hyperlordotic posture (Figure 14.4b)

There is an increase in the lumbosacral angle, producing an increased lumbar lordosis, increased anterior tilt to the pelvis and hip flexion. There is often a compensatory increase to the thoracic kyphosis – this is termed 'kypholordosis'. It is often associated with slack gluteal and abdominal muscles. The commonest causes are pregnancy, obesity, weakness of the abdominal musculature and sustained poor posture.

Flatback posture (Figure 14.4c)

There is a decreased lumbosacral angle, decreased lumbar lordosis and a posterior tilting of the pelvis. The commonest causes are long-term slouching or maintenance of flexed postures.

Scoliosis (Figure 14.4d)

Most backs have some degree of slight lateral curve or physiological scoliosis. The main part of the curve tends to be concave to the dominant side and there may be compensatory curves above or below. The cause may be due to leg-length inequality or asymmetry of the pelvis. This is a benign variation of normal which may cause problems because of muscular fatigue of stress, but should not be confused with pathological scoliosis leading to more serious orthopaedic pathology.

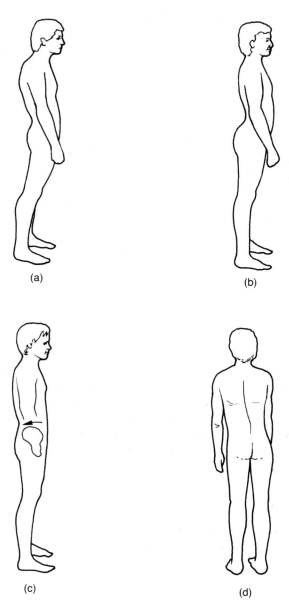

Fig. 14.4 Some postural abnormalities: (a) swayback; (b) hyperlordosis; (c) flat-back; (d) scoliosis

Increased kyphosis (dowager's hump) (Figure 14.5a)

This is characterized by an increase in the thoracic kyphosis, round shoulders through protraction of the scapulae and a forward position to the head, leading to a poking chin and compensatory cervical lordosis. Again it is caused by relaxation of the muscles necessary to

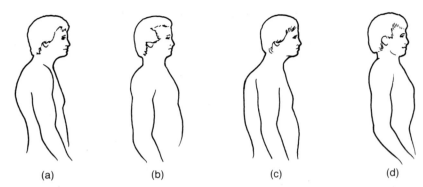

(a)	(b)	(c)	(d)

Fig. 14.5 More postural abnormalities: (a) increased kyphosis; (b) flat upper back; (c) forward head; (d) flat neck

counteract gravity, slouching and sustained flexed postures. It may cause pain due to abnormal muscle stresses and by decreasing the space for nerves to travel, as in thoracic outlet syndrome.

Flat upper back (Figure 14.5b)

There is a decrease in the thoracic kyphosis, depression of the scapulae and clavicles and a flat neck position, i.e. loss of the cervical lordosis. It is close to an exaggerated military posture. This may lead to pain due to fatigue of the anti-gravity muscles working to maintain such a posture and muscle imbalance. It is commonly caused by over-emphasis on maintaining what is thought of as a good posture, endeavouring to make the body as upright as possible.

Forward head (Figure 14.5c)

At the cervical area there is increased flexion of the lower cervical and upper thoracic regions and increased extension of the upper cervical vertebrae and the occiput on the first cervical vertebra. Problems may arise due to stress of the anterior or posterior longitudinal ligaments, muscle tension and fatigue or impingement of the cervical plexus. It is commonly caused by occupational working postures that require forward leaning or due to muscle weakness and fatigue, e.g. computing.

Flat neck (Figure 14.5d)

There is a decreased cervical lordosis with increased flexion of the occiput on the atlas. It is often seen with an exaggerated military posture, as in the flat upper back posture described above.

Effect of postural malalignment on back pain

Biomechanical changes that occur in the spine from postural malalignment arise primarily from the alterations of moments produced by the new posture (Norkin and Levangie, 1985). In standing, the main stresses caused by an abnormality in the sagittal plane are increases in shear and compressive stresses. For example, normally the primary function of the apophyseal joints is to guide movement; in hyperlordosis they may have to resist up to one-sixth of the compressive force of the spine (Adams and Hutton, 1983). Shear stresses at the anterior of the annulus fibrosis also increase with hyperlordosis (Farfan, 1978). During *et al.* (1985) suggest that postural aberrations may act by disturbing the even distribution of stress across the structures of the spine, causing overloading, strain and pain.

Bullock-Saxton (1988) points out that it is difficult to be sure whether posture abnormality has been a cause of low back pain, or whether the low back pain itself has led to a change in posture.

Those authors who relate postural alignment to biomechanical changes offer the therapist a rationale for the cause of low back pain. Others state that postural malalignment can lead to undue tension on muscles and ligaments (Kendall and McCreary, 1983).

However, many postural abnormalities are seen in subjects who are asymptomatic and a clear relationship between hyperlordosis and back pain has not been conclusively established. Nevertheless, the assumption of a hyperlordotic posture as a protective mechanism against existing back pain, and the frequency with which hyperlordosis is seen in back pain sufferers, make it essential for the therapist carefully to assess posture in patients with back pain and to determine the influence of postural modifications on the behaviour of that pain (Bullock-Saxton, 1993).

The assumption of an asymmetrical posture for a prolonged period of time may lead to multiplication of the fibroblasts in the muscles along the lines of stress and the formation of extra collagen (Editorial, 1979).

Causes of postural problems

In addition to the obvious causes of postural faults listed earlier, the therapist needs to consider other possible causative factors and how they may interact to influence postural health. Lee-Jones (1988) suggests that causative factors fall into five categories: genetic, environmental, psychosocial, physiological and idiopathic.

Genetic factors

- gender
- body type

- congenital birth defects
- intrinsic disability and disease
- eyesight
- joint flexibility.

Environmental factors

- nutrition
- trauma, postural strain
- extrinsic disability and disease
- ageing
- clothing
- physical adaptation
- occupation
- physical exercise
- climate.

Psychosocial factors

- self-esteem
- body image
- mental health
- learned postural habits
- peer influence
- lifestyle
- motivation.

Physiological factors

- age
- growth
- pregnancy
- physiological processes
- fatigue
- body weight
- tissue degeneration
- muscle tension
- flexibility
- pain.

Idiopathic factors

- paralysis
- bone malformation
- vestibular system function.

In some categories it will be easier to influence change than in others. Phelps *et al.* (1956) stated that environmental circumstances are among the chief influences in producing man's postural dispositions. Manipulation of the patient's environment may be an important strategy in helping the patient to regain postural health. Many patients will show combined causal factors rather than a simple cause–effect relationship. For example, age-related osteoarthritis of the joints, poor nutrition, low self-esteem, post-menopausal osteoporosis and poor postural habits may combine to give a patient poor postural health. The therapist may identify the relationship between the different causative factors of faulty posture and manipulate them to try to achieve efficient intervention strategies to regain postural health for the patient.

OCCUPATIONAL POSTURAL PROBLEMS

A good working posture is one that allows the task to be accomplished in an efficient and effective manner, but with minimal muscular effort, in a relaxed fashion using the muscles in a dynamic rather than a static posture. Pheasant (1986) lists the basic guidelines which should be adopted when considering working postures:

- avoid forward inclination of the head and neck
- avoid forward inclination of the trunk
- avoid work that requires the upper limbs to be used in a raised manner
- avoid twisted or asymmetrical postures
- where possible, keep joints within the middle third of their range of motion
- provide a backrest to all seats and design the seat and workstation so that the backrest can be utilized
- where muscular force must be exerted, the limbs should be in the position of greatest strength.

Ramazzini (1713) described the consequences of poor working postures: '. . . other workers in whom certain morbid affectations gradually arise from other causes, i.e. from some particular posture of the limbs or unnatural movements of the body called for while they work'.

Grandjean (1988) cites the relative energy expenditures related to workers' posture. Compared to the energy consumption of lying down, sitting increased energy consumption by 3–5%, standing by 8–10%, kneeling by 30–40% and stooping by 50–60%. Thus if a good working posture is one that is metabolically efficient, the posture will be an important factor in the overall efficiency of the work.

Work and upper limb postural disorders

Upper limb work disorders, including repetitive strain injury, are becoming a more common problem which the physiotherapist may have to treat. Although opinion differs as to the relative importance of ergonomic, psychological and physiological factors, recognition of poor postural patterns and their relationship to the patient's symptoms is important.

These problems may relate to poor head and neck posture; for example, the person working at a visual display unit (VDU) will try to view the screen straight on, i.e. with a horizontal line of sight. If there is not sufficient scope for adjustment of posture available, by adaptations to the desk and chair height or angle the computer operator will tend to slouch in the seat, flexing the spine, except for the upper cervical spine which will be extended to view the screen.

Grandjean (1988) surveyed the incidence of musculoskeletal complaints among a group of sedentary office workers and found that 14% complained of head pain, 24% neck and shoulder discomfort and 57% back discomfort. This tendency for people to complain of upper limb problems related to posture has increased with the introduction of technology to the workplace. Hunting *et al.* (1981) compared workers engaged in VDU tasks and traditional office jobs or typists. He found that 38% of those engaged in VDU tasks complained of tendomyotic pressure pains in their shoulders and necks, compared with 11% of those performing traditional office tasks. He postulated that this was mainly due to the constrained postures necessary to work at the VDU which did not afford much opportunity for change in posture, compared with the variations of posture that are found in traditional work.

Other work may induce problems by forcing the arms to be held in a raised position so that the shoulders are elevated and the arms flexed or abducted. Such a position will be assumed if the work surface is too high, increasing the static loading on trapezius and deltoid.

Prolonged work with the arms raised above shoulder height, e.g. in decorating, filing, electrical work, etc., increases the loading on the muscles of the shoulder girdle. It will put the cervical spine in an awkward posture, tending towards extension and side flexion, and is metabolically more costly to the cardiovascular system.

Griegel-Morris *et al.* (1992) examined the incidence of common postural abnormalities in the cervical, shoulder and thoracic regions and related these to complaints of pain in people who were otherwise completely healthy. They found that the following postural abnormalities were prevalent: forward head position 66%; kyphosis 38%; right rounded shoulder 73%; left rounded shoulder 66%.

While no relationship was found between the severity of the postural abnormalities and the severity of pain, severe postural abnormalities were related to an increased incidence of pain, suggesting a relationship between the presence of some postural abnormalities and the incidence of pain.

Recommended posture for VDU tasks

Grandjean (1987) recommends the following as being an ergonomically sound design for VDU workstations which will encourage the adoption of a good working posture. The furniture should be as flexible as possible and adjustable in the following dimensions:

● keyboard height (floor to home row), 700–850 mm
● screen centre height above floor, 900–1150 mm
● screen inclination to horizontal, 88–105°
● keyboard (home row) to table edge, 100–260 mm
● screen distance to table edge, 500–750 mm
● seat height, 420–500 mm
● desk height, 740–780 mm

These recommendations were made in an attempt to prevent constrained seating postures and restricted body positions.

EXAMINATION AND MEASUREMENT OF POSTURE

Traditionally, posture is measured by attempts to align significant bony anatomical landmarks. Normal alignment is usually used following the guidelines of Kendall and McCreary (1983) in which the lobe of the ear, seventh cervical vertebra, acromial process, greater trochanter, anterior to the midline of the knee and anterior to the lateral malleolus, form a theoretical line around which the body is balanced in skeletal and muscular alignment. There is, however, little standardization of methods of measuring posture. It is important that the comfortable, erect posture assumed by a patient on request at the time of examination is representative of their true postural alignment. Researchers have used a range of tools to assess changes in posture. These include:

● rating scales for subjective assessment of any observed deviations from normal
● flexirule (Burton, 1986)
● photography (Keegan, 1953)
● lordisometry (Tichauer *et al.*, 1973)
● inclinometer (Bullock-Saxton, 1993).

Bullock-Saxton (1993) found that on any one day a consistent postural alignment is assumed and that the perception of posture and, therefore, postural alignment, remains constant for 2 years. Thus a patient's awareness of what is, for them, a comfortable posture does not vary and any deviations from so-called normal posture would appear to be part of the individual's perception of posture and not due to chance positioning at the time of measurement.

PRINCIPLES OF POSTURE RE-EDUCATION

The importance of re-educating posture and the importance of maintaining normal postural alignment has not been clearly established. Deviations from a position usually considered to represent normal alignment suggests musculoskeletal imbalance which may be causally implicated in the development of pain and dysfunction. It is based on this assumption that posture correction has been used as a treatment approach for alleviating pain thought to have a postural cause. However, there is a paucity of published material to substantiate this approach.

It is vital that careful observation of the subject's posture is made and attempts made to see that the posture assumed is consistent over time. Where possible the therapist should try to identify any underlying cause for that posture existing, e.g. leg-length inequality, poor range of joint motion or muscle imbalance. Once the causative problems can be identified, therapeutic interventions can be directed towards correcting them. However, the patient must also be helped to recognize that there is a problem with their posture and be motivated to attempt to change this and recognize the benefits of a balanced, upright posture (Oliver, 1994).

Re-education of postural abnormalities may be achieved using the following treatment principles:

- relief of pain and abnormal tension
- increasing the range of motion of specific tight structures
- increasing muscle strength and balance
- retraining neuromuscular control and postural awareness
- reinforcing the link between posture and pain
- prevention of recurrence of postural problems.

Relief of pain and abnormal tension

Most patients who are seen with postural abnormalities present with pain and discomfort. Pain may be generated by abnormal tension due to muscle imbalance or due to holding a muscle group in a static position. The patient should be taught to feel the difference between the muscle being held in a tight, tense manner and in a relaxed manner. This may be demonstrated by using the opposite muscle group in an active manner to induce reciprocal relaxation and improve circulation to the area (Chapter 13).

Other therapeutic interventions may be made to relieve pain and decrease the muscle tension, e.g. massage, hot packs, ice, etc. However, the patient must be taught to feel the difference between abnormal tension in the area resulting from faulty posture and relaxation in the muscle when that posture is not present.

Increasing the range of motion of specific tight structures

If the cause of the postural abnormality is thought to be due to limitation of range of specific structures, then appropriate stretching exercises should be taught. The patient should be taught to apply a slow, sustained stretch for approximately 10 seconds to the tight structure, relax and then repeat the stretch. For example, limitation to the range of pectoralis major may cause neck and shoulder discomfort.

The therapist can passively stretch the muscle for the patient by holding the patient's arms with their hands behind the head and their elbows horizontally abducted. Alternatively, the patient could be taught to stretch the muscle themselves by holding a stick or piece of Cliniband/Theraband in front of them and then elevating the shoulders and bringing the band behind the head with the shoulders abducting the scapulae. Thus the patient can be taught to increase the range of motion and stretch tight structures.

Increase muscle strength and balance

In patients who have lost muscle strength and flexibility, muscle strengthening exercises need to be taught to train and strengthen the muscles necessary to maintain postural balance. The exercises should be progressed from an easy start point, concentrating on localized contraction of anti-gravity muscles and relaxation of antagonistic groups. Once the patient has sufficient strength to control the motion, the exercises can be progressed until the patient has sufficient strength and muscle balance to maintain the body in proper alignment.

Retraining neuromuscular control and postural awareness

For the patient to regain full postural health, they must re-establish both conscious awareness of a correct posture and its automatic control. The patient must be taught awareness of normal, balanced posture and its effects, awareness through improved proprioception of where the body is in space and learn to control that positioning. Through practice and correction, together with visual reinforcement, the patient's kinaesthetic awareness will increase. Correct postures and movements should be reinforced by verbal, tactile and visual means. Because a faulty posture has often become established and become entrenched in the patient's mind as normal, it may be difficult for them to grasp the concept of a correct posture.

Visual reinforcement will assist in this – a lateral photograph will often come as a shock to the patient who had not realized that they habitually assumed such a posture. Active training can be greatly helped by visual cues such as videotaping the patient from a side view and displaying the image via a television monitor directly in front of them. They will thus be able to monitor the effect that minor changes to their body configuration make, until unaccustomed positions feel more comfortable and normal (Figure 14.6). The use of mirrors can similarly assist in visually reinforcing the adoption of a correct posture.

Reinforcing the link between posture and pain

The patient will be much more motivated to attempt to develop correct postural alignment if they believe that there is a connection between the development of their pain and their posture. The patient should take up their habitual posture and maintain it until they feel discomfort. When this occurs, the posture should be corrected and the patient able to see the relief of discomfort that accompanies this change. Once the patient has accepted that a simple relationship does exist between the onset of pain and their posture, they will be better motivated to use the strategies taught to them to maintain correct postural alignment.

Additionally, the therapist should provide positive feedback as the patient begins to corrrect their postural faults, until they have gained an increased awareness of their postural state and can detect and correct their own faulty posture.

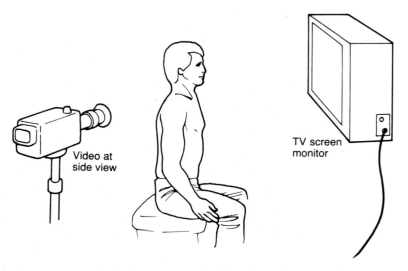

Fig. 14.6 Video arrangement for visual reinforcement of posture correction

Prevention of recurrence of postural problems

Once the patient has been taught to maintain good body alignment and re-education of posture is complete, they must be assisted to avoid recurrence of the problem. This will mean teaching of good body mechanics when carrying out activities of daily living such as walking, driving, stooping and lifting. Preventive exercises or recreation may be suggested to keep the anti-gravity muscles strong, e.g. swimming. The patient may also be encouraged to review his daily working environment pattern to identify if his work habitually calls for the adoption of static postures, or if office furniture necessitates stooped postures.

If the patient can identify such risk areas, they should seek to have them changed. Additionally, they should be made aware of the hazards and taught to employ preventive strategies such as frequent changes of posture, using a lumbar support and adopting common sense and good safety habits.

ALTERNATIVE THERAPY TECHNIQUES TO RE-EDUCATE POSTURE

In addition to the physiotherapeutic techniques outlined earlier, there exist a number of alternative therapy disciplines based upon the recognition and correction of faulty body posture. The most well known of these are the Alexander technique, the Feldenkrais method and functional integration. A brief description of each is given.

The Alexander technique

The Alexander technique was formulated almost 100 years ago by the Australian actor Frederick Matthias Alexander. He suffered from chronic vocal hoarseness when performing and was little helped by medical treatment or vocal training. Alexander decided to try to solve the problem himself by observing himself in specifically positioned mirrors. He noticed that he tended to pull his head back and down and was convinced that this was the cause of his problems. Alexander believed that the head, neck and body were all interconnected and that whatever was done with one area of the body inevitably affected other areas. He forced himself to dissociate from what felt natural and right and to learn new habits by conscious effort.

Alexander developed his theory to include the concept that the spine is where 'primary control' lies. In order to function properly, the spine must be lengthened not compressed, as shortening of the spine puts undue strain on all limbs and organs. The neck should always be free of tension, as tension will spread from there throughout the body. The head should be held forward and up and the torso should be allowed to lengthen and widen out. With practice, the new ways of holding the body will lead to new sensory experience and unconscious bad habits will be replaced. Alexander also preached the need for proper breathing, but never formulated a set of breathing exercises as such. The capacity of the chest should be increased and to this end patients were taught to breathe out on a whispered 'ah' sound.

Essentially the Alexander technique is a way of unlearning how the body is used incorrectly and substituting positive habits in order to achieve full potential mentally, physically and emotionally. Using the Alexander technique, the influence of the mind and body and how they are inextricably linked is stressed. The technique also contends that the process is more important than the end result and teaches the subject to become consciously aware of the ways that they use their body in everyday tasks. Thus the Alexander technique is primarily about the correction of faulty posture by conscious control. For further information, see Maisel (1974) and Hodgkinson (1988).

The Feldenkrais method

Dr Moshe Feldenkrais developed a method of self-observation and learning following a serious leg injury. He studied his habitual way of moving and became aware of how limiting these were. He began to explore alternatives in order to minimize his physical disability and ultimately improve the general functioning of his body. Throughout his life (1904–1984) Feldenkrais worked on methods to teach people how to learn about their bodies, developing two techniques, 'awareness through movement' and 'functional integration'.

Awareness through movement

A class setting is usually followed and verbal instructions are given for movements, which lead to a heightened sensory awareness of the whole body. Each lesson follows a theme either based on developmental sequences, such as crawling or rolling, or on functional movements, such as standing and walking. Attention is focused on how the joints and muscles move and interact with each other. The subject becomes aware both of each individual part of the body and also of how each individual part contributes to the overall functioning of the body as a whole.

Learning through sensory experience, improved body alignment and mechanics are recognized and superfluous exertion is eliminated. Initially the person has to become sensitive to the feedback that they receive from their body and, to this end, small movements are made and the verbal instructions do not give any indication of a goal to work towards. The same movement is repeated many times until the subject is sensitive to and aware of that movement.

Functional integration

This is based on the same principles that have been outlined above, but is taught on an individual basis, with tactile feedback replacing verbal instructions. The subject's attention is drawn to the parts touched and, as the teacher moves these parts, to the connections between different parts of the body (Feldenkrais, 1975).

SUMMARY

- Posture is concerned with the relative alignment of adjacent body parts, in which the body is in a state of balance, requiring minimal muscular effort to maintain that position.
- The concept of an ideal posture and normal alignment is defined, together with the effects of small variations from that norm.
- The differing postural attitudes that may be found are standing, sitting and lying.
- The concept of poor posture is explored, together with a description of characteristic postural abnormalities and their effect on back pain.
- The interaction of many causative factors – environmental, psychosocial, genetic, idiopathic and physiological – to give a model of postural health is described.
- The role of workplace factors in producing occupational postural problems, particularly in the upper limb, is discussed.
- Methods to examine and measure posture include observation, photography and inclinometry.
- The concept of posture re-education together with therapeutic interventions are outlined.
- Finally, the role of complementary therapies in the treatment of postural problems is discussed.

REFERENCES

Adams, M.A. and Hutton, W.C. (1983) The mechanical function of the lumbar apophyseal joints. *Spine*, **8**, 327–330

American Academy of Orthopaedic Surgeons (1947) *Posture and Its Relationship to Orthopaedic Disabilities*, AAOS,

Andersson, G.B.J. and Ortengren, R. (1974) Lumbar disc pressure and myoelectric back muscle activity during sitting. *Scandinavian Journal of Rehabilitation Medicine*, **3**, 104 –114

Bullock-Saxton, J.E. (1988) Normal and abnormal postures in the sagittal plane and their relationship to low back pain. *Physiotherapy Practice*, **4**(2), 94–104

Bullock-Saxton, J.E. (1993) Postural alignment in standing: a repeatability study. *Australian Journal of Physiotherapy*, **39**, 25–29

Burton, K. (1986) Measurement of regional lumbar sagittal mobility and posture by means of a flexible curve. In *Ergonomics of Working Postures* (eds E.N. Corlett, J.R. Wilson and I. Manenica), Taylor and Francis, London

During, J., Goudfrooji, H., Keeson, W., Becker, W. and Crowe, A. (1985) Towards standards for posture. Postural characteristics of the lower back system in normal and pathological conditions. *Spine*, **10**, 83–87

Editorial (1979) Stay young by good posture. *New Scientist*, **82**, 544

Farfan, H.F. (1978) The biomechanical advantage of lordosis and hip extension for upright activity: man as compared with other arthropods. *Spine*, **3**, 336–342

Feldenkrais, M. (1975) *Awareness Through Movement*, Penguin, London

Grandjean, E. (1987) *Ergonomics in Computerised Offices*, Taylor and Francis, London

Grandjean, E. (1988) *Fitting the Task to the Man*, 4th edn, Taylor and Francis, London

Griegel-Morris, P., Larson, K., Mueller-Klaus, K. and Oatis, C.A. (1992) Incidence of common postural abnormalities in the cervical, shoulder and thoracic regions and their association with pain in two age groups of healthy subjects. *Physical Therapy*, **72**(6), 425–431

Hodgkinson, L. (1988) *The Alexander Technique*, Piatkus, London

Hunting, W., Laubli, T. and Grandjean, E. (1981) Postural and visual loads at VDT workplaces. Part I: Constrained postures. *Ergonomics*, **24**, 917–931

Joseph, J. (1960) *Man's Posture*, Charles C. Thomas, Illinois

Keegan, J.J. (1953) Alterations to the lumbar curve related to posture and seating. *Journal of Bone and Joint Surgery*, **35-A**, 589–603

Kendall, H.O., Kendall, F.P. and Boynton, D.A. (1952) *Posture and Pain*, Williams and Wilkins, Baltimore

Kendall, F.P. and McCreary, E.K. (1983) *Muscle Testing and Function*, Williams and Wilkins, Baltimore

Kisner, C. and Colby, L. (1990) *Therapeutic Exercise: Foundations and Techniques*, 2nd edn, F.A. Davis, Philadelphia

Lee, W.A. (1989) A control systems framework for understanding normal and abnormal posture. *American Journal of Occupational Therapy*, **43**(5), 291–301

Lee-Jones, S. (1988) Issues to consider for postural health. *Physics in Canada*, **40**(3), 172–174

Maisel, E. (ed.) (1974) *The Alexander Technique: Essential Writings of F.M. Alexander*, Thames and Hudson, London

Nachemson, A. and Elfstom, G. (1979) Intravital dynamic pressure measurements in lumbar discs. *Scandinavian Journal of Rehabilitation Medicine*, Suppl. 1, 1–40

Norkin, C.C. and Levangie, M.S. (1985) *Joint Structure and Function: A Comprehensive Analysis*, F.A. Davis, Philadelphia

Oliver, J. (1994) *Backcare – An Illustrated Guide*, Butterworth–Heinemann, Oxford

Pheasant, S.T. (1986) *Bodyspace – Anthropometry, Ergonomics and Design*, Taylor and Francis, London

Pheasant, S.T. (1991) *Ergonomics, Work and Health*, Macmillan Press, London

Phelps, W., Kiputh, R. and Goff, C. (1956) *The Diagnosis and Treatment of Postural Defects*, 2nd edn, Charles C. Thomas, Springfield, Illinois

Ramazzini, B. (1713) *De Morbis Artificum*, Trans. W.C. Wright (1940) *Diseases of Workers*, University Press, Chicago

Tichauer, E.R., Miller, M. and Nathan, I.M. (1973) Lordisometry: a new technique for the measurement of postural response to materials handling. *American Indian Industrial Hygiene Journal*, **34**, 1–12

Woodhull, A.M., Maltrud, K. and Mello, B.L. (1985) Alignment of the human body in standing. *European Journal of Applied Physiology*, **54**, 109–115

15.   *Gait analysis*

INTRODUCTION

This chapter is intended to give the reader a comprehensive overview
of gait. The basic nomenclature is defined and the gait cycle described.
The study of gait analysis techniques are covered, varying from simple
observational techniques to the primary areas of gait study: distance
and temporal factors, kinematics and kinetics. The joint range of
motion and muscle activity of normal gait are outlined. Thus the broad
outline of the study of normal gait is covered. The latter part of the
chapter covers the changes in gait that occur with ageing or due to
pathological abnormalities. This chapter will outline:

- the gait cycle
- range of motion at the joints in walking
- muscle activity while walking
- clinical methods of gait analysis
- gait changes with age and disease
- guidelines for the correction of gait
- gait re-education.

DEFINITIONS

For the purposes of this chapter, the terms 'gait', 'locomotion' and
'walking' are all considered to be interchangeable. Gait is the movement of the body from one point to another. It is a cyclical activity, the
location of the feet alternating repetitively, representing a movement of
the body centre of mass along a horizontal trajectory in a certain
direction and at a certain velocity. Thus walking is a smooth, well-
co-ordinated reciprocal activity which encompasses a vast range of
normal states in terms of speed and style pattern. Strictly speaking,
gait refers to the manner of progression, whereas walking is used to
describe the process of locomotion.

As a cyclical activity gait can be analysed in one cycle to represent a
whole period of activity. It must be stressed that each cycle will show

variation within one individual due to distractions, changes in the ground surface, alterations in speed or changes in the level of concentration. Walking is a learned feature, gained during infancy until in the young adult it is almost automatic. In walking, one foot is alternately placed in front of the other while multiple body segments move around their joint articulations, resulting in numerous combinations about the feet and the movement forwards of the entire body.

In normal gait, the basic qualitative movements are the same for all individuals, but small variations in the degree to which each movement is performed result in very characteristic gait patterns among individuals. At its most extreme this can be represented by the contrasting styles of a John Cleese style walk, with vast fluctuations in the forwards velocity and the centre of mass going backwards in some instances, to the old films of Groucho Marx with a crouched posture and very little vertical displacement of the body.

Gait analysis can be divided into time and distance factors. In normal walking, the start of the cycle begins with heel strike on one side and finishes when the same foot returns to heel strike. Thus a step is half the stride length. Figure 15.1 shows the basic distance and temporal factors.

Gait cycle. The interval between successive contacts with the ground and one foot.

Step length. The linear distance between two consecutive contralateral foot contacts.

Stride length. The linear distance between two consecutive ipsilateral foot contacts.

Foot angle. The angle between a line representing the long axis of the foot and the line of walking.

Step width. The lateral distance between consecutive mid-heel placement positions.

Cadence. The number of steps per unit of time, e.g. steps/min.

THE GAIT CYCLE

This has been comprehensively described by Inman *et al*. (1981) and many other authors. The gait cycle is initiated with the first contact of one foot with the ground and terminates when the same foot again makes contact with the ground. Each leg will go through a period when it is in contact with the ground (*stance phase*) and a period where it is not in contact (*swing phase*).

Within each phase there are certain time instants which are of great importance for the analysis of gait (Figure 15.2). If the gait cycle starts when the first contact is made with the ground, then this time instant can be called '*heel strike*'. If the contact does not occur with the heel,

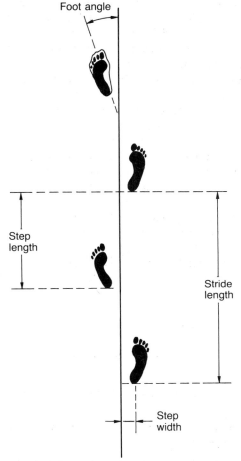

Fig. 15.1 Distance and temporal factors in gait

which is the normal pattern, then this can be redefined as *'ground contact'*. As the foot leaves the ground at the end of the stance phase it is usual for the toe to be the last point of contact and the instant of the toe leaving the ground can be called *'toe-off'*.

Stance phase

Between the time instant of heel strike and toe-off there are three important time instants. As the heel touches the ground the foot will be carefully controlled to come down towards the ground and provide a stable base of support for the rest of the body. When the whole foot is in contact with the ground this reaches the point of *'foot flat'*. The body will then roll over the foot, with the ankle joint acting as a pivot

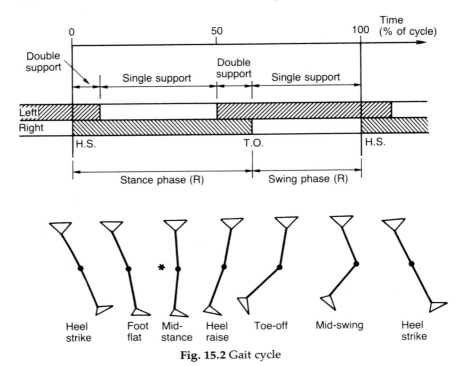

Fig. 15.2 Gait cycle

point. The hip joint extends until it is directly over the ankle joint and the knee is in slight flexion to give a stable position called 'mid-stance'. Thereafter, the purpose of the lower limb is to propel the centre of gravity forward and this can be called the 'push-off' phase, with an important time instant occurring when the heel loses contact with the ground at 'heel-off'. As the heel is raised, the hip is extended, as is the knee. As the toe is lifted from the ground in toe-off, the stance phase is completed and the swing phase begins.

For normal subjects walking at their customary walking pace on a flat, even surface, the whole stance phase takes approximately 62% of the gait cycle.

Double support phase

If the assumption is made that gait is symmetrical, it means that the heel strike of the left foot should occur exactly halfway between the heel strike of the right foot at the beginning of the cycle and heel strike of the right foot at the end of the gait cycle. If each foot is in contact with the ground for 62% of the cycle, not just 50%, there must obviously be 12% of the cycle where both feet are in contact with the ground at the same time. This is called the 'double support phase' and represents the changeover and fine control for the smooth transition from limb support on the left and limb support on the right. If the walking speed is increased, this double support phase is decreased until there is no double support phase and this represents the Olympic classification of running.

Swing phase

The swing phase can be split into three periods. The first involves the foot leaving the ground and accelerating forward and is called the 'acceleration phase'. As the ankle goes below the hip joint this can be called '*mid-swing*', as the swinging limb passes the stance limb. Thereafter, the foot is controlled ready for the next heel strike and the limb undergoes deceleration.

Although swing phase does not involve large forces of contact between the ground and the foot, two important points must be considered. First, the limb must be shortened in vertical length so that the foot can clear the ground during the forward swing. This is achieved by flexing the hip and knee. Secondly, the masses of the lower limb will be subjected to accelerations and decelerations which will, of course, be represented as force actions requiring muscular effort at the joints. As the swing phase is speeded up, the accelerations become larger and this becomes more demanding for musculature.

RANGES OF MOTION AT THE JOINTS IN WALKING

Many different researchers have performed studies of gait and produced figures for the ranges of motion for the ankle, knee and hip joints in walking. The ranges described here represent the average value for normal subjects (Chao and Cahalan, 1990). The ranges of motion are described according to the plane of movement in which they occur (Chapter 2).

Truncal rotation

As the left leg is taken forwards the pelvis rotates clockwise as viewed from above. The trunk rotates in the opposite, anticlockwise direction. With the next step, this pattern of rotation is reversed so that the pelvis rotates anticlockwise and the trunk clockwise. In normal subjects walking at their own chosen pace, truncal rotation has been measured as $6.8 \pm 2.1°$ (Murray *et al.*, 1971).

Hip joint

At heel strike, viewed in a sagittal plane, the hip joint is at approximately 30° of flexion. It extends through foot flat and is at 0° flexion at the time of heel-off. The hip continues to extend throughout the stance phase, until maximum extension is reached just before toe-off.

At heel strike, viewed from a coronal plane, the hip is in neutral and adduction occurs throughout foot flat, reaching maximum adduction at 80% of the stance phase. In the swing phase the hip is abducted.

In the transverse plane the hip is at neutral at heel strike, rotating to a position of maximum internal rotation at toe-off. In the swing phase the hip is externally rotated until just before heel strike, when it internally rotates ready for weight bearing:

stance: extension adduction and internal rotation
swing: flexion abduction and external rotation

Knee joint

In the sagittal plane the knee is extended at heel strike, moving to around 20° of flexion in mid-stance and further flexion occurring throughout toe-off.

In the coronal plane, adduction of approximately 5° occurs at heel strike and throughout the foot flat phase. During the swing phase the knee abducts and returns to neutral.

In the transverse plane the knee internally rotates at heel strike and starts to externally rotate at heel-off until mid-swing when internal rotation starts:

stance: extension, adduction and internal rotation
swing: flexion/extension, abduction and external rotation

Ankle joint

At heel strike the ankle is usually neutral. From foot flat to heel-off there is dorsiflexion as the body weight transfers over the foot followed by rapid plantar flexion at toe-off. During the swing phase, dorsiflexion brings the foot back to neutral to clear the ground and prepare for the next step. Normal ranges of motion are from 10° dorsiflexion to 20° plantar flexion.

The subtalar joint is in supination at heel strike, pronates at foot flat and supinates again for heel-off.

DISPLACEMENT OF THE CENTRE OF MASS

Efficient locomotion requires that the centre of gravity is kept as near to the midline position as possible. Figure 15.3 shows the overall trajectory of the centre of mass. From the side view the sagittal motion of the centre of mass goes along an approximate sinusoidal curve. This has repercussions, because if the centre of mass is moved upwards the forward velocity must decrease. Likewise when the centre of mass

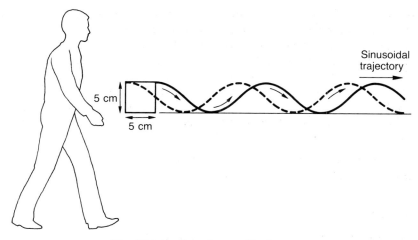

Fig. 15.3 Centre of mass displacement

descends, the velocity increases. There is also a lateral displacement, again in a sinusoidal trajectory, as the body weight is moved from side to side. These mediolateral accelerations produce mediolateral inertial forces which help to give balance while walking by generating forces which maintain the body's centre of gravity over a narrow base.

The overall vertical and lateral displacements in normal gait are ± 50 mm in each direction. In normal walking, much of the mechanism of walking is devoted to moving the centre of gravity as little as possible to save energy. In pathological gait conditions, this figure of 50 mm may vary greatly.

MUSCLE ACTIVITY WHILE WALKING

The muscle activity in walking is highly variable and there is only partial agreement about what constitutes the normal pattern of muscle activity during walking. This is partially because of individual variability of muscle usage and partly because of the difficulties of measurement, which is usually performed using electromyography (Whittle, 1991).

The hip

The gluteus maximus is the primary extensor of the hip. It is active in extending the hip at heel strike and is consistently active during the transition from the stance to the swing phase. In early swing phase the hip flexors, primarily iliopsoas, act to flex the hip and carry the limb forward. This is controlled in the deceleration phase by gluteus maximus and the hamstrings acting eccentrically. Gluteus medius and minimus are also active to stabilize the pelvis and prevent lateral trunk flexion, which would result in a waddling or Trendelenburg gait (see later). In normality, gluteus medius and minimus contract to prevent the pelvis from dipping on the unsupported side.

The knee

At heel strike there is immediate knee flexion produced by the hamstrings, the degree of which is controlled by the quadriceps acting eccentrically. Shortly afterwards at foot flat, quadriceps is active in extending the knee until mid-stance. During swing phase the hamstrings actively flex the knee to achieve ground clearance, until mid-swing when the knee extends under its own momentum, eccentric hamstring activity controlling the deceleration ready for heel strike.

The ankle

At heel strike the dorsiflexors work eccentrically to control foot placement. At mid-stance when the body translates over the supporting base of the foot, the plantar flexors work eccentrically to prevent forward shifting of the tibia on the foot. At toe-off, dorsiflexion is active, holding the ankle joint in a neutral position throughout the swing phase.

ENERGY EXPENDITURE IN WALKING

Average adults walk at speeds between 60 m/min and 100 m/min. At such speeds the rate of oxygen uptake averages $12 \, \text{ml} \, \text{kg}^{-1} \text{min}^{-1}$ (Waters *et al.*, 1983).

As a subject ages, the speed of walking tends to fall. Elderly subjects also have a lower rate of oxygen consumption at the customary walking speed than do young adults. Walking uses less than 50% of the VO_2max in normal subjects irrespective of age. However, it is significant that as a subject ages they have small aerobic reserves (i.e. a smaller VO_2max) to accommodate any decreases in physiological efficiency that pathological gait disorders may bring.

CLINICAL METHODS OF STUDYING GAIT

Gait analysis is important both in order to understand basic movement principles and to be able to spot the abnormal. Gait analysis allows:

- the measurement of the degree and extent of departure from the norm
- documentation of changes due to therapeutic interventions
- evaluation of the results of rehabilitation compared with the initial injury or disability.

There are many methods for studying gait analysis, varying from simple observational, qualitative techniques which require no specialist apparatus to sophisticated biomechanical gait laboratory analysis. In the clinical setting, quantitative data may be obtained via temporal and distance measurements or electromyography. For greater depth of study, kinematic and kinetic analysis will have to be performed, usually in a specific gait laboratory. The methods available include:

- qualitative: observations
- quantitative: stopwatch/footswitch
- kinetic
- kinematic: cinematography/Selspot.

Observational techniques

The subject must be clothed suitably to allow a clear view of the body segment configurations. Gait should be studied with the footwear on and off.

While watching the subject, the following points need to be compared to normal:

- Is the weight evenly distributed between the feet?
- Can weight transference occur without loss of balance?
- Are the stance phases even?
- Is the step width normal?
- Is the step length even?
- Does the foot point straight ahead?
- Is the gait rhythmic and efficient?
- Can the patient change direction, speed, surface, etc?
- Is there a limit to the total walking distance?

After gaining an overall impression of the subject's gait pattern, the observer needs to observe the body parts systematically and observe if deviations from normal are occurring. For example, at the pelvis is there:

- lack of forward rotation
- ipsilateral drop
- contralateral drop
- excessive rotation.

At the hip:

- flexion – limited or excessive in motion
- rotation – excessive/lacking
- abduction – where present.

At the knee:

- flexion – limited or excessive
- extension – inadequate

- hyperextension
- valgus/varus (adduction/abduction).

At the ankle:

- foot flap
- excessive plantar flexion
- excessive dorsiflexion
- no heel-off
- contralateral vaulting.

At the toes:

- clawed
- hyperextending.

Each joint must be observed and the point at which phenomena occur noted, i.e. whether it is in swing or stance phase and where in each phase it occurs. By sequentially relating what is observed at each joint with the adjoining body segments, primary gait deficits may be recognized and differentiated from secondary compensatory gait problems.

As detailed gait analysis is time consuming and the patient with pathological gait deformities may have poor exercise tolerance or compliance, it is often useful to videotape the subject's gait and record it later.

The reported reliability of observational gait techniques is questionable. Krebs *et al.* (1985) reported that observational gait analysis appears to be a convenient but only moderately reliable technique. Other authors rate its reliability as slight to moderate (Eastlack *et al.*, 1991). However, it is a convenient method and a skill every clinician should have mastered prior to using more sophisticated analysis methods.

Quantitative gait analysis

Time–distance measurements

These include the following methods:

- use of stopwatch
- use of footswitches.

Time–distance factors underlie the need to relate joint forces, muscle activity and ranges of motion to different phases of the gait cycle. Gait changes with the speed of walking, thus most analysis is performed with the subject walking at their most comfortable or customary walking speed (Andriacchi *et al.*, 1977). The subject's customary walking speed will reflect overall efficiency and correlates to abnormalities such as pain and limping. Walking speed is the combination of step length and step frequency (cadence), so alterations may occur through lengthening or shortening the step length or varying the pace of walking.

Using a measured distance, e.g. 100 m over a gym floor with a stopwatch, it is easy to gather basic quantitative data on:

- average walking velocity
- cadence
- stride length.

Cadence
This is measured by counting the number of steps taken over a given period of time. Measurements should be taken at the subject's customary walking speed.

$$\text{Cadence (steps/minute)} = \text{Steps counted} \times 60/\text{Time (sec)}$$

Velocity
This is measured by timing the patient as they walk a known distance:

$$\text{Velocity (m/s)} = \text{Distance (m)}/\text{Time (sec)}$$

Stride length
This may be measured by counting the trace of footprints left by the patient which may be measured to show left and right step lengths, stride length and base of support (see Figure 15.1). Alternatively, stride length may be calculated if the cadence and velocity are known:

$$\text{Stride length (m)} = \text{Velocity (m/s)} \times 120/\text{Cadence (steps/min)}$$

(the number 120 is used as cadence and is measured in half strides not full strides).

Observational analysis is often facilitated by the patient leaving a trace by standing on an ink pad and walking along a paper trail, or the patient walking after standing in talcum powder so that they leave a clear trace of their footprints. The percentage of time spent in the swing and stance phases may also be analysed, although it must be remembered that as the walking speed increases the percentage of these in the gait cycle will decrease. It is useful to note if the percentage of time spent in the double support phase has increased, since with a unilaterally painful condition, double support time may increase unilaterally due to hesitancy about weight-bearing on the affected side.

Such basic quantitative data may readily be obtained by the clinician and will be useful in documenting how effective a therapeutic intervention has been. The test–retest reliability for time–distance measures has been reported as high in normal subjects (Boenig, 1977) and those with hip disease (Wadsworth *et al.*, 1972).

Footswitches

Footswitches are an inexpensive method for analysing gait. They consist of a flexible insole which is placed inside the subject's shoe. The insole contains microswitches which are sensitive to pressure. Commonly, a switch is placed under the heel, under the first and fifth metatarsal heads and on the great toe. Thus when pressure is applied to a given area of the foot, the switch is activated and the duration of activation, together with the sequence of activity, is displayed via a voltmeter (Figure 15.4). Thus without being unduly intrusive to the patient or disturbing their customary walking pattern, quantitative information about overall speed, step length, cadence and lengths of stance and swing phase may be obtained as well as an indication of the pattern of weight transference through the foot.

Kinetic techniques

Kinetic analysis is based on the study of the ground reaction forces between the foot and the floor and the internal forces within the joints caused by motion. These are usually measured by force platforms, e.g. the Kistler force plate (Kistler Instrumente AG, Switzerland). The force platform consists of a top plate supported on a base plate by four legs, each of which is fitted with transducers and instrumented to measure the forces applied in three planes of movement. As yet, no standard exists for co-ordinated systems in gait laboratories, each laboratory

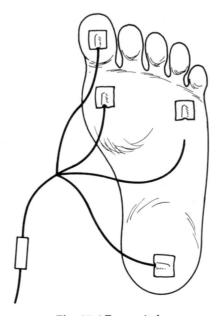

Fig. 15.4 Footswitch

devising their own. In the Oxford laboratory the parameters used are vertical forces (Fz), anteroposterior forces (Fx) and mediolateral forces (Fy).

The maximum vertical force and the vertical impulse for each leg will highlight problems with weight-bearing. A characteristic double peak pattern will be obtained if the vertical forces transmitted from the force plate are studied for a normal subject. Figure 15.5 illustrates the expected pattern of reading that would be obtained for normal subjects.

A disadvantage of the force platform is that it must be unobtrusive to avoid the subject deliberately trying to alter their gait in order that they hit the centre of the force plate while walking. Numerous repetitions from different starting positions are often required before the subject successfully strikes the centre of the force plate. A large space is required to accommodate the force platform, as there must be enough space to initiate gait up to a normal walking speed and also to slow down without running into a wall or other obstacle, as the fear of impending collision will alter the subject's gait style. Generally a room 25–30 metres long is required to house the force platform (Whittle, 1991).

The data gathered can be displayed as an electrical signal such as by an oscilloscope or by paper chart recorder. The paper will move through the recorder at a set rate and voltage changes will be recorded by the pen moving at right angles to the direction of paper movement. Alternatively, the data may be collected by dedicated computer link.

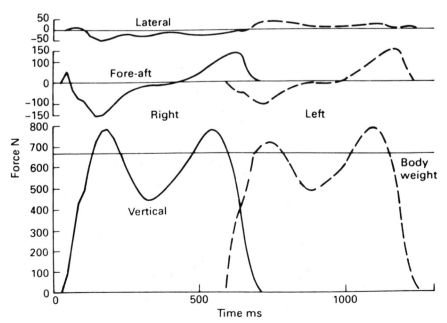

Fig. 15.5 Normal kinetic ground reaction forces

Human Movement Explained

Kinematic analysis

Kinematics is the study of the motion of the body without reference to force, i.e. it is concerned with the relative movement of different body segments. The methods available for this type of gait analysis are:

- stroboscopy
- cinematography
- electrogoniometry
- video-taping
- semi-automated systems, e.g. Selspot, Vicon, Coda.

Stroboscopy

To enable the collection of data for a series of time instants, one photograph may be produced with multiple exposures on it. Stroboscopy involves a subject walking in front of a still camera taking multiple exposures. This produces the whole motion as one picture as a stick diagram, with the position of the limbs shown at equal time intervals (Figure 15.6).

Cinematography

More sophisticated than using multiple exposure of one single photograph is to use exposures of several photographs taken in sequence throughout time, i.e. cine-photographic techniques. Cine-film effectively multiplies up the effect obtained from still photographs. As up to 50 frames per second can be obtained, it must be remembered that the amount of analysis required for each sequence of data can quickly become excessive. The film may either be studied to give a general impression of the gait pattern, or anatomical landmarks may be spotted on each frame of cine-film and digitized against a grid to give co-ordinates of each joint position. From this the relative displacements

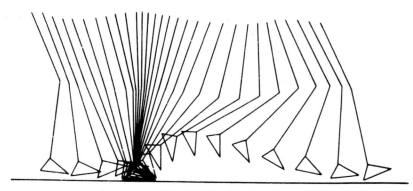

Fig. 15.6 Stroboscopy

of, for example, the hip, knee and ankle joints may be calculated. Commercial systems exist to speed up this process.

The advantage of this type of system is that no apparatus has to be attached to the patient, allowing complete freedom of movement. Much information can also be obtained in a very short period of time.

Electrogoniometry

Electrogoniometers may be attached to bony landmarks on the skeleton to describe the position of one body segment relative to another. They are particularly well suited to studying knee movements. At other sites, such as the ankle or hip, it is difficult to measure exactly at which joint the movement is occurring or in which plane, as combinations of hip and spine and ankle and subtalar joints are not easy to eliminate.

The advantage of the system is that it is relatively inexpensive and easy to apply to the subject. The disadvantage is that great accuracy is needed to place the electrogoniometers consistently. Electrogoniometers tend to be more accurate and reliable at detecting changes in flexion/extension than abduction, adduction or rotational forces.

Video-taping

Simple video-recording equipment may be used to gather information. The information collected in this way is useful for overall descriptive analysis, but care must be taken if it is used quantitatively, as it is difficult to measure co-ordinates due to poor picture resolution and the low frame rate (25 Hz).

Vicon/Selspot/Coda

With increasing sophistication in electronics, systems exist which semi-automate the process of marker detection and co-ordinate measurement. A scanning beam is used to pick up anatomical landmarks and this generates co-ordinates relative to the beam. Thus three-dimensional co-ordinates for each camera image can be generated electronically. There are a variety of semi-automatic optoelectronic devices available. However, all are expensive and require a high level of training before they can be used with ease.

Vicon. This uses television scanning to produce co-ordinates of selected landmarks.

Selspot. This uses infra-red detection which incorporates marker numbering.

Coda. This utilizes rotating mirrors and reflective prisms.

Each system has its own strengths and relative advantages and weaknesses compared with the others.

Combined kinetic and kinematic analysis

Motion systems such as those described above, where fixed detectors in the laboratory measure the relative positions of marked anatomical points, are frequently used in combination with kinetic measurements from the force platform, giving a much more comprehensive picture of the gait pattern occurring. This is illustrated in Figure 15.7.

Gait analysis is an expanding field as technology advances. While sophisticated equipment will produce a vast amount of very detailed data, this requires large resources in terms of finance, space and technical back-up. An overview of the techniques available has been given in order that the reader is aware of the range available, even though many clinically based therapists will have little opportunity to come into contact with such apparatus. However, the reader should

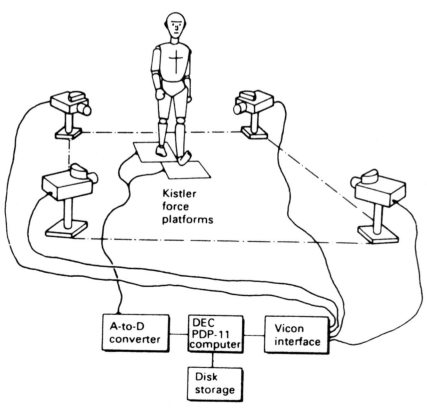

Kistler
force
platforms

| A-to-D converter | DEC PDP-11 computer | Vicon interface |

Disk storage

Fig. 15.7 Vicon

not discount the great wealth of useful information that can be obtained from simple observational techniques or from temporal–distance analysis. The physiotherapist is well used to accurate observation of patients' functional activities and as such is well suited to documenting gait in a clear, scientific manner, pertinent to the clinical setting.

GAIT CHANGES WITH THE NORMAL AGEING PROCESS

Age may affect gait either due to age itself or due to pathological processes. If pathology is excluded, gait in the elderly is a slowed down version of that of younger people. Age-related changes typically occur between 60 and 70 years of age. Most studies report that the elderly:

- have shorter step and stride lengths than younger people
- have lower average walking speeds
- demonstrate a greater variation in their stride width
- have a wider base of support (Hageman and Blanke, 1986).

Most of these changes are as a result of trying to improve stability; thus decreasing the stride length and increasing the base of support make it easier to maintain balance. Reducing the cadence leads to a longer period of double support time and hence greater stability. These are generalized findings and encompass a large variation among both old and young subjects.

Many of the patients who have walking difficulties and on whom clinical gait correction will have to be practised will be elderly. It is therefore necessary to understand the normal changes that occur to gait with age. Rehabilitation of walking in the elderly is often the most important intervention that can be made in terms of improving quality of life, minimizing disability and maintaining independence (Buchener *et al.*, 1992). The elderly frequently complain of difficulties with walking and it has been estimated that around 15% of older people experience disability through gait disturbances in the absence of any other specific diagnostic cause (Koller *et al.*, 1985).

Gait re-education of the elderly

This needs to be targeted at the patient's home environment, so that if they live in an institution without carpeting on the floor, they are confident about walking on shiny lino floors. Likewise, those accustomed to carpets should receive their walking training on such a surface, so that the training received is readily translated to the home environment.

The average distances that the patient will need to walk has to be considered in the re-education programme, e.g. how far is the nearest shop? The speed of walking required must also be relevant to the patient's needs, e.g. could the patient cross a pelican crossing in the street within the time allowed by the automatic mechanism? Such factors will need to be taken into consideration when setting the treatment goals of the gait rehabilitation programme.

Falling

In the elderly, an admission to hospital is frequently precipitated by a fall at home. Poor gait may cause falls, or they may be the cause of the gait problem developing. In comparison with elderly non-fallers, those elderly who fall demonstrate slower walking speeds, shorter step lengths, a wide variation in their step lengths and a wide range of step frequencies.

Smidt (1990) identifies the main implications for gait training in the elderly to be that the elderly patient will emphasize cadence rather than step length when walking at faster speeds. However, overall gait differences between elderly and young are small and rehabilitation should follow the same principles as for any age group, with added emphasis on making it representative of the conditions that the patient will meet at home.

GAIT IN CHILDREN

The main features of gait in children that differ from that of the adult gait pattern are:

- the base is wider
- there is no heel strike – contact is made with a flat foot
- stride length and velocity are less
- cadence is greater
- there is reduced knee flexion in the stance phase
- there is no reciprocal arm swing.

Most children will have attained heel strike and knee flexion by 2 years of age and a narrow walking base and arm swing by the age of 4. Cadence, stride length and velocity are related to height and usually reach adult values at around 15 years of age (Sutherland *et al.*, 1988; Whittle, 1991).

GAIT CHANGES IN DISEASE

In this section the common, classic gait abnormalities that a physiotherapist should be able to recognize due to musculoskeletal or neuromuscular disorders will be described. These include:

- antalgic gait
- Trendelenburg gait
- osteoarthritis and rheumatoid arthritis involving the hip
- osteoarthritis and rheumatoid arthritis involving the knee
- gait deformities from chondromalacia patellae
- hemiplegic gait
- Parkinsonian gait.

Antalgic gait

The antalgic gait is very typical of unilateral pain. There is lateral displacement of the head and trunk towards the involved weight-bearing side. With an antalgic gait the patient is subconsciously trying to decrease the forces through the joint by shifting his weight over the affected joint; thus there is only vertical force through the joint as the horizontal lever arm between the joint and the body's line of gravity has been eliminated (Figure 15.8a).

Trendelenburg gait

A Trendelenburg gait typically shows excessive pelvic tilt on the side opposite the involved stance limb. When standing on one leg, the centre of gravity of the body is brought over the weight-bearing leg by

(a)

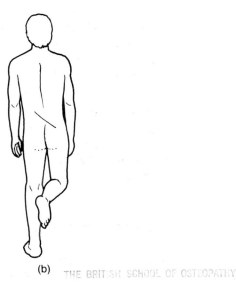

(b)

Fig. 15.8 Gait abnormalities: (a) antalgic gait; (b) Trendelenburg gait

action of the gluteal muscles on that side tilting the pelvis. Thus the pelvis is raised on the swing (unsupported) phase leg. A patient's gait is described as being Trendelenburg positive if the buttock on the non-weight-bearing leg fails to rise.

If a patient exhibits a bilateral Trendelenburg gait, there will be a waddling gait with excessive shoulder swing. This swing is due to the inability of weak gluteal muscles to prevent pelvic rotation, so that the pelvis will dip on the side of the leg which is off the ground (Figure 15.8b).

Osteoarthritis and rheumatoid arthritis of the hip

The characteristic features of arthritis of the hip, as manifested by changes in gait, have been described by many authors including Murray *et al*. (1971). These include:

- decreased walking speed
- longer time spent in single leg stance – approximately 70% of cycle compared with normal value of 62%
- in unilateral disease, least time will be spent on the affected hip
- decreased step length
- decreased cadence
- increases in lateral displacement of head and shoulders
- presence of anterior tilt of the pelvis
- loss of symmetry in walking.

The extent to which these features are exhibited will obviously vary with the severity of the hip disease and pain.

Gait following total hip replacement

When the gait of patients who have undergone unilateral total hip replacement is compared with their preoperative state they characteristically show:

- increased walking speed
- increased step and stride lengths
- increased cadence
- less displacement of the body's centre of gravity
- greater overall walking distances
- a more efficient gait in terms of energy consumption.

However, while the postoperative result is considerably better than preoperatively, the gait characteristics are still altered when compared to normal subjects.

Osteoarthritis and rheumatoid arthritis of the knee

Walking ability will again be determined by pain and disease severity. Pain usually results in a decrease in knee flexion. Rheumatoid arthritis patients use only 25% of normal knee flexion while walking, compared with 46% in normal subjects (Kettlekamp *et al.*, 1972). Brinkmann and Perry (1985) found average walking velocities of 27 m/min for rheumatoid arthritis patients and 38 m/min for those with osteoarthritis, compared with normal subjects' values of 80 m/min. They also found decreased knee flexion values of 36° for rheumatoid arthritis and 39° for osteoarthritis, compared with normal values of 50–55°.

Other gait changes are:

- shorter step lengths
- increased cadence
- decreased walking speed
- asymmetrical gait
- decreased knee range of motion
- decreased stance time on affected side.

Gait following total knee replacement

Postoperatively, total knee arthroplasty patients exhibit:

- increased walking speed
- increased walking distance
- increased step and stride lengths
- greater symmetry
- increased range of knee motion.

However, even when patients exhibited no symptoms of pain or disability they did not perform as well as age-matched normal subjects (Jevsevar *et al.*, 1993).

Chondromalacia patellae/anterior knee pain

Compared with normal subjects, such patients exhibit:

- decreased knee flexion in stance phase
- increased internal rotation of the femur just before heel strike
- increased external rotation of the femur during swing phase
- excessive knee hyperextension.

Hemiplegic gait

Gait abnormalities arising from hemiplegia will be very variable according to the cause of the condition, but clinicians often refer to a hemiplegic gait. Many authors, including Musa (1986), describe the hemiplegic patient exhibiting all or some of the following features:

- decreased walking speed
- decreased stride length
- abnormal swing/stance phase ratio
- shortened stance phase on affected side
- decreased hip movement in sagittal plane
- poor co-ordination
- poor weight transference between the lower limbs
- hip hitching
- circumduction of the leg.

The kinematics of the affected leg tends to give a picture of hip flexion, knee extension, ankle plantar flexion and lower limb circumduction during the swing phase. During stance there is characteristically knee hyperextension and a lack of toe-off.

While these are common features exhibited in a hemiplegic gait, subjects will demonstrate a wide variety of patterns according to the location and extent of the original brain injury and to the development of compensatory movements. It is imperative that careful individual assessment occurs to identify gait problems and plan appropriate rehabilitation programmes.

Holden *et al*. (1984) reported temporal distance analysis to be reliable when used to study neurological conditions. They recommend that velocity, cadence, step length and stride length are the most valid parameters when studying the neurological patient.

Gait changes in Parkinson's disease

The common features of Parkinson's disease include rigidity, tremor, bradykinesia (slowness of movement) and postural instability. Bradykinesia results in a diminution or loss of spontaneous and automatic movements, so the gait characteristically is:

- slow
- greatly reduced arm movements
- difficulty in weight transference between legs
- lack of trunk rotation
- weight tends to be too far forward, so the trunk is tilted forwards
- short, shuffling steps
- wide base of support
- difficulty in turning quickly
- difficulty in initiating walking or ceasing it
- if the patient loses balance there is an increase in speed with small steps (festinant gait).

Most of these characteristic features of a Parkinsonian gait arise from disturbances of the balance, as the patient has difficulty in weight transference and fine motor control (Chapter 4).

GUIDELINES FOR CORRECTION OF GAIT ABNORMALITIES

Once an assessment of gait has been completed and the problems causing it identified, a treatment plan must be devised aimed at correcting those faults. The following are a few suggestions for interventions to correct some of the more common gait abnormalities. They are not intended to be a comprehensive guide to treatment, but a guide as to how to tackle the often complex problem of gait correction.

Decreased/unequal step length

This frequently occurs in painful conditions such as arthritis or following trauma or surgery to the lower limb. A set of marks placed on the floor between parallel bars may assist the patient to attain an even step length by giving a visual cue. As the patient progresses, the distance between the marks may be increased until a satisfactory step length has been achieved and the step lengths are equal on each side.

Walking with flexed hips

Once it has been established that no fixed flexion hip deformity exists (Thomas's test – see Chapter 18), gait correction should be concentrated on the stance phase. Standing erect in front of a long mirror and extending the lumbar spine will help the patient to stand taller, as will providing counter-pressure in the direction required, i.e. in front of the hips and behind the head. This will provide good sensory feedback of the required posture to the patient. If fixed hip flexion contractures exist, soft-tissue stretching and manual mobilizations of the joints where range is limited may be required. For example, a patient may walk with a flexed hip and hyperextended knee when there is limitation of dorsiflexion at the ankle. The ankle joint may either be mobilized or, if this is not possible, a small heel raise supplied.

Walking with flexed knees

The therapist needs to establish if full range of motion is present at the knees, particularly extension. If the range is limited, attention should be paid to increasing it by mobilization and stretching techniques. The therapist should also check that the gait abnormality is not due to muscle weakness, and provide appropriate strengthening exercises should this be the case.

Should examination reveal no abnormalities in joint range or muscle strength, then gait re-education will concentrate on the patient attaining a good heel strike with the knee in an extended position, followed by efficient transfer of weight through the foot. If the patient is encouraged to develop an exaggerated heel–toe action, flexion at the knees can often be eliminated and a more normal gait pattern established.

Walking with an anterior pelvic tilt

This frequently occurs in arthritic conditions or where a patient has been chair-bound for a prolonged period. Efforts should be made to mobilize the lumbar spine and lumbosacral junction by pelvic tilting exercises or by manual mobilizations. In the parallel bars, the patient should be encouraged to observe the position of the iliac crests of the pelvis in a long mirror during walking, and to notice the effect of altering the lumbar lordosis.

Walking with a lateral shift of the pelvis

This is frequently seen following a period of non-weight-bearing. The trunk is usually flexed towards the affected leg and there is a failure to shift the body weight over the affected leg in the stance phase of walking. Correction of gait may be achieved by encouraging the patient to practise weight transference from side to side while standing between the parallel bars, until confidence is increased in the affected leg's ability to bear weight. Walking sideways towards the affected leg will also encourage correct pelvic alignment, as will stepping sideways onto a small step or block.

Vaulting

This occurs when the patient makes it easier to bring the swing phase leg through by going up on to the toes of the stance phase leg, a manoeuvre referred to as 'vaulting'. It results in excessive vertical displacement of the body and is an inefficient gait style.

Circumduction

This usually occurs due to leg-length inequality or in the presence of increased tone (spasticity). The swing phase leg moves in an arc, swung outwards, to increase the ground clearance for the swing phase leg.

Flat-footed gait

This is frequently seen in arthritic conditions where there is often limitation of ankle dorsiflexion, resulting in a gait that is flexed at the hips and knees and flat footed. Mobilization of the ankle joint and encouragement of a heel–toe pattern with exaggerated push-off from the toes will often restore a more bouncy gait pattern. Other causes of a flat-footed gait may be torn Achilles tendon, weakness of the posterior tibialis, weakness of the intrinsic foot muscles, pain under the metatarsal heads or a painful forefoot.

GAIT RE-EDUCATION

An ideal gait style will be:

- symmetrical and rhythmic
- metabolically efficient using minimum energy
- asymptomatic
- efficient.

There are authors who consider that walking is an innate, automatic process which does not require specific training. This approach assumes that with a reduction in the symptoms and pathology responsible, gait abnormalities will automatically correct themselves. The first step in good gait re-education will be a thorough and accurate assessment of what the abnormalities in the gait pattern are. Trying to push a patient to achieve a standardized 'ideal' gait will be unsuccessful if the reasons for that pattern being adopted have not first been established. Thus some patients may prefer to walk with a limp because it is metabolically less stressful than attempting to learn to use a standard walking style.

However, in many patients the adoption of a certain gait style becomes habitual and is continued long after the physical reason for it has been eliminated. For example, patients with painful unilateral arthritis of the hip will often demonstrate an antalgic gait to reduce pain, a Trendelenburg gait because of muscle atrophy and vaulting to compensate for leg-length inequality. Following total joint arthroplasty, gait re-education is mandatory to remove these habitual movements and replace them with a smooth, symmetrical style.

The key to successful gait re-education is to identify the factor that is responsible for a faulty pattern of movement and to target the rehabilitation programme to that specific area. For example, if the gait problem is due to weak hip abductors, the treatment programme should be directed at muscle strengthening followed by walking training.

While gait assessment is important, it should be remembered that it only reveals information about a single motor function and that such information should be combined with that from clinical examination and the subjective history.

Benefits of clinical gait assessment

The following are the principal benefits of clinical gait assessment:

- it may act as an adjunct to physical dysfunction problem-solving
- it may help determine deviations from the normal
- it may act as an outcome measure to monitor the efficiency of therapeutic interventions.

An outline of a clinical gait assessment

The patient should be dressed in clothing that will not inhibit them from moving normally or prevent the therapist from accurately visualizing the relevant body parts. The assessment should include:

1. Age (years).
2. Height (metres).
3. Weight (kilograms).
4. Sex.
5. Medical diagnosis category (if known).
6. Footwear – ideally a patient will be assessed in good footwear; slippers give the feet no support and may slip off. The type of shoe and its heel height should be noted. The footwear should be examined for wear patterns which will give information about the usual pattern of weight transference through the foot.
7. Barefoot – although walking barefoot is not usual for most people, it is important to examine the patient briefly while they are unshod. Some patients, particularly the elderly, may find walking barefoot cold and uncomfortable.
8. Foot examination. The foot should be examined to determine its flexibility or rigidity. Pressure areas, callosities or corns should be noted.
9. Orthoses or prostheses. If these are worn, the type of device should be identified together with which leg it is worn on.
10. Static posture. The posture in standing should be noted, as an abnormal gait is frequently an extension of postural musculoskeletal or neurological abnormalities.
11. Assistive devices used. Any walking aids habitually used should be noted, together with the hand they are held in and the style of gait used.
12. Patient walking. The patient should be observed walking slowly, at their customary walking pace and fast. Initially this should be done in an environment that is safe, e.g. in parallel bars, and without any verbal instructions or cues. While the patient is walking, temporal–distance factors such as walking velocity, cadence and step length can be recorded. If the patient is able, they should be viewed walking on carpets, in open spaces, on slopes, outdoors and on

steps and stairs to give a comprehensive picture of their walking abilities.

13. Pain. The location and degree of intensity of any pain experienced during the walking examination should be noted.

14. Cardiovascular factors. The patient's resting heart rate and post-exercise heart rate should be noted, together with any respiratory symptoms such as shortness of breath, wheezing, etc.

SUMMARY

- An understanding of normal walking provides a basis for the systematic analysis, treatment and management of patients with pathological gait.
- Gait is a series of rhythmical and alternating movements of the extremities which results in forward movement. Gait is described in terms of the component parts of the walking cycle, i.e. stance and swing phases.
- At different stages of the gait cycle, specific patterns of movement will be observed at the joints, which are representative of normal.
- Efficient locomotion results in the centre of gravity of the body describing a sinusoidal curve in both the vertical and lateral directions.
- Qualitative methods of gait analysis are simple, require no apparatus and are an essential skill for the clinician. However, they are hard to repeat accurately in a research setting.
- Quantitative analysis by distance–time factors is easily achieved in the clinical setting by a variety of methods with the need for minimal expense or equipment.
- Sophisticated gait analysis in a biomechanical laboratory setting can be obtained both in terms of kinetics, i.e. ground reaction forces, or kinematics, i.e. relative movement of the body parts. The two may be combined to give cumulative data of both the kinetics and kinematics of walking. These techniques involve expensive equipment and post-graduate training in their use and thus are often not readily available to clinical therapists, but remain a vital research tool.
- It is important to understand the normal changes that occur in gait with age, in order to set appropriate treatment goals for therapeutic intervention in the elderly.
- An understanding of the commonest gait deformities due to musculoskeletal or neurological disorder will assist in the correct identification of the causative factors of pathological gait and help the clinician to devise appropriate treatment programmes.
- The goals of gait education and an outline of a clinical gait assessment are described to assist the reader in performing gait analysis in the clinical setting.

REFERENCES

Andriacchi, T.P., Ogle, J.A. and Galante, J.O. (1977) Walking speed as a basis for normal and abnormal gait measurements. *Journal of Biomechanics*, **10**, 261–268

Boenig, D. (1977) Evaluation of a clinical method of gait analysis. *Physical Therapy*, **57**, 795–798

Brinkmann, J.R. and Perry, J. (1985) Rate and range of knee motion during ambulation in healthy and arthritic subjects. *Physical Therapy*, **65**(7), 1055–1060

Buchener, D.M., Cress, E.M., Wagner, E.H. and De Lateur, B.J. (1992) The role of exercise in fall prevention. In *Falls, Balance and Gait Disorders of the Elderly* (ed. B. Vellas), Elsevier, Paris

Chao, E.Y.S. and Cahalan, T.D. (1990) Kinematics and kinetics of normal gait. In *Gait in Rehabilitation* (ed. G.L. Smidt), Churchill Livingstone, New York

Eastlack, M.E., Arvidson, J., Snyder-Mackler, L., Danoff, J.V. and McGarvey, C.L. (1991) Interrater reliability of videotaped observational gait analysis assessments. *Physical Therapy*, **71**(6), 465–472

Hageman, P.A. and Blanke, D.J. (1986) Comparison of gait of young women and elderly women. *Physical Therapy*, **66**(9), 1382–1387

Holden, M.K., Gill, K.M., Magliozzi, M.R., Nathan, J. and Piehl-Baker, L. (1984) Clinical gait assessment in the neurologically impaired. *Physical Therapy*, **64**, 35–40

Inman, V.T., Ralston, H.J. and Todd, F. (1981) *Human Walking*, Williams and Wilkins, Baltimore

Jevsevar, D.S., Riley, P.O., Hodge, W.A. and Krebs, D.E. (1993) Knee kinematics and kinetics during locomotor activities of daily living in subjects with knee arthroplasty and in healthy control subjects. *Physical Therapy*, **73**(4), 229–239

Kettlekamp, D.B., Leaverton, P.E. and Misol, S. (1972) Gait characteristics of the rheumatoid knee. *Archives of Surgery*, **104**, 30–34

Koller, W.C., Glatt, S.L. and Fox, J.H. (1985) Senile gait. *Clinical Geriatric Medicine*, **1**, 661–669

Krebs, D.E., Edelstein, J.E. and Fishman, S. (1985) Reliability of observational kinematic gait analysis. *Physical Therapy*, **65**, 1027–1033

Murray, M. P. (1967) Gait as a total pattern of movement. *American Journal of Physical Medicine*, **46**, 290

Murray, M.P., Gore, D.R. and Clarkson, B.H. (1971) Walking patterns of patients with unilateral hip pain due to osteoarthritis and avascular necrosis. *Journal of Bone and Joint Surgery*, **53-A**, 259–274

Musa, I.M. (1986) Recent findings on the neural control of locomotion: implications for the rehabilitation of gait. In *Stroke, International Perspectives in Physiotherapy* (ed. M. Banks), Churchill Livingstone, Edingburgh

Smidt, G.L. (1990) *Gait in Rehabilitation*, Churchill Livingstone, New York

Sutherland, D.H., Olshen, R.A., Biden, E.N. and Wyatt, M.P. (1988) *The Development of Mature Walking*, MacKeith Press, London

Wadsworth, J.B., Smidt, G.L. and Johnston, R.C. (1972) Gait characteristics of subjects with hip disease. *Physical Therapy*, **52**, 829–837

Waters, R.L., Hislop, H.J., Perry, J. *et al.* (1983) Comparative cost of walking in young and old adults. *Journal of Orthopaedic Research*, **1**, 73–76

Whittle, M. (1991) *Gait Analysis: An Introduction*, Butterworth–Heinemann, Oxford

16. *Walking aids and orthotics*

This chapter is intended to give the reader an overview of the types of walking aid available to the patient. The selection, measurement and use of walking aids are described together with particular gait patterns. An overview of the common types of orthoses to be seen in clinical practice is also given.

This chapter will outline:

- types of walking aid
- the importance of height of walking aid
- the relative use of sticks on the contralateral and ipsilateral side
- the metabolic cost of using a walking aid
- the common gait patterns with walking aids
- the commonest types of orthoses available and their clinical indications.

OVERVIEW

Walking aids such as crutches and sticks are commonly recommended to assist with problems of pain, balance, weakness or musculoskeletal abnormality. The walking aid will transfer to the upper limbs some of the load that normally goes through the pelvis and lower limbs thus relieving weight-bearing from a painful or weak leg. They also increase balance and confidence, both by biomechanical support and by psychologically giving the user a feeling of greater stability and confidence (Figure 16.1).

The main supply of walking aids in the UK is from the NHS via the accident and emergency department, orthopaedic or geriatric clinics and from the physiotherapy departments. The selection of walking aid, adjustment and training in its use are usually the function of the physiotherapist, but in practice aids are often issued by a variety of people sometimes without adjustment or training. Additionally, the individual may select and purchase their own equipment direct from a large number of consumer outlets. The Department of Health funds research into collating information about suitable equipment for the

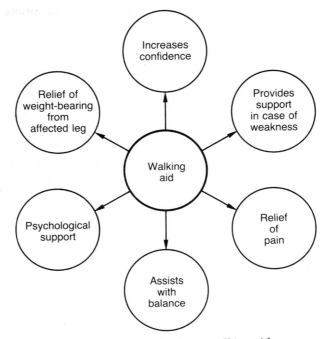

Fig. 16.1 Effects of using a walking aid

disabled. The comprehensive guide produced by the Disability Information Trust (1991), *Walking Aids*, gives details of a wide variety of equipment, indications for its use and from where it can be purchased. There are British Standards for wooden walking sticks, wooden axillary crutches, elbow crutches, tripods, tetrapod walkers and walking frames which ensure the design, materials and construction of the walking aid are satisfactory for the normal use to which they will be put.

TYPES OF WALKING AID

The type of walking aid chosen will depend on the age of the patient, their disability, their general physical condition, e.g. strength, and the duration for which the walking aid is likely to be used. The aid may be temporary, e.g. after fracture or surgical procedure, or more permanent, e.g. in the elderly or after severe injury such as spinal cord trauma.

The broad subdivision of walking aids would be between frames, crutches and sticks. There are many different designs and levels of support available within each category. Generally it is accepted that walking frames give the most support for the more heavily dependent patients and sticks the least, but there are always exceptions to this rule and each patient needs to be assessed for the most appropriate style of walking aid.

Walking frames

Walking frames give a large amount of support as they have a very wide base and are thus very stable. They are most commonly used for the elderly, although they will be used transiently in younger age groups when particularly high levels of support are required, e.g. when the patient with a total hip arthroplasty first starts to mobilize. They are also frequently used in children with neurological or musculoskeletal dysfunction.

The types of walking frame available include:

- rigid frames
- folding frames
- reciprocal frames
- forearm supporting frames
- wheeled frames
- rollators.

Rigid frames

The frame is primarily prescribed to assist in balance and to reduce weight-bearing. It needs to be adjusted to the correct height, allowing the patient to stand upright with the elbows flexed at approximately 15°. The frame should not be too heavy for the user to propel forwards easily; different weight frames are available. As the walking frame disrupts a normal gait pattern, only allowing a 'walk-to' gait, other walking aids should be used if possible.

As with other walking aids, regular maintenance checks should be carried out to check for loose screws and that ferrules have not worn smooth or become damaged.

Frames may either be fixed in height, in which case there are four sizes, small, medium, large and extra large varying from 685 to 915 mm, or adjustable in height and sized small, medium and large varying from 640 to 915 mm. They are covered by BS4986:1990 and BS5104:1990, respectively.

The frames are usually made from a lightweight aluminium, with four legs and plastic or foam rubber hand grips (Figure 16.2a). The British Standard specifies that even when the height adjustment for the front and rear legs is incorrectly set so that the one is at its shortest and the other at its longest setting, it must remain stable. Modifications to the standard frame consists of lightweight models and narrow-based models.

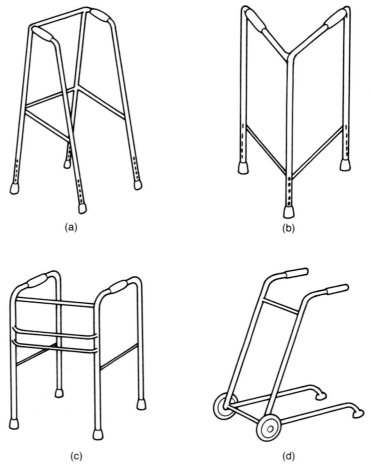

Fig. 16.2 Types of walking frame: (a) standard type of frame; (b) folding frame; (c) reciprocal frame; (d) rollator

Folding frames

Folding frames are particularly useful if the user is regularly trans-ported by car, if space is limited or if there is little room for storage. Care must be taken in reassembling the frame correctly and, because the frame is less stable than a standard non-folding model, warning must be given to the user because folding frames may tip or skew to one side if pushed too far forwards. Folding frames may either be a three-legged V-shaped or traditional four-legged collapsible design. Many different models and manufacturers exist (Figure 16.2b).

Reciprocal frames

These are particularly useful for those patients who find it difficult to lift a traditional frame. The advantage of the reciprocal frame is that it allows each side to be moved independently, as it is hinged at the front (Figure 16.2c). A normal gait pattern is thus achievable, one leg being propelled forwards together with the opposite side of the frame. Learning to do this is difficult and in practice the patient who can achieve a satisfactory reciprocal gait with such a frame could often cope with a less supportive walking aid, such as tripod or crutches. Their main indication is in certain neurological conditions where ataxia is a dominant feature, as they allow one side to be firmly braced while the other side is moved forwards.

Forearm supporting frames

These may also be called pulpit or gutter frames. They allow walking training of the patient who has difficulty weight-bearing through the upper limbs because of poor hand function, e.g. the patient with rheumatoid arthritis.

Wheeled frames

Most standard adjustable height walking frames can be converted by removing the front extension legs and replacing them with small wheeled legs. A frame with wheels at the front and ferrules at the back is less disruptive to a normal walking pattern than a standard walking frame, as it does not have to be picked up between steps. It also helps the patient who finds the traditional frame too heavy or tiring to use.

Alternatively it is possible to buy three-wheeled walking frames with two fixed rear wheels and a swivelling front wheel. These are advantageous as they encourage a more normal gait pattern and are manoeuvrable, but they need to be used with care as they lack stability.

Rollators

These are more like traditional frames in design, with two fixed wheels at the front and two ferrules at the rear. A continuous walking pattern is possible as the rear ferrules will slide over a smooth floor. The style of use is similar to that of a traditional frame, but it negates the need to lift the frame forward. The rollator has disadvantages because it is stable but not very manoeuvrable, can be awkward in tight spaces and corners, and may deviate from a straight course if unequal weight-bearing to one side occurs. It may also stick or become caught on uneven floor surfaces (Figure 16.2d).

Walking pattern with a frame

With a standard walking frame the patient lifts the frame forwards, transfers their weight onto it and takes two steps up to the frame. The patients should aim to take even steps, keep the frame well forwards, but ensure that they place all four legs of the frame on the ground at the same time before taking a step. A common fault is to use the frame tipped back onto just its rear legs at the beginning of the step. A patient who tends to lean backwards can be encouraged to bring their weight forwards by using a frame that is slightly too low and placing it well forwards.

Selection of walking frames

Hall *et al.* (1990) conducted a trial of 28 different walking frames with a wide variety of design features and from a range of manufacturers. Their principal findings were that the frame had to be considered in relation to the user's home environment and lifestyle, as opposed to their ability to use it in a physiotherapy department setting. Otherwise they did not show any one frame to be superior to another, but advise that prescription should be based on the user's disability and home environment.

Crutches

There are three main categories of crutches:

- axillary (underarm) crutches
- elbow (forearm) crutches
- forearm (trough/gutter) crutches.

Axillary crutches

Axillary crutches (Figure 16.3a) are usually prescribed when non-weight-bearing gait is required. The axillary top is rested against the chest wall stabilizing the crutch, rather like a ladder resting against a wall, and the bulk of the patient's weight is borne through the hands. As such they give good lateral stability. The top of the crutch should be approximately 50 mm below the axillary fold when the patient is standing in normal footwear. The largest disadvantage with axillary crutches is the tendency for patients to lean on the axillary bar, risking neurovascular damage. Goh *et al.* (1986) found that in incorrectly adjusted axillary crutches up to 34% of the body weight was placed on the axilla, with commensurate risk of nerve palsy. Axillary crutches are also less efficient in terms of energy costs than elbow crutches (Stallard *et al.*, 1978).

The principal indication for axillary crutches is for those patients who require the extra lateral stability that they offer. However, in North America they are issued as the standard model of crutch to most patients. The crutches are usually made of wood and adjusted via

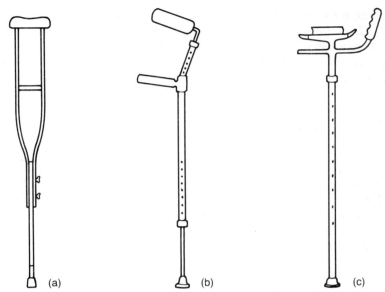

Fig. 16.3 Types of crutch: (a) axillary crutch; (b) elbow crutch; (c) forearm/gutter crutch

wingnuts, although aluminium models do exist. A further disadvantage of axillary crutches is that they inhibit functional activities such as opening doors, carrying a cup of tea, etc.

Elbow crutches

These are the most functional type of crutch and are particularly indicated for longer term use. They are suitable for both non- and partial weight-bearing gaits. In the UK they are most commonly issued with a closed cuff, which allows the use of the hand for such tasks as opening doors without the crutch falling off. On the Continent, open cuff crutches are the norm. Downwards pressure of the body weight through the handle of the crutches creates counter-pressure against the cuff on the forearm and increases stability. The handle of the elbow crutch is usually fixed at 97° to the shaft; thus they may not be suitable for patients with a fused wrist. Elbow crutches will also be contraindicated if there is more than 45° of elbow flexion contracture. Long-term use of elbow crutches may create stress on the hands and wrist, or on the shoulders if they are not correctly adjusted for the user.

Elbow crutches have the advantage of being easy to handle on stairs, for opening doors and getting in and out of cars, etc. Modifications to the standard elbow crutch design include lightweight models, moulded anatomical handgrips and differing levels of height adjustment. They may be either singly adjustable in height, by altering the length of the main shaft, or doubly adjustable, by adjusting the height between the handle and cuff as well as the shaft length (Figure 16.3b).

Forearm/gutter crutches

Forearm crutches are particularly useful for the patient who is unable to use normal handgrips due to pain or deformity or where there are flexion contractures of the wrist or elbow. Velcro straps fix the forearm into the trough and weight is applied via the forearm.

They can cause problems if they are not lightweight, imposing stress on the shoulder joints both during loading and when swung forwards. They should be adjusted so that the shoulders are not hunched or so that the patient has to lean forwards into them. Most forms of crutch have an adjustable height shaft, padded trough with Velcro strap closure and may be adjusted for forearm length and angle rotation (Figure 16.3c).

Walking sticks

Walking sticks provide effective support for the patient with good grip and sound joints in the upper limb. They are only suitable for partial weight-bearing, with loads of up to 20% of body weight. In most cases the stick should be used in the contralateral hand and be of the correct height for the user. Handles may be rounded, square, offset or with anatomically moulded handpieces, in which case right- and left-handed sticks need to be identified and correctly used. Variations in appearance, materials and weight are found. Standard hospital issue sticks are either wooden, which are cheap and easily cut to length, or more commonly nowadays metal, which are quicker to adjust and more likely to be returned when they are no longer needed. British Standards exist for wooden sticks (BS5181:1975) and metal sticks (BS5205:1990).

Other walking sticks may be foldable sticks which collapse into four or five sections and ergonomically designed sticks with shaped hand-pieces. The ferrules on sticks should be checked and the patient instructed to check and replace them regularly, either by returning to the hospital department that issued the stick or from local sources such as ironmongers or shoe repair shops.

Tripods/tetrapods

These are three- or four-legged sticks which give greater stability than a traditional stick (Figure 16.4). The tetrapod is the more stable of the two. They may be indicated where a patient has particularly poor balance or lacks confidence. Again they should be correctly adjusted and ferrules checked regularly. The size of the base between the legs can vary and selection will depend on how much extra stability the patient needs and on the room available to manoeuvre the walking aid. They are frequently used in the progression of the gait training of

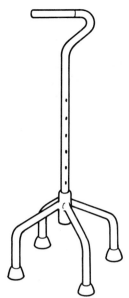

Fig. 16.4 Tetrapod

bilateral above-knee amputees following training in the parallel bars. They are also commonly seen used by hemiplegic patients, although this does not reflect good practice and should be discouraged. The main disadvantage of such devices is their relative heaviness compared with a single stem stick and that they cannot be used safely on stairs.

WALKING AID HEIGHT

The height of a walking aid will vary with the reason for which an aid is being used. Most guidelines, such as Mulley's guidelines, refer to the height of a walking aid that is being used to support body weight. An aid that is primarily designed to alleviate weight-bearing may well be shorter than one whose main purpose is balance.

Most physiotherapists adopt Mulley's guidelines for measuring the height of walking aids, measuring from the ulnar styloid to the ground, with the patient standing erect, the shoulders relaxed and the elbows flexed to 15° (Mulley, 1988). If the walking aid is too long, the elbow will be flexed too much and the shoulders hunched. If too low, the hip, knee and ankle will need to flex more than normal to compensate during the swing phase of gait.

Bauer *et al.* (1991) studied seven different crutch-length estimation techniques based on adjusting axillary crutches. They found that the best predictors were minor arithmetical adjustments to the subject's self-reported height, setting the crutches at either 77% of reported height or height minus 16 inches. However, these methods are de-signed for axillary crutch measurement and require that the patient has an accurate knowledge of their own height. If the walking aid is primarily going to be used to provide balance, then it is usually adjusted to be higher in order to provide both support and pro-prioceptive feedback.

THE IMPORTANCE OF THE HEIGHT OF WALKING AIDS

Although many authors have advocated specific height guidelines for walking aids, there is some controversy over how critical walking aid height actually is. Crosbie *et al.* (1992) studied walking aid height in healthy subjects using a non-weight-bearing gait with elbow crutches. They set the crutches 2 cm and 4 cm above and below the convention-ally recommended settings and analysed the results by kinetic and temporospatial data. They concluded that it was only at 4 cm above the recommended height setting that performance deteriorated with greater shoulder abduction. They suggested that greater attention be paid to the subjective impression of the user regarding comfort and ease of use. As this study was performed on normal subjects it is probably useful to continue to follow the conventional wisdom for height setting as well as paying attention to the expressed preference of the user.

Dean and Ross (1993) also concluded that the notion of a correct length and stick hand may be of less importance in providing safe usage of sticks than comfort and following that which the patient is most confident using. Conversely, Sainsbury and Mulley (1982) found that the majority of elderly people who fell at home were using walking aids that were incorrectly adjusted by the criteria of Mulley (1988).

CONTRALATERAL VERSUS IPSILATERAL STICK USAGE

Theoretically a walking aid should be utilized in the opposite hand to the affected leg. In this position a normal gait pattern may be achieved, as it will automatically be brought forward at the same time as the affected leg in a reciprocal pattern. Blount (1956) demonstrated that a moderate force applied to a stick on the contralateral side greatly reduces the force on the femoral head of the affected extremity. Use of a contralateral stick reduces abductor muscle force on the ipsilateral hip by creating a force that will act in the same direction as the abductor muscle pull and thus make it easier to balance the pelvis. Basmajian (1990) recommends that the patient be taught to bear down on the stick held in the contralateral hand as the unaffected leg begins the swing phase to prevent lurching and tilting of the trunk. Edwards (1986)

Table 16.1 Contralateral v. ipsilateral stick usage

Advantages of contralateral gait	Advantages of ipsilateral gait
Reduces the force through the affected leg	If used in the dominant hand, feels more natural
Prevents tilting of the pelvis	May limit hip and knee flexion
Facilitates a reciprocal gait pattern	Subjectively feels to offer more support as it is adjacent to the affected leg
Provides stability as it has a greater base of support	

studied the effect of contralateral and ipsilateral stick usage on patients who had undergone lower limb arthroplasty and concluded that unless limited hip movement was desired, contralateral usage was preferable.

However, many patients prefer to use the stick ipsilaterally, especially when they have hip or knee instability due to pain or weakness. They tend almost to use the stick as a splint, advancing it simultaneously with the affected leg. Patients will often use a stick ipsilaterally unless they are instructed otherwise. If a patient is very dominant towards one hand, they may have great difficulty using a stick in their non-dominant hand. In this case it may be better to teach them proper use of the stick in the dominant hand regardless of which side the affected leg is on. Overall, contralateral stick usage should be taught as the technique of choice, especially when balance is poor, as this gives the greatest base of support (Table 16.1).

Ipsilateral stick usage creates an increased lateral shift of the trunk and limits the reciprocal gait pattern. It gives less stability as the base of support is smaller, but may have advantages if limited hip or knee motion is required. Ipsilateral stick usage will also require greater pressure through the hand to produce the same amount of pressure relief at the hip joint. When a walking aid is solely used for balance, laterality is simply a matter of convenience or, in the case of upper limb dysfunction, of necessity.

METABOLIC COST OF WALKING USING WALKING AIDS

A swing-through gait with crutches requires a very high rate of physical effort compared with normal walking. Waters and Lunsford, (1986) compared the heart rate and oxygen costs for amputees, fracture patients and paraplegics using swing-through crutch walking at their most comfortable speed. They found much higher energy consumption than for corresponding values for normal subjects during fast walking. Kathrins and O'Sullivan (1984) compared unilateral non-weight-bearing gait with touch-down, partial weight-bearing gait using crutches. They found that both heart rate and myocardial oxygen consumption were significantly higher in non-weight-bearing ambulation which they attributed to the greater need for isometric exercise of both upper and lower extremities in non-weight-bearing gait.

Subjects also perceived a partial weight-bearing gait style to be subjectively less stressful than non-weight-bearing ambulation. There is also evidence that, with time, crutch users become adapted so that their energy expenditure and heart rate decrease as they become habitual walking aid users, suggesting the presence of both upper limb conditioning and a training response. There is little evidence that usage of a single stick imposes any metabolic costs on the user.

Forces through the upper limbs when using walking aids

If a person is utilizing a walking aid in a non-weight-bearing or partial weight-bearing manner, then most of the body weight will be transmitted through the upper arms, via the walking aid to the ground. Opila *et al.* (1987) estimate that such a gait style creates joint moment forces on the shoulder of a similar magnitude to those on the hip joint during non-aided gait. This is further accentuated when the walking aid does not give much lateral stability, so that mediolateral forces may be higher than fore–aft ones. This further increases the loading of the shoulder and the energy costs of achieving a forward motion. This is particularly true of the swing-to and swing-through styles of gait. It is important to realize that if a patient is issued with a walking aid for long-term use, there may be a risk of increased degenerative joint disease in the joints of the upper extremities.

Pre-walking exercise programmes

Many patients, particularly those with orthopaedic problems, will have had a period of prolonged bedrest prior to the commencement of walking training. As crutch walking is a learned skill, the patient must demonstrate adequate muscle strength, balance and co-ordination if they are to become proficient. Basmajian (1990) advocates that exercises are started while the patient is still confined to bed to increase strength, balance, co-ordination and timing. The strength of the upper extremities can be increased by weight-resistive exercises, graduated springs, the use of theraband, and proprioceptive neuromuscular facilitation techniques, etc. Balance exercises can occur in the bed or by mat work. Finally the patient may exercise in parallel bars to regain the feeling of being upright, but in an environment with greater stability and security than can be provided by a walking aid alone. When these stages have been followed, the patient will be stronger and have a better sense of balance to assist in learning whatever gait style is most appropriate for their condition and capability.

GAIT PATTERNS WITH WALKING AIDS

Ipsilateral two-point gait with one stick (Figure 16.5)

The stick in the ipsilateral hand is moved forward, together with the affected leg. This is followed by the non-affected leg.

Contralateral two-point gait with one stick (Figure 16.6)

The contralateral hand and stick are moved, together with the affected leg. The weight is shared between the stick and affected side as the non-affected leg is brought through.

Three-point gait (Figure 16.7)

A three-point gait requires two walking aids, either crutches or sticks, followed by the affected leg then the unaffected leg. If a minimal weight-bearing gait is required, e.g. toe touching only, then a delayed three-point gait must be utilized where the walking aid makes contact with the ground before the affected leg touches the floor. Partial weight-bearing is often prescribed in orthopaedic conditions, with a gradual progression of weight-bearing over time. Some procedures, e.g. uncemented hip arthroplasty, require long periods of partial weight-bearing ambulation.

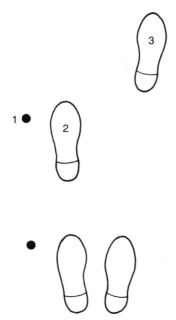

Fig. 16.5 Ipsilateral two-point gait with one stick (left foot, left hand)

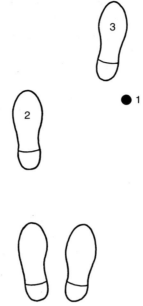

Fig. 16.6 Contralateral two-point gait with one stick (left foot, right hand)

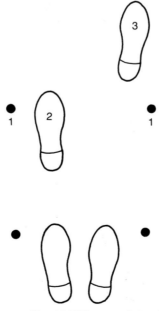

Fig. 16.7 Three-point gait

The patient may be given a clear idea of how much weight to take through the partially weight-bearing side by standing on a set of scales and feeling what the resulting pressure is like, when a set amount of weight, such as 15 kg, is applied through the affected leg. The faster the patient walks, the greater is the weight that they will transmit through the partial weight-bearing leg. Therefore, walking speeds should be kept down to around a quarter of normal free walking speeds. Repeated training will be needed for the patient to maintain the desired amount of loading through the leg.

Four-point gait (Figure 16.8)

In a four-point gait, two walking aids are used, one for each leg. The right walking aid is put forward, followed by the left leg, the left walking aid and the right leg. A four-point gait is very stable, as at no time are both feet or both walking aids off the ground at the same time. However, this also means that they produce slower walking speeds than swing-gait styles. A four-point gait is only appropriate where near full weight-bearing is allowed, as ground reaction forces will be around 95% of body weight, because when each walking aid is moved forwards the contralateral leg bears almost all of the weight. However, a four-point gait is ideal for balance and as a step to relearning a normal reciprocal gait pattern.

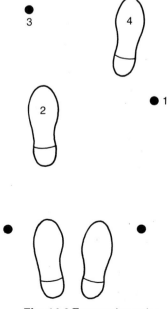

Fig. 16.8 Four-point gait

Reciprocal two-point gait (Figure 16.9)

The reciprocal two-point gait uses two sticks, right leg and left stick being placed on the ground together, followed by left leg and right stick. Again it provides a style of walking that allows fast walking speeds to be achieved.

Swing-to gait (Figure 16.10)

In the swing-to gait, both crutches are brought forwards together. The trunk and lower extremities lean forwards, weight is transferred to the upper limbs and walking aids and both lower limbs are lifted and swung forwards to the level of the crutches.

Swing-through gait (Figure 16.11)

This gait requires the best balance, upper limb strength and timing and is also least efficient in terms of energy consumption. However, it is the fastest means of ambulation with crutches. Both crutches are taken forwards, then both lower limbs are lifted and swung past the crutches, so that the crutches are left behind the point where the feet land on the floor. Therefore, the crutches must rapidly be brought forwards before hip and trunk balance is lost, and the next step started. A swing-through gait is most commonly used by those with no lower limb control such as spinal cord injury patients. It is unsuitable for

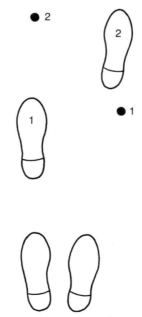

Fig. 16.9 Reciprocal two-point gait with two sticks

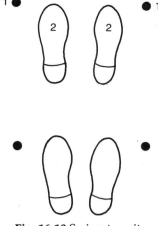

Fig. 16.10 Swing-to gait

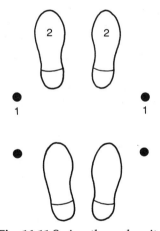

Fig. 16.11 Swing-through gait

those with painful lower limbs, as considerable ground reaction forces are generated when the feet land on the floor.

COMMON TYPES OF ORTHOSES

An orthosis is defined by the DHSS as 'a device applied direct and externally to the patient's body with the object of supporting, correcting or compensating for an anatomical deformity or weakness, however caused. It may be applied with the additional object of assisting, allowing or restricting movement of the body'.

An orthosis augments the body by preventing unwanted motion or by replacing the action of weak or dysfunctional muscles, hopefully enabling the wearer to move efficiently and comfortably by reducing impairment. As an orthosis is worn over the limb, the underlying articular and motor functions of the body will continue to influence the patient's gait. Orthoses vary from the simple foot orthosis (FO) such as a medial arch support insole, to complicated multi-articular devices, such as the Parawalker HKAFO (hip, knee, ankle, foot orthosis). A brief outline of the range of orthoses for the lower limb are described, together with some of the orthotic gait deviations that the physiotherapist may observe and need to correct in clinical practice. Prosthetic gait is not described in this book; for information on this the reader is referred to Engstrom and Van de Ven (1985).

Foot orthoses

The foot is the point at which contact is made with the ground and ground reaction forces are generated and take effect. In normal gait the foot is in a neutral position at heel strike; the arch of the foot is raised just before stance by active extension of the great toe tightening the plantar fascia and supinating the foot. As the foot is lowered to the ground the arch descends and the foot pronates. At mid-stance this process is reversed, with supination of the foot, raising of the arch and external rotation of the tibia. Thus the foot moves in three planes.

Bowker *et al.* (1993) describe the three different types of foot pathology that are amenable to orthotic management. These are:

- foot instability or deformity due to muscle weakness or imbalance
- foot instability or deformity due to structural malalignment
- foot instability or deformity arising due to loss of structural integrity within the foot.

Foot instability due to muscle weakness or imbalance

Weak supinators
On weight-bearing, such a weakness will result in a pronated foot, which may exhibit a calcaneovalgus deformity where the line of pull of the tendo Achilles has been shifted laterally. There is a general tendency for the bones on the medial side of the foot to be distracted and those on the lateral side to be compressed. This change in foot shape causes the medial longitudinal arch to be depressed and the deltoid ligament stretched. If the foot is pronated at heel strike, a medial flare (Figure 16.12a) will tend to cause the ground reaction force to turn the foot towards its correct functional position. This will be achieved more accurately by a wedge built into an insole worn inside the shoe (Figure 16.12b). Similar biomechanical corrections may be made by a heel cup (Figure 16.12c) or a flexible insole (Figure 16.12d), if the deformity is

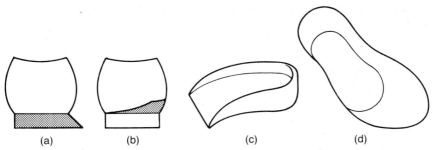

Fig. 16.12 Foot orthoses: (a) medial heel flare; (b) medial wedge insole; (c) heel cup; (d) flexible insole

mild. All will be manufactured from a cast of the foot taken with the subtalar joint in a neutral position and non-weight-bearing.

Weak pronators

A foot with weak or absent pronators will adopt a supinated position at foot contact. It is often associated with dysfunction of the peronei. The talus becomes dorsiflexed, the plantar fascia tightened and a high arched pes cavus deformity results. The angle at the metatarsal joints becomes accentuated, leading to tightness of the long extensors and possible clawing of the toes. In order to correct this, a valgus moment is required, usually by a lateral flare or wedge.

Weak toe extensors/flexors

These are often found in association with problems at the ankle joint. However, digital deformities may occur alone. If flexor muscle activity is unopposed, clawing of the toes may occur, with subluxation of the metatarsophalangeal joints and pressure on the metatarsal heads. Hammer toe, i.e. plantar flexion deformity of the proximal interphalangeal joint, or mallet toe, the abnormal plantar flexion of the distal phalanx, may also occur. As the lever arms with which to correct such deformities by three-point pressure make it near impracticable, moulding using polyurethane or silicone materials may be most effective.

If weakness of the flexors or extensors exists it may lead to malalignment of the metatarsal heads and the development of metatarsalgia. Metatarsalgia is pain at the front of the foot under the metatarsal heads, usually due to poor distribution of the load across the metatarsal heads. The metatarsal heads may become more prominent due to migration of the fat pad covering them distally and to clawing of the toes. The pressure on the metatarsal heads may be relieved by an insole with either a metatarsal dome or bar (Figures 16.13a,b). However, these may cause pressure problems if the toes have clawed, and if there is insufficient room between the tops of the toes and the shoe upper to allow the insertion of an insole. Alternatively, a metatarsal bar

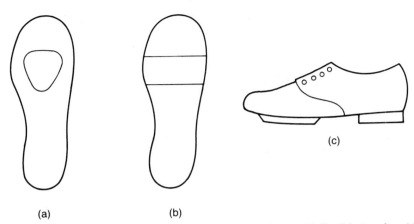

Fig. 16.13 Foot orthoses: (a) insole with metatarsal dome; (b) flexible insole with metatarsal bar; (c) external metatarsal bar rocker

may be fixed to the bottom of the shoe so that the patient rocks over the metatarsal heads without directly loading them (Figure 16.13c).

Foot instability or deformity due to structural malalignment

Structural malaligments are often congenital and generally result in a foot in which the joints are mobile but function about abnormal positions. In children, the foot can readily be realigned, usually using a moulded heel cup.

Foot instability or deformity due to loss of structural integrity within the foot

Structural integrity may be compromised by trauma, repeated strains or arthritic changes. Pain may result from joint instability or excessive motion. The patient will try to avoid this pain by changing the portion of their foot that they present to the ground, which may cause secondary pressure problems. The foot joints need to be stabilized using a stiff material for wedges or flares. If most pain is felt on weight-bearing, shock must be absorbed either by shock-absorbing insoles, such as Sorbothane or Poron insoles, or by lessening the pronation that occurs at foot contact by a flexible medial arch support.

Heel pain may be due to a calcaneal spur, plantar fascitis or a feature of some rheumatoid diseases. The aim of the orthosis is to relieve the pressure on the inflamed tissue of the heel by supporting the mid-section of the foot and rolling the calcaneum inwards. This is achieved

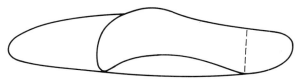

Fig. 16.14 Rose-Parker insole

with a Rose-Parker insole, made of firm Sorbo rubber in a convex wedge higher medially than laterally, extending from behind the metatarsal heads to the end of the medial arch (Figure 16.14).

Ankle–foot orthoses

This is the commonest type of orthosis to be prescribed. There are numerous designs and trade names but they all fulfil the same basic purpose: to support the foot in a stable, neutral position and provide stability to the ankle, while also having the option of limiting the range of motion at the ankle joint. They are commonly prescribed for:

- neurological conditions causing weakness, e.g. multiple sclerosis, polio
- muscular paralysis, e.g. muscular dystrophy
- spastic paralysis, e.g. cerebrovascular accident, head injury
- pain relief in arthritis.

The ankle–foot orthosis (AFO) comes in two main categories – metal and leather, and plastic-moulded.

Metal and leather
For example, single leg iron. These have a leather-covered cuff band, with one or two metal bars inserting into the heel of the shoe. According to the mechanism of insertion, e.g. square socket or round socket, the range of motion attainable at the ankle joint can be varied or limited altogether. Variation in the amount of resistance offered by the AFO can be provided by using springs or fixed stops on the bar (Figure 17.15a).

Plastic-moulded AFO
For example, the Yates splint (Figure 16.15b). These are thermoplastic splints moulded to fit the limbs closely and to be inserted inside the patient's footwear. They are made by taking a plaster cast of the patient's foot in a pain-free, functional position. They are lighter than a traditional AFO and more acceptable cosmetically. The patient may need to wear shoes a size larger than their usual size to accommodate the AFO.

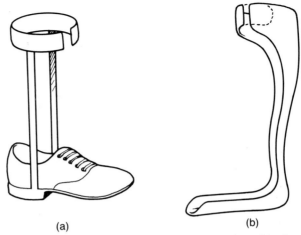

Fig. 16.15 Ankle–foot orthoses: (a) metal and leather; (b) plastic-moulded

Knee–ankle–foot orthoses (long leg orthoses)

The main functions of the knee–ankle–foot orthoses (KAFO) are to:

- reduce the stress within the leg
- stabilize the knee in the sagittal or coronal plane
- limit movement within the knee.

Clinical usage includes cases of:

- delayed union of lower limb fractures
- valgus/varus instability due to arthritis
- to limit range of motion at the knee
- cast braces during normal fracture repair.

KAFOs may be made of traditional leather and metal construction or moulded plastic (Figures 16.16a,b). The traditional KAFO will usually have a leather cuff around the thigh, a hinge to allow a prescribed amount of knee flexion, which may be lockable or free swinging, and an insertion into the shoe as previously described for ankle–foot orthoses.

Thermoplastic moulded KAFOs have a moulded thigh support, a plastic conforming AFO and a hinge joining the two sections. These have the advantage of being lighter, more cosmetic and easier to keep clean.

Cast braces

These are used to maintain normal limb function as far as possible while fracture healing occurs. Fracture union is more rapid if fixation is

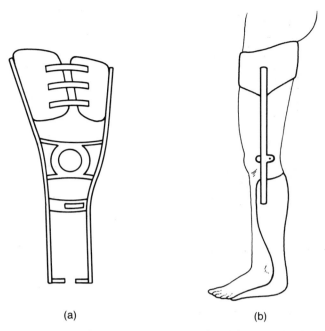

(a)   (b)

Fig. 16.16 Knee–ankle–foot orthoses: (a) metal and leather; (b) thermoplastic

less, provided that shortening and malalignment do not occur. Cast braces have gained popular approval as they support fracture sites while allowing normal limb and knee function.

Knee orthoses (braces)

A knee orthosis (KO) is commonly prescribed for two reasons:

● relief of pain and to provide stability in arthritis
● sports bracing for ligamentous damage.

KOs are biomechanically difficult as mechanically they have to act with a short lever arm, over areas of soft tissue, making it difficult to stop them sliding down the leg from gravitational forces or becoming twisted. They may provide support to the knee, but often their use is as much warmth and psychological support as on sound biomechanical orthotic principles. France *et al.* (1987) concluded that the majority of prophylactic knee braces available appear to be biomechanically inadequate.

KOs exist to try to control hyperextension, such as the Swedish knee cage commonly used in hemiplegic patients (Figure 16.17a).

Sports braces will attempt to control the joint axis changes that occur in the knee joint with sudden movements, usually to compensate for a

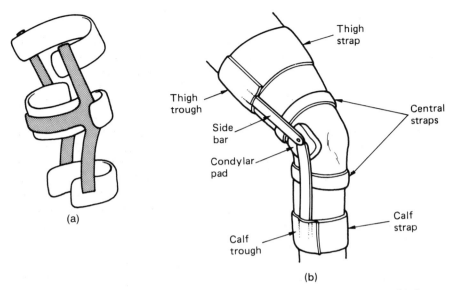

Fig. 16.17 Knee orthoses: (a) Swedish knee cage for hyperextension; (b) frame brace

damaged or absent cruciate ligament. They consist of a firm metal frame which acts as an anchor for straps positioned to resist tibial displacement on the femur. Such braces include the Donjoy and Lennox Hill braces (Figure 16.17b).

Trunk and limb braces

Hip–knee–ankle–foot orthoses (HKAFOs) are most commonly used in neurological conditions such as:

- traumatic paraplegia
- congenital paraplegia – spina bifida
- flaccid paralysis from polio
- some types of cerebral palsy.

They aim to provide stability to the skeleton and control to allow other working muscle groups to produce locomotion. The main types are the hip guidance orthosis (HGO) and the reciprocating gait orthosis (RGO).

The general principles of both types of orthosis are the same in that the body is braced from mid-trunk to the feet with the hips and knees immobilized. The hips can flex and extend in the sagittal plane, but adduction is prevented when the leg is lifted from the ground. They are both used with either crutches or a rollator but have differences in design, gait characteristics and weight.

Hip guidance orthosis

The HGO, also called the Parawalker, was devised by Gordon Rose (1979) at the Orthotic Research and Locomotor Assessment Unit (ORLAU) at Oswestry, UK (Figure 16.18a). It has free hip joints between stops at the limit of flexion and extension. The patient's shoes fit onto metal plates with rocker soles. The HGO is usually worn outside of the patient's clothing. The patient walks by using the arms and walking aids to move the trunk forwards, the weight being taken on the forward leg. Pushing down on the contralateral arm, the patient leans towards the supporting leg and gravity causes the other leg to swing forwards and a step to be taken. Once the leg has swung forward, weight is transferred onto it, the trunk is moved forward and the cycle repeated. Thus there is a pendular action in the swing phase of gait and a vaulting action in the stance phase. The advantages of the HGO are as follows:

- it has relatively low energy consumption requirements
- it allows a competent user to achieve walking speeds of approximately 50% of normal

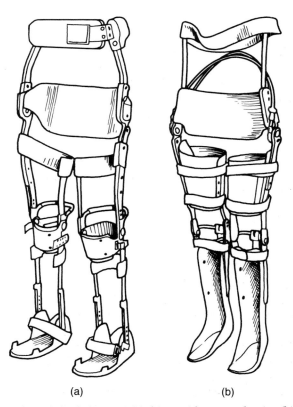

(a) (b)

Fig. 16.18 Trunk and limb braces: (a) hip guidance orthosis; (b) reciprocating gait orthosis

- it can go up steps of approximately 6 in (150 mm) and down slopes
- it is independent to put on and take off.

Reciprocating gait orthosis

The RGO was developed by Douglas *et al.* (1983) at Louisiana State University, USA. The RGO (Figure 16.18b) has hip joints linked by a cable so that extension on one side causes flexion at the other. The lower section of the RGO consists of a moulded plastic AFO which fits inside the patient's shoes, and the device can be worn under normal clothing. Forward motion is achieved by pushing down through the crutches with both arms and pulling the pelvis forward, elevating on one side. As the trunk advances, the leg in contact with the floor which has body weight on it is left behind and the cable linkage automatically advances the other leg. Once this leg is in a forward position, weight is transferred onto it and the cycle is repeated for the other side.

Relative advantages of the HGO and RGO

The DHSS commissioned a trial to compare the HGO and RGO at the Nuffield Orthopaedic Centre, Oxford, the results of which were published in 1989 (Whittle and Cochrane, 1989). The trial compared the two devices on clinical, psychological, biomechanical and ergonomic criteria. Each patient in the trial used each type of orthosis, following full training, for 4 months. There was a random order of which orthosis was given first and at the end of the trial the subjects were able to choose which orthosis they kept. Table 16.2 summarizes the results of the trial.

Reasons for choosing HGO (4 subjects)
- speed to put on and take off
- worn only as a form of exercise
- instantly understood by observers.

Reasons for choosing RGO (12 subjects)
- cosmetically acceptable
- lighter
- ability to stand unsupported.

Table 16.2 Comparison of the HGO with the RGO

Walking aid with HGO	Walking aid with RGO	Success with HGO	Success with RGO	Final choice
Crutches 14	1	Yes 17	Yes 15	RGO 12
Rollator 5	19	No 2	No 5	HGO 4
				None 6

Reasons for choosing neither (6 subjects)
• pressure area problems (3)
• too tiring to use (2)
• too embarrassed to wear (1).

Other studies question the methodology used in this study and have also compared the two devices (Banta *et al.*, 1991).

EVALUATING ORTHOTIC GAIT PATTERNS

In evaluating the patient's gait it is imperative that the therapist distinguishes between gait abnormalities that are imposed by the orthosis which are necessary for its successful function, e.g. a fixed knee position, and those that result from poor use of the device by the patient. Alternatively, the gait abnormality may be due to the orthosis failing to compensate for the anatomical problem or to its introducing new abnormalities.

Sagittal plane deviations
These include inadequate dorsiflexion, knee hyperextension or trunk flexion. They may be caused by an orthosis failing to maintain adequate dorsiflexion during the swing phase causing the patient to stumble, such as by an inadequate spring-loading system. Knee hyperextension may occur if a KAFO does not have a stop on the knee mechanism.

Coronal plane deviations
For example:

• abducting the lower limb
• trunk flexion laterally
• circumduction of the lower limb
• vaulting of the lower limb.

Transverse plane deviations
For example:

• internal/external hip rotation
• tibial torsion
• excessive medial/lateral foot contacts.

The therapist must distinguish between those abnormalities caused by a fault in the orthosis and those caused by the patient, e.g. flexing the trunk over the hips. Once the cause of the fault has been correctly identified, appropriate interventions to correct it can be initiated either through exercise programmes or referral back to the orthotist.

SUMMARY

- Walking aids such as frames, crutches and sticks are commonly used to assist with problems of pain, weakness, balance or musculo-skeletal abnormality.
- The walking pattern used with a walking aid will depend on the patient's strength, balance, disability and age.
- Methods for measuring the correct height of walking aids are described, but how important this factor is in successful walking aid use is controversial.
- Walking aids may be used either contralaterally or ipsilaterally, but contralateral use is generally recommended, except where limited lower limb joint range is desired.
- Using a walking aid such as crutches imposes a metabolic cost on the patient, especially when used in a non-weight-bearing gait.
- Accepted gait styles using walking aids are point (two-, three- and four-point) and swing gaits (swing-to and swing-through). Point gaits tend to be more stable but swing gaits are faster.
- Orthoses may be prescribed to correct an anatomical deformity giving rise to a gait disorder. The orthoses are described according to the anatomical structures that they support.
- The need to differentiate between failure of an orthosis to perform its task adequately and other anatomical gait abnormalities is high-lighted in the evaluation of orthotic gait.
- It should be remembered that gait with either a walking aid or an orthosis is not normal walking, but the therapist should aim to achieve optimal performance within the restrictions imposed by the patient's disability.

REFERENCES

Banta, J.V., Bell, K.J., Muik, E.A. *et al*. (1991) Parawalker energy cost of walking. *European Journal of Paediatric Surgery*, **1** (Suppl. 1), 7–10

Basmajian, J.V. (1990) Crutch and cane exercises and use. In *Therapeutic Exercise* 5th edn (eds J.V. Basmajian and S.L. Wolf), Williams and Wilkins, Baltimore

Bauer, D.M., Finch, D.C., McGough, K.P. *et al*. (1991) A comparative analysis of several crutch-length-estimation techniques. *Physical Therapy*, **71**(4), 294–300

Blount, W.P. (1956) Don't throw away the cane. *Journal of Bone and Joint Surgery*, **38-A**, 695–708

Bowker, P., Condie, D.N., Bader, D.L. and Pratt, D.J. (1993) *Biomechanical Basis of Orthotic Management*, Butterworth–Heinemann, Oxford

Crosbie, J., Armstrong, E. and Kempson, J. (1992) Is walking aid height critical? *Australian Journal of Physiotherapy*, **38**(4), 261–266

Dean, E. and Ross, J. (1993) Relationships amongst cane fitting, function and falls. *Physical Therapy*, **73**(8), 494–504

Disability Information Trust (1991) *Walking Aids – Equipment for Disabled People*, Disability Information Trust, Oxford

Douglas, R., Larson, P.F., D'Ambrosia, R. and McCall, R.E. (1983) The LSU reciprocation-gait orthosis. *Orthopaedics*, **6**, 834–838

Edwards, B.G. (1986) Contralateral and ipselateral cane usage by patients with total knee or hip replacement. *Archives of Physical Medicine and Rehabilitation*, **67**(10), 734–740

Engstrom, B. and Van de Ven, C. (1985) *Physiotherapy for Amputees: The Roehampton Approach*, Churchill Livingstone, Edinburgh

France, E.P., Paulos, L.E., Jayaraman, G. *et al*. (1987) The biomechanics of lateral knee bracing. Part II: Impact response of the braced knee. *American Journal of Sports Medicine*, **15**, 420–438

Goh, J.C.H., Toh, S.L. and Bose, K. (1986) Biomechanical study on axillary crutches during single leg swing–through gait. *Prosthetics and Orthotics International*, **10**, 89–95

Hall, J., Clarke, A.K. and Harrison, R. (1990) Guidelines for prescription of walking frames. *Physiotherapy*, **76**(2), 118–120

Kathrins, B.P. and O'Sullivan, S.D. (1984) Cardiovascular responses during non weight bearing and touchdown ambulation. *Physical Therapy*, **64**, 14–18

Mulley, G.P. (1988) Everyday aids and appliances – walking sticks. *British Medical Journal*, **296**, 475–476

Opila, K.A., Nicol, A.C. and Paul, J.P. (1987) Forces and impulses during aided gait. *Archives of Physical Medicine and Rehabilitation*, **68**(10), 715–722

Rose, G.K. (1979) The principles and practice of hip guidance articulations. *Prosthetics and Orthotics International*, **3**, 37–43

Sainsbury, R. and Mulley, G.P. (1982) Walking sticks used by the elderly. *British Medical Journal*, **284**, 1751

Stallard, J., Sankarankutty, M. and Rose, G.K. (1978) A comparison of axillary, elbow and Canadian crutches. *Rheumatology and Rehabilitation*, **17**, 237–238

Waters, R.L. and Lunsford, B.R. (1986) Energy cost of paraplegic locomotion. *Journal of Bone and Joint Surgery*, **67-A**, 1245–1250

Whittle, M.W. and Cochrane, G.M. (1989) *A Comparative Evaluation of the Hip Guidance Orthosis (HGO) and the Reciprocating Gait Orthosis (RGO)*, Health Equipment Information, 192, DHSS, London

17. *Ergonomic approach to lifting and handling*

AIMS OF THE CHAPTER

This chapter will outline:

- back pain in the health care professions
- the causative factors in lifting and handling injuries
- prevention of lifting and handling injuries
- the ergonomic approach to lifting and handling
- risk assessment
- legislation
- principles of lifting
- practical patient handling techniques.

BACK PAIN IN THE HEALTH CARE CONTEXT

Lifting and handling of patients in hospital has been identified as an activity capable of causing occupational back pain (Videman *et al.*, 1989; Klaber Moffett *et al.*, 1993). Nurses have nearly 30% more days' sickness absence each year from back pain than the general population (Pheasant and Stubbs, 1992).

Scholey and Hair (1989) found that physiotherapists, who share many of the occupational risk factors with nurses, had a similar prevalence of back pain to the general population, but the initial onset of pain was more likely to be attributed to a work-related incident. Overall, the epidemiological data seem to point to a relationship between lifting and handling and back pain problems in health care workers.

CAUSES OF LIFTING AND HANDLING INJURIES

It is difficult to establish a clear cause–effect relationship between individual risk factors and injury. A combination of many vocational risk factors may contribute to lifting and handling injuries. These include:

- weight on the load
- frequency of lifting and handling
- lifting while the spine is rotated
- lifting while the spine is flexed
- lifting while the spine is side flexed
- prolonged stooped postures
- lack of variation in posture

All of these occupational risk factors may be found while handling patients. There is an overall association between heavy lifting and back problems in nursing. This is reflected by the higher incidence of back pain in nurses working in areas traditionally regarded as heavy, such as orthopaedics and geriatrics. Baty and Stubbs (1987) observed that a nurse may spend over 20% of the working day working in a stooped posture. The psychological stress associated with health care occupations may also act as a risk factor for back pain (Hawkins, 1987).

PREVENTION OF LIFTING AND HANDLING INJURIES

There are three main preventive strategies to deal with the risks posed by lifting and handling at work:

- selection (pre-employment screening)
- training
- ergonomic interventions.

Selection

Pre-employment screening attempts to identify those individuals who are particularly at risk of developing back pain, and tries to avoid recruiting them into the workforce.

A large study in American industry carried out by Snook *et al.* (1978) failed to show any relationship between the health screening selection policies and a decrease in back injury at work. This strategy also eliminates staff from the workforce based on one criterion, while ignoring their other abilities and qualities which may be more relevant to the job.

Training

Most health care staff will receive some training in lifting and handling. Training is based on the assumption that some techniques are less hazardous than others and that these techniques can be taught. There has been little evidence to show that traditional methods of training in lifting and handling have succeeded in reducing the incidence of manual handling injuries (Girling *et al.*, 1988; Videman *et al.*, 1989).

Videman *et al.* (1989) showed that skills could be acquired by a suitable training programme, but they were inadequate when faced with poor environmental conditions and ergonomically unsound working areas.

Ergonomic interventions

It has already been stated that training in lifting and handling will be ineffective if the worker has to work in an intrinsically unsafe working environment. Stooping and poor posture at work can be eliminated by the redesign of everyday equipment, such as drug trolleys, height-adjustable treatment couches, etc.

Changes in the workplace layout may also be needed to avoid unnecessary stretching, reaching or awkward and constrained postures. Snook *et al.* (1978) demonstrated that low back injuries could be reduced by 30% when ergonomic approaches for the reduction of work stress and effort were implemented.

Much equipment had been designed to assist in patient-handling tasks, such as patient-handling slings, sliding boards and hoists. The availability of such suitable and appropriate lifting equipment, together with an environment where there is room to use it, will help to reduce the incidence of back injuries.

In a comparison of the three preventive strategies, Snook *et al.* (1978) concluded that ergonomic intervention was likely to be the most effective.

THE ERGONOMIC APPROACH

Ergonomics attempts to optimize the fit between the individual worker and the task that they are undertaking. It lays its emphasis on 'fitting the job to the person', as opposed to workers having to adjust to suboptimal working conditions, i.e. 'fitting the person to the job'. Using the knowledge from four main, pure, scientific disciplines of anatomy, physiology, biomechanics and psychology, the ergonomist tries to ensure that the task performed is safe, efficient, comfortable and will not give rise to musculoskeletal disorders or other problems with health. Ergonomics is a concept that takes into consideration human abilities and behavioural characteristics and tailors the workstation, equipment and environment to make best use of the full potential of the worker.

Ergonomics tends to use a systems approach, which recognizes that changes or problems in one part of the working environment will affect the others.

At the centre of this approach is the individual worker and all the factors which may interact and impinge on him (her) to affect his efficiency and well being are considered. Thus the whole organization is considered in terms of concentric rings spreading from the worker, his interface with equipment at work, his immediate working environment and the environment around him both in terms of physical, organizational and social factors. Figure 17.1 shows a conceptual ergonomic model upon which risk assessment may be based, the main aim of the assessment being to eliminate unsafe situations that may cause injury, rather than to take corrective actions after an injury has occurred.

The objective of an ergonomic job assessment is to study the relationships that exist between the job demands, environmental conditions and human functional characteristics (Khalil *et al.*, 1993).

At the centre of the assessment is the task that needs to be performed, e.g. moving a patient up the bed. The load, i.e. the patient, needs to be considered in terms of their weight, size, physical condition and level of co-operation, etc. The carer needs to be considered in terms of their spinal posture, age, health and physical capability. The equipment used needs to be evaluated for suitability, ease of use, etc. In the general working environment such factors as flooring – is it uneven or slippery? – lighting, temperature and humidity need to be considered.

The work organization will affect this approach by such factors as

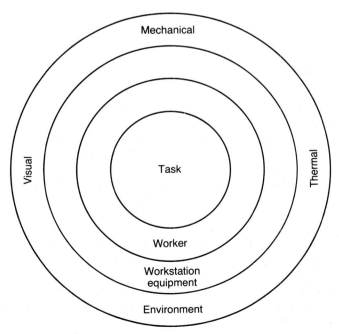

Fig. 17.1 Conceptual model of ergonomics

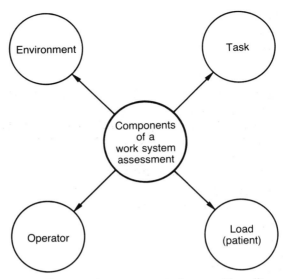

Fig. 17.2 Ergonomic assessment of manual handling

management policies which are in place across the organization, or routines such as always bathing patients in the mornings.

The social environmental factors will include the psychosocial characteristics of individuals as they interact with each other, negative attitudes to good working practices and the influence of pressure and poor communication.

An ergonomic assessment of manual handling requires consideration of the risks under four main areas (Figure 17.2) which must then be considered and weighted to give an overall assessment of the risk. These factors are:

- *Task*, e.g. what actions are performed, what is the overall workload?
- *Load/patient*, e.g. weight, size, stability, condition
- *Environment*, e.g. workspace layout, general working conditions
- *Operator*, e.g. age, sex, strength, fitness, etc.

RISK ASSESSMENT CHECKLIST FOR MANUAL HANDLING TASKS

The following questions need to be asked. A positive answer indicates that a potential problem may exist.

Task

1. Does the task require the person to support a load or exert sustained force while:

    (a) lifting

(b) pushing
(c) lowering
(d) carrying
(e) pulling
(f) other, e.g. throwing?

Such a sustained force may lead to prolonged stress on the spine, increasing the potential for back injury.

2. Does the task involve holding the load at a distance from the trunk? The stress on the spine increases as the distance that a load is held from the body increases.
3. Does the task involve force exertion while:

(a) stooped forwards
(b) leaning or reaching sideways
(c) twisting
(d) using one hand
(e) sitting
(f) reaching upwards?

Each of these positions will increase the stress on the spine by creating an asymmetrical posture or by constraining posture.

4. When handling, is the body held in one posture for a long time while:

(a) stooped
(b) crouched
(c) the arms are held above mid-trunk height
(d) twisted
(e) bent sideways?

Prolonged periods spent in disadvantageous postures will increase the stress on the spine and increase the risk of an injury occurring.

5. Does the task stop easy changes in posture? If the task does not allow ready changes of posture, the stresses will be concentrated on one area of the spine rather than being spread more evenly.
6. Does the task involve a sudden movement of the load? Sudden movements require a burst of muscular effort and do not give the spine time to prepare to receive loading.
7. Does the task involve inadequate rest periods? If the time allowed for recovery is inadequate, the lifter will become fatigued and more prone to injury.
8. Do workers have to perform handling activities as a team? Team lifting requires extra training as there may be problems with inequality in capabilities, sizes and strength and a good system for communicating instructions needs to be established.
9. Does the task involve frequent or prolonged effort? Consideration of the effect of fatigue needs to be given.
10. Does the task involve excessive lifting or lowering distances, handling above shoulder height or below knee height? Lifting below knee height and above shoulder height will be more stressful as the

load will not be kept as close to the trunk and it will be possible to keep the spine in an optimal position.

Load/patient

11. Is the load:

    (a) heavy
    (b) bulky
    (c) difficult to grasp
    (d) unstable
    (e) sharp, hot or potentially damaging
    (f) contents likely to shift during movement?

    Any of the above may make the lifter more likely to drop the load or to have to adjust their grasp on it while lifting.

12. In the case of a patient, assess:

    (a) co-operation level
    (b) ability to help themselves
    (c) medical condition, e.g. frail skin, blood pressure
    (d) presence of other equipment, e.g. drips and drains.

Environment

13. Are there obstructions in the working area which necessitate the worker:

    (a) reaching over obstructions
    (b) reaching into something
    (c) failing to adopt a good foot position?

    These factors will tend to increase the distance of the load from the spine, thus increasing the stresses on the spine.

14. Are there space constraints preventing good posture, e.g.:

    (a) headroom
    (b) narrow passages
    (c) limited leg-room?

15. Are there uneven, slippery or loose floors?
16. Are there variations in level of floors or work surfaces? Both of the above may increase the likelihood of tripping.
17. Are there problems with:

    (a) temperature
    (b) humidity
    (c) noise
    (d) lighting
    (e) ventilation
    (f) air circulation?

    Working in such environmental conditions may cause suboptimal performance, causing the lifter to fatigue more quickly or lose concentration.

Operator

18. Are workers under 18 or over 55? Statistically such people have an increased risk of sustaining injury.
19. Do the workers have medical problems affecting their suitability for handling work, e.g.:

    (a) previous back injury
    (b) hernia
    (c) cardiovascular or respiratory problems?

    Previous injury is the best predictor of who will sustain a lifting injury.
20. Are any workers pregnant?
21. Are there problems with personal or protective clothing? Such clothing may restrict the lifter's posture or ability to get close to the object being lifted.
22. Does the task require special information or training for its safe performance?

Steps to reduce risk of injury

Once the manual handling task and workplace have been assessed, an action plan can be produced to reduce the risks. Possible solutions include changing the workplace layout to allow better access, provision of mechanical lifting devices or hoists or training the workforce in safer working practices. The overall aim is to eliminate manual handling tasks wherever possible. The following are the steps to be taken to reduce the risk of injury:

- provision of equipment to reduce need for lifting, e.g. hoists, trolleys
- redesign of the task to minimize manual handling operations
- staff rotation to avoid the same person carrying out repeated manual handling tasks
- even distribution of work between staff and over the working day; for example, all patients should not be walked in the morning and do bed exercises in the afternoon, but have a mix of each throughout the day
- change of workplace layout to enhance free body movement and variety of posture.

LEGISLATION

From 1 January 1993 the Manual Handling Operations Regulations 1992 came into force. The regulations are made under the Health and Safety at Work Act 1974. They implement the European Directive 90/269/EEC (Commission of the European Communities, 1990) which sets out the minimum health and safety requirements for the manual handling of loads. The impetus for the regulations was the European Directive. A Directive sets out the provisions that the Member States are required to incorporate into their legal framework, but leaves the exact way in which this is done to the individual Member States. The regulations supplement the general duties placed upon the employers by the Health and Safety at Work Act 1974 as well as replacing a number of earlier and outdated legal provisions. The regulations establish a clear hierarchy of measures:

1. *Avoid* hazardous manual handling operations as far as is reasonably practicable, e.g. by redesigning the task or use of equipment.
2. *Assessment* – make a suitable and sufficient assessment of any hazardous manual handling operations that cannot be avoided.
3. *Action* – take action to reduce the risk of injury from manual handling as far as is reasonably practicable, using the assessment as a basis for action.

The regulations seek to prevent injury not only to the back but to other areas of the body. Like the European Directive, the regulations set no safe weight limits. The ergonomic approach has shown that this is too simplistic an approach and that weight is just one of a large number of factors that must be assessed.

Unlike many of the earlier health and safety regulations, the Manual Handling Operations Regulations require the employer to take a positively active role in making health and safety decisions, not just to implement what the Health and Safety Executive (1992) has decided. They do not seek to lay down what is or is not safe in any given working environment.

Avoiding manual handling

Consideration must be made as to whether the lift can be avoided altogether, e.g. can the treatment be brought to a hospital patient rather than taking the patient to the treatment? While many handling procedures can be avoided, e.g. by the use of a bed pan rather than helping to transfer a patient onto the toilet, this may be appropriate. The regulations state that lifting should be avoided 'as far as is reasonably practicable'. In many hospital settings it will not be possible or desirable, in view of the overall goals of patient management, to avoid manual handling but the risks may be lessened by the use of aids, etc.

Assessment

If manual handling cannot be avoided the employer must undertake an assessment of manual handling. Under Regulation 6 of the Management of Health and Safety at Work Regulations 1992, each employer may appoint one or more competent people to help him undertake and implement assessments. The guidance in the regulations points out that there is a difference between a final risk assessment and day-to-day judgements. The aim of the assessment is to establish the range of risk and to form a basis for preventive action. Generic assessments which draw together broadly similar operations are acceptable. Unless the assessment is very simple or the risk very low, a record of the assessment should be kept either in writing or on a computer.

Action

Once the assessment has been carried out, the employer must take steps to reduce the risks to employees as far as is reasonably practicable. This may be by ergonomic interventions which are practical and effective. The effectiveness of the action should be evaluated and if necessary reappraised.

'Reasonably practicable'

The regulations contain the phrase 'as far as is reasonably practicable'. The test of that which is reasonably practicable is satisfied if an employer can show that the cost of any further preventive steps would be grossly disproportionate to the benefit that would be gained from them.

Training

Section 2 of the Health and Safety at Work Act 1974 and Regulations 8 and 11 of the Management of Health and Safety Regulations 1992 require employers to provide all employees with health and safety information and training.

Enforcement

The Manual Handling Operations Regulations are implemented under the Health and Safety at Work Act, and as such attract the same penalties as imposed for breach of the Act. The Health and Safety Executive may therefore impose an improvement or prohibition notice on an employer who is in breach. As a breach of the Health and Safety at Work Act is a criminal offence, the employer may be prosecuted. The fines at present in force for offences under the Health and Safety at Work Act are up to a limit of £20 000 for each offence.

PRINCIPLES OF LIFTING

Preparation for lifting

Planning

There are many points that must be considered before any actual movement of the object or patient to be lifted occurs. Most important is what one is trying to achieve by the handling procedure.

If the objective, for example, is to move a patient up the bed, how much the patient can assist, whether the use of mechanical aids is appropriate and the space available would need to be considered.

Once a lifting technique has been decided upon, the number of lifters required should be agreed and they should be gathered.

Posture of lifter

It is important that the lifter assumes a good posture, supporting their own back and using good body mechanics. The strong leg muscles should provide most of the force for the lifting movement. Their efficiency will be increased by working over bent hips, knees and ankles. Thus the lift will start with the hip and knee joints bent and they will straighten during the lifting action.

The transference of weight from one leg to the other as the lift occurs should allow a smooth movement to occur.

The spine should be bent as little as possible while lifting; either heights should be adjusted, or the legs bent and a squat position assumed.

With the back in an optimal position, the arm position can be considered. Bent elbows will bring the patient closer to the lifter's body, but the arms should not be used to perform the lifting movement – the back should be moved first, then the arms.

Bracing

This will maintain correct posture during the lift. The lower back is using the intra-abdominal pressure which is raised and acts to splint the spine by pressing against it. The rest of the body is braced by slightly raising the head and elongating the body so that the body feels firm, but not stiff and rigid.

Biomechanics of lifting

In upright standing, the load on the spine arises from the effects of gravity transmitted down the vertebral column, requiring little muscular effort. On bending forwards, the tension in the muscles increases until, when the back is horizontal, the load on the spine is almost entirely from back muscle tension.

The stress on the spine may be calculated as a series of levers, with the combined weight of the load acting at a distance from the spine. The further away the load is from the body, the greater the effort is to lift it (Figure 17.3).

Foot position

A large stable base should be attained. Lifting strength is greatest when the feet are placed directly under the load. It decreases as the horizontal distance between the feet and the load increases.

The position of the feet also affects the stability of the lifter; the base of support consists of the lifter's feet and the area between them. Thus a base will be larger and safer if the feet are apart. The lift and all associated movements should be inside the area dictated by the foot position. If lifting occurs outside this area, efficiency is reduced and the risk of injury increases.

The leading foot should be pointed in the direction in which the lift is going to occur (Barker *et al.*, 1992) (Figure 17.4).

Grip

A firm grasp should be taken of the object to be lifted before any effort is applied. If the object is difficult to grasp, it may be necessary to tip it to allow a good grip to be attained.

Preparation of the lifting area

The immediate environment around the object to be lifted should be cleared, e.g. a patient's bed must be arranged to allow the maximum of space, such as chairs moved, lockers cleared away. The brakes should

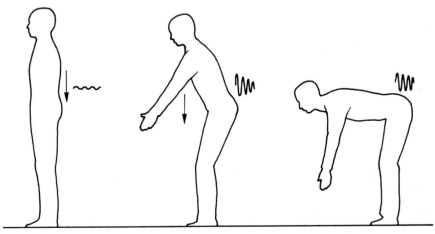

Fig. 17.3 Lifting – leverage and distance from spine

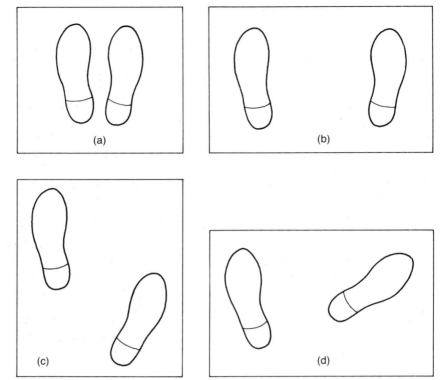

Fig. 17.4 Lifting – foot position: (a) unstable; (b) stable; (c) lifting forwards – stable; (d) lifting sideways – stable

be applied to the bed and, if applicable, chair or wheelchair. If the bed is height adjustable, it should be adjusted to the most suitable height for the lifters.

When all this has been prepared, clear instructions must be given by the lifter in charge of the procedure and, on a given command, the lift can occur.

The main rules for lifting

The following are the main rules to be adopted when lifting:

- get close to the object to be lifted
- put the feet in a wide stable position for balance
- get a good firm hold on the object or patient
- bend the knees at the start of the lift
- use your own body weight to assist in the lift
- lift slowly and smoothly.

Lifting style – squat v. stoop

'Squat' lifting

Conventional wisdom since the 1930s has advocated the lifting of heavy loads by a squat posture (back straight, knees bent) (Figure 17.5a). Many authors claim that this form of lifting is least hazardous to the lower back (Edgar, 1979; Asmussen *et al.*, 1965). The rationale for this is that this position maintains the back in its natural alternating curves position. The forces on the cartilaginous end-plates of the vertebrae and discs will thus be evenly distributed and it will be the muscles not the ligaments or bone that will counteract gravitational forces (Poulsen and Jorgensen, 1971). In the squat position the load is close to the body, reducing the lever arms and decreasing stress (Bendix and Eid, 1983).

However, industrial workers rarely use the squat position. Davis *et al.* (1965) suggested that this is because the leverage exerted by the quadriceps in this position is ineffective and the average worker does not have sufficiently strong leg muscles to lift in this way. Alternatively, degenerative changes at the knees may make it too painful to lift with bent knees. A squat style of lifting also requires a greater expenditure of energy as the weight has to be lifted through a greater vertical height than with other lifting styles (Garg and Herrin, 1979). This results in the lifter becoming tired and adopting a poor lifting technique (Oliver, 1994).

'Stoop' lifting

Stoop lifting involves a bent back, legs straight posture. Generally, stoop lifting generates low compressive forces in the spine but they are

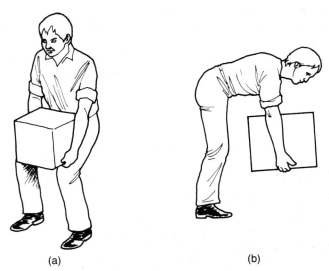

(a)   (b)

Fig. 17.5 Lifting styles: (a) 'squat' lifting; (b) 'stoop' lifting

distributed unevenly over the disc, concentrating pressure on one area and possibly creating an area of weakness. Although it involves significantly lower energy expenditure than other styles, stoop lifting is usually condemmed as representing poor technique as the load is held a long way from the spine.

In practice, most workers will adopt a freestyle technique, lifting in the manner they perceive as easiest that is efficient in terms of energy and time. Davis *et al*. (1965) found that as a load became heavier workers tended to change from a crouched squat lift to a stooped lifting posture.

PRACTICAL PATIENT-HANDLING TECHNIQUES

Lifting patients manually is a last resort and should be avoided. Lifting aids or hoists should be used by preference whenever possible.

Holds

There are a number of ways in which two lifters may hold each other and the patient. It is important that the grip shall be comfortable and secure.

If two people are lifting the patient, the technique of choice would be to use a patient-handling sling (Figure 17.6). The patient-handling sling is comfortable, it extends each lifter's reach and reduces the amount that the lifter has to stoop. If a handling sling is not available the double wrist hold or finger hold should be used.

Double wrist hold

Each lifter grasps the wrist or forearm of the other (Figure 17.7a). The wrist grasp requires one lifter to have the arm fully pronated and the other lifter having the arm fully supinated.

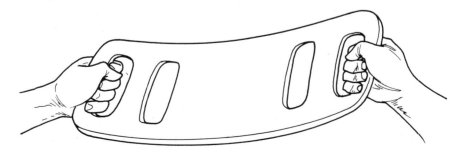

Fig. 17.6 Patient-handling sling

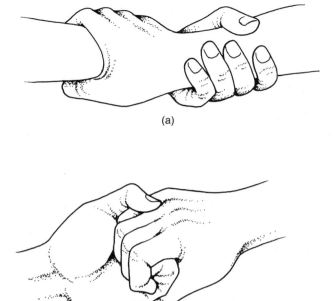

(a)

(b)

Fig. 17.7 Types of hold: (a) double wrist hold; (b) finger hold

Finger hold

The two lifters hook their fingers in each other's (Figure 17.7b).

Axillary hold

In the axillary hold (Figure 17.8a) the lifter stands facing the patient. The lifter's forward foot is across the feet of the patient, their knee blocking the patient's knee. The lifter's other foot is beside the patient. The patient is leaned forward with their feet tucked in close to the chair so that their centre of gravity is over their feet.

The lifter puts their hand nearest the patient's body over the patient's shoulder blade. On the side away from the lifter, the lifter inserts her fingers as far as possible under the axilla and around the chest wall.

Elbow hold

This is particularly useful when transfers have to occur in a confined space. The lifter must be tall enough to reach over the seated patient. The lifter stands with the forward foot blocking the patient's feet and the lifter's leg blocking the knees (Figure 17.8b). The other foot is at right angles to the side of the patient. The patient is leaned forward so

Human Movement Explained

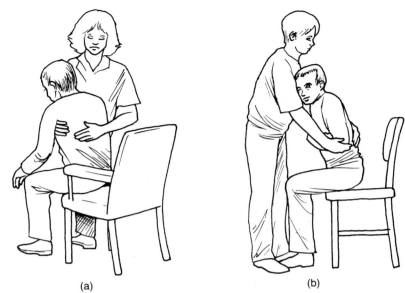

Fig. 17.8 Types of hold: (a) axillary hold; (b) elbow hold

that their near shoulder rests on the lifter's trunk. The lifter then leans across the patient's back, grasps the outsides of the patient's elbows with their fingers beneath them and holds both elbows tucked in to the patient's waist and slightly back from the mid-axillary line.

This grip gives good control of the patient's movement.

Waist-belt hold

The lifter stands in front of the patient, one foot blocking the patient's feet and knees and the other beside them. The lifter's thumbs are inserted into the side of the patient's trousers and grasped ready to lift up and forwards (Figure 17.9).

Alternatively, a patient-handling belt is strapped firmly around the patient's hips and grasped by the lifter. Many commercially available belts with hand-holds are available for this purpose.

Use of rocking/body weight

This is based upon the principle that a moving body will be easier to move, as it will have momentum and kinetic energy. In rocking techniques, the patient rocks themselves rhythmically forwards and backwards in the chair to build up momentum prior to standing up. The lifter can help the patient to rock forwards and backwards before using momentum to stand up.

Rocking transfers are particularly useful for sitting to standing prior to walking or sitting somewhere else, or in a pivot transfer where a

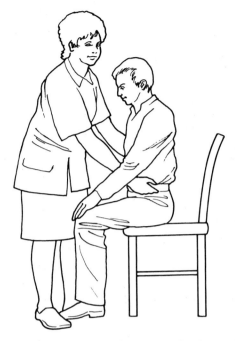

Fig. 17.9 Waist-belt hold

patient is horizontally transferred from one seated position to another through 90°, e.g. transferring from wheelchair to toilet.

Counterbalance

The lifter may use their own body weight to counterbalance the patient's weight. When raising a patient to standing, the lifter can drop their own weight backwards to counterbalance that of the patient as they start to come up into the standing position.

Both of these techniques require the co-operation and understanding of the patient.

Techniques for moving the patient in the bed

A height-adjustable bed is essential for ease of handling.

Sitting the patient up in bed

Ideally the patient should be encouraged to do this for themselves by using a monkey pole or rope ladder fixed to the bottom of the bed and pulling themselves up.

The patient may be able to push themselves up the bed by bending one knee and digging the foot into the bed while holding onto a monkey pole.

Alternatively, blocks may be used, the patient pushing on the blocks to raise themselves clear of the bed and push backwards. Manual lifts are very much a last resort. Four techniques may be used.

- two-sling lift
- shoulder or Australian lift
- combined lift
- through-arm lift.

Wherever possible, patient-handling slings should be used as these extend the lifter's reach and decrease stooping.

Two-sling lift

The two-sling lift (Figure 17.10) is suitable when a patient cannot sit up, but has head control. Two slings are used: one under the patient's upper thighs and the other placed at the upper waist. The bed height should be adjusted so that it is midway between the lifter's knees and hips when they are standing. The lifters stand facing each other on opposite sides of the bed. The lifters should take up the slack on the slings, lean slightly backwards to counterbalance the patient's weight and lift them slowly and smoothly up the bed. It is important that the lifters' bodyweight has moved the patient and the weight is not all taken on the lifters' arms. If patient-handling slings are not available, a draw sheet may be substituted.

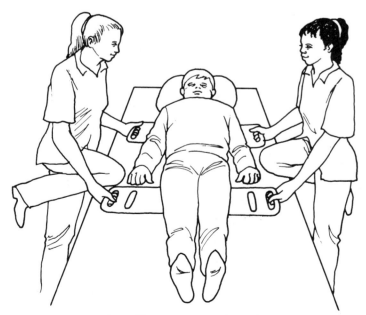

Fig. 17.10 Two-sling lift

Shoulder/Australian lift

This lift (Figure 17.11), which is known by the two names, the shoulder lift and the older name of the Australian lift, may be used for lifting a patient up the bed or transferring from the bed to chair. It requires that the patient can sit up and does not have pain in their shoulder, upper back or chest or wall.

The lifters stand on each side of the bed, facing the top of the bed with their inside knee placed on the bed level with the patient's hips. From behind the patient the lifters press their near shoulders into the patient's chest wall under his axilla, while the patient rests his arms along their backs. The lifters' near arms are then passed under the upper thigh and grasped using either a wrist or finger grip (see earlier). Alternatively a patient-handling sling may be used under the upper thighs.

The lifters' other hands are placed on the bed behind the patient with the elbows flexed, alternatively they may place them on the bed head. The lifters keep their heads raised to maintain good cervical posture and on a clear command they kneel up, straightening the elbows of their supporting arms. Their standing legs provide the power for the lift, and the patient is lifted clear of the bed and then lowered by bending the standing leg and supporting elbow.

The patient should only be moved a short distance at a time so that their hips stay behind the lifters' hands on the bed. If a large distance needs to be covered, the procedure will have to be repeated. As neither lifter can observe the patient's face, great care is needed to ensure that

Fig. 17.11 Shoulder or Australian lift

the patient has full understanding of the procedure and that clear instructions as to the timing of the lift are given.

Combined lift

This lift (Figure 17.12) may be used where the shoulder lift is inappropriate due to the patient having a painful shoulder or chest wall on one side, or where the patient has limited range of motion at the shoulder joint.

The lifters stand on opposite sides of the bed, one facing either direction. The lifter who faces the top of the bed places a knee next to the patient's hip and uses a shoulder or Australian technique as described above.

The other lifter places a knee by the patient's hips but faces the

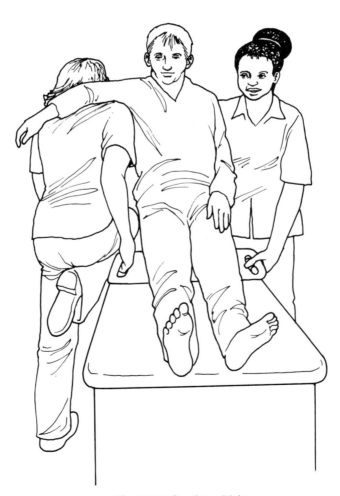

Fig. 17.12 Combined lift

bottom of the bed. That lifter places one hand under the patient's sacrum and uses the patient-handling sling under the patient's upper thighs with the partner.

This lift is more stressful than the shoulder lift, but has the advantage that the patient can be observed. This is useful for hemiplegic or arthritic patients who lack the range of motion in their shoulders to allow a full shoulder lift to occur.

Through-arm lift
The through-arm lift (Figure 17.13) is useful as the patient assumes a compact body configuration. It is also useful for transferring a dependent patient from the bed to the chair.

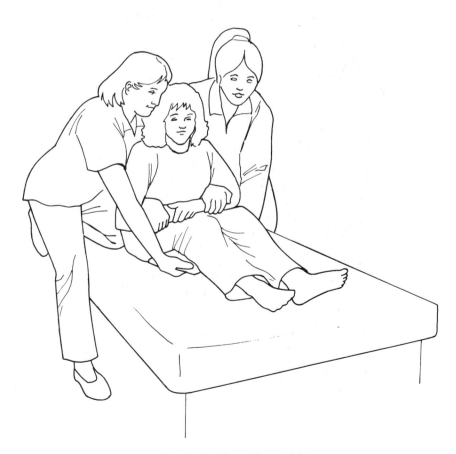

Fig. 17.13 Through-arm lift

The two lifters stand on opposite sides of the bed, facing the foot of the bed and with their inside knees on the bed behind the patient's hips. With the patient's arms crossed, the lifters place a hand between the patient's chest and arms from behind and grasp the patient's forearms as near to the elbow as possible. With the outside hand they then hold a patient-handling sling placed under the upper thigh. The patient is lifted towards them, the lifters rocking back on their heels using their body weight to assist with the lift. Only a short distance should be covered by the lift.

Single handed. If only one person is available, the through-arm lift may be performed using an identical through-arm grip, provided that no attempt is made to lift all of the patient's weight. The patient should assist by bending their knees and digging their heels into the bed. However, unless the patient is very light this will still be very stressful for the lifter.

Transfers from bed to chair

Patients who cannot take any of their own weight should use a hoist for these transfers. Alternatively, a sliding board may be used.

Shoulder lift
The RCN (1992) recommends that for a patient weighing up to 8 stone (51 kg) and who can sit up, a shoulder lift may be used. The patient should be helped to sit up with their feet hanging over the side of the bed and should be told where they are going and understand the sequence of events that will occur.

The lifters join hands under the patient's upper thighs and place their shoulders into the patient's axilla, pressing against the chest wall, with the patient's arms along the lifters' backs. Their other hands are placed on the bed next to the patient's hips with their elbows bent. On an agreed command, they lift by straightening their kness and pushing through their supporting elbows to come into an upright position. Once upright, the supporting hands are placed behind the patient's back while the patient is carried to the chair that they are being transferred into. To lower the patient, the lifters will place their supporting hands on the arms or seat of the chair (not the back) and bend their knees and supporting elbows to lower the patient into the chair.

Through-arm lift
This lift (Figure 17.14) is useful, provided that there is not a high backrest to the chair. If the lifters are not of the same height, the taller should be at the patient's head.

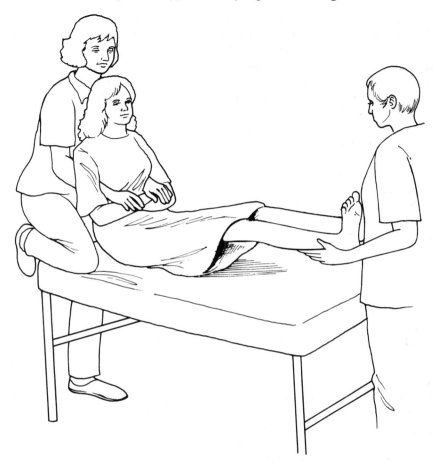

Fig. 17.14 Through-arm lift – bed to chair

The patient is moved to the side of the bed. The taller lifter applies a through-arm hold from behind the patient. The smaller lifter places a chair adjacent to the bed, faces the patient and passes their hands under the patient's legs to support them. On a given command, the legs are pulled over the side of the bed and the taller nurse lifts the patient's trunk from the bed, bending the knees to lower the patient into the chair.

Transfers from sitting to standing

If the patient is fairly independent, transferring from sitting to standing may be facilitated by a chair that is the correct height. If the feet are tucked under the patient, who can push through his arms onto the arms of the chair, all the help that may be required is verbal encouragement.

Axillary lift

In this lift (Figure 17.15), the lifter stands in front of the patient with the knees either side of the patient's knees and one foot slightly in front of the other. The lifter thus has a wide stable base and can easily transfer weight from front to back foot during the sit-to-stand transfer. The patient's feet are placed directly underneath the front edge of the chair seat ready to weight-bear. The lifter assumes an axillary hold, putting one hand over the patient's shoulder blade. The other hand is placed around the patient's chest wall from the front and under the axilla.

The patient's hands are placed on the arms of the chair and the patient is asked to push down as the lift begins.

Elbow hold transfer

This utilizes the elbow lift grasp. The nurse stands in front of the patient with one leg and foot blocking the patient's feet and the other foot slightly to the side. The lifters should feel stable and find it easy to transfer weight from one foot to the other. The nurse leans the patient forward and reaches across their back, grasping the outside of their elbows with the fingers beneath them. In order to stop any trunk movement of the patient, the far shoulder of the patient is locked by the lifter bringing an arm in front of it.

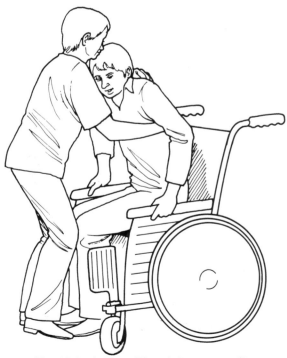

Fig. 17.15 Axillary lift – sitting to standing

The nurse rocks the patient, then moves them into a forward raised position.

Rocking pivot transfers

For patients who can co-operate and control the position of their heads and arms, the stress of lifting can be eased by gently rocking them forwards and backwards while they are seated, as a preliminary to the move (Figure 17.16). The rocking gives the patient momentum and, by using this, the lifter is relieved of much of the effort of lifting.

The patient is assisted to the front edge of the chair by rocking them from side to side and pulling alternate legs forward while the weight is supported by the contralateral side. Then, with the patient's knees at 90° and their feet and knees together, the lifter stands with one foot beside the patient and the other in front, blocking their feet and knees. The nurse then holds the patient by an axillary grasp, elbow grasp or belt grasp, depending on the patient's condition and personal preference. The lifter begins the rocking movement by rhythmically swaying backwards and forwards and, keeping the patient held close to the body, sets the patient rocking backwards and forwards also. By using

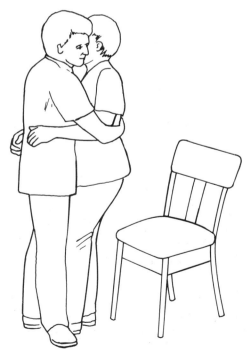

Fig. 17.16 Rocking pivot transfer – sitting to standing

body weight, the lifter conveys enough kinetic energy to the patient to achieve the transfer.

After the rhythm has been set up, an increase in the movement is given and, with a command such as 'ready steady up', the transfer is completed.

There are many other patient-handling techniques adapted to particular medical conditions or environmental circumstances. A fuller description of such manual handling procedures, together with a description of best practice, is available from *The Guide to the Handling of Patients* published by the National Back Pain Association and the Royal College of Nursing (RCN, 1992).

Banned lifts

There are a number of techniques of moving patients which were formerly taught but are now considered dangerous and unacceptable by the Royal College of Nursing (Figure 17.17). Unfortunately they are still in use in some areas. They have been condemned not just by the Royal College of Nursing but by other lifting experts.

Orthodox or cradle lift
This lift (Figure 17.17a) was traditionally used to raise a patient, e.g. prior to inserting a bedpan. Here two lifters face each other, clasp hands under the patient's thighs and behind the patient's back. They must then lift the patient up and transfer sideways.

This is banned because:

• it involves excessive stooping
• the load is a long way from the lifter's body
• the lift occurs sideways, adding twist to the spinal compression.

Drag lift
One lifter stands on either side of the patient and hooks their arm into the patient's armpits (Figure 17.17b). The patient is then lifted up the bed. Hollis (1991) states that it risks injuring the patient.

It is banned because:

• it risks injury to the patient's shoulders
• the lifter has to lift asymmetrically
• it is uncomfortable for both patient and lifter.

Neck hold
When a lifter tries to transfer a patient from sitting to standing, the risk is greatly increased if the patient puts their arms around the neck (Figure 17.17c).

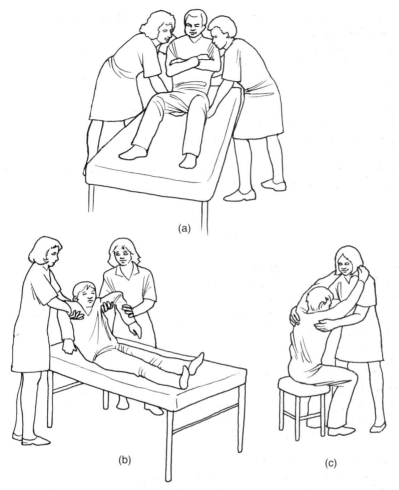

Fig. 17.17 Lifts banned as dangerous practice by the Royal College of Nursing: (a) orthodox or cradle lift; (b) drag lift; (c) neck hold

This leads to:

- stooped posture for the lifter
- patient at long distance from the lifter
- excessive stress on the lifter's neck.

All of these techniques are biomechanically dangerous and should not be used.

Experimental comparison of lifting procedures

Stubbs *et al*. (1983) and Pheasant and Stubbs (1992) have attempted to determine the relative biomechanical stress of the different lifting procedures. They based their analysis on experimental studies using intra-abdominal pressure as a measure of biomechanical stress during lifting.

It has been demonstrated that an intra-abdominal pressure in excess of 45 mm Hg in women is associated with an increased risk of back injury (David, 1987). Pheasant and Stubbs (1992) experimentally investigated different nursing procedures. They calculated that for a 70 kg patient (an average male adult), this safe level of 45 mmHg would be breached 20% of the time for the shoulder or Australian lift, but 61% for the orthodox or cradle lift. Using a drawsheet (patient-handling sling) to lift, the safe limit was only breached on 8% of occasions. With pivot turns, the axillary hold pivot turns through 90° exceeded the safe limit of 45 mmHg 31% of the time and the elbow pivot turn through 180° 48% of the time.

Although these results are based on small numbers in a largely experimental setting, they do allow judgements to be made about how hazardous the procedures are relative to one another. Table 17.1 gives the risk assessment suggested by Pheasant and Stubbs (1992).

While this research helps to determine which of the patient handling procedures are safest, it still shows that there is considerable stress associated with patient handling. Thus the use of hoists and mechanical aids remains the safest way of lifting and handling patients.

Table 17.1 Risk assessment of various lifting procedures

Manoeuvre	Risk
Cradle (orthodox) lift	High
Pivot turn – elbow hold through 180°	Moderate/high
Pivot turn – axillary hold through 90°	Moderate
Shoulder	Low
Drawsheet (patient-handling slings)	Very low

SUMMARY

- The lifting and handling of patients is recognized by most authors as a primary cause of low back pain.
- Wherever possible, movement of loads or patients by manual handling techniques should be avoided.
- The relative advantages of preventive strategies has demonstrated that pre-employment staff selection and traditional technique-orientated training have been ineffective at reducing occupational back injury. Adding in an ergonomic approach to such training has been found by some authors to increase its effectiveness.
- Current legislation in effect since 1992 requires that all lifting and handling activities should be considered to see if they can be *avoided*; where this is not possible, an *assessment* of the risks should be made

and an *action* plan made based on this assessment to reduce the risks.

● Manual handling techniques are very much a last resort and the use of appropriate mechanical aids and hoists should be used wherever possible.

● Where a manual handling technique has to be performed, consideration should be given to the relative risks of the procedures, to their suitability for the task being performed and to optimizing the help a patient can give.

REFERENCES

Asmussen, E., Poulsen, E. and Rasmussen, B. (1965) Quantitative evaluation of the activity of the back muscles in lifting. *Communication of the Danish National Association for Infant Paralysis*, No. 21

Barker, K.L., Bell, R.E., Green, W.H. and Klaber Moffett, J. (1992) *Trainers' Manual For Prevention of Back Injury*, Nuffield Orthopaedic Centre, Oxford

Baty, D. and Stubbs, D. (1987) Postural stress in geriatric nursing. *International Journal of Nursing Studies*, **24**, 307–316

Bendix, T. and Eid, S.E. (1983) The distance between the load and the body with three bio-manual lifting techniques. *Applied Ergonomics*, **14**(3), 185–192

Commission of the European Communities (1990) Council directive on the minimum Health and Safety Requirements for the Manual Handling of Loads, 90/269/EEC. *Official Journal of the European Communities* 21.6.90, No. L156/9–13

Davis, P.R., Troup, J.D.G. and Burnard, J.H. (1965) Movements of the thoracic and lumbar spine when lifting: a chronocyclophotographic study. *Journal of Anatomy*, **99**, 13–26

Edgar, M. (1979) Pathologies associated with lifting. *Physiotherapy*, **65**, 245–247

Garg, A. and Herrin, D.B. (1979) Stoop or squat; a biomechanical and metabolic evaluation. *AIIE Transactions*, **11**(4), 293–301

Girling, B., Birnbaum, R. and Pheasant, S.T. (1988) Prevention of musculo-skeletal stress in hospital employees: an ergonomic approach to training. In *Contemporary Ergonomics* (ed. Megoul), Taylor and Francis, London

Hawkins, L. (1987) An ergonomic approach to stress. *International Journal of Nursing Studies*, **24**, 307–316

Health and Safety Executive (1992) *Manual Handling – Guidance on Regulations*, HMSO, London

Hollis, M. (1991) *Safer Lifting for Patient Care*, 3rd edn, Blackwell, Oxford

Khalil, T.M., Abdel-Moty, E.M., Rosomoff, R.S. and Rosomoff H.L. (1993) *Ergonomics In Back Pain*, Van Nostrand Reinhold, New York

Klaber Moffett, J.A., Hughes, G.I. and Griffiths, P. (1993) A longitudinal study of low back pain in student nurses. *International Journal of Nursing Studies*, **30**(3), 197–212

Oliver, J. (1994) *Back Care – An Illustrated Guide*, Butterworth–Heinemann, Oxford

Pheasant, S.T. and Stubbs, D. (1992) Back pain in nurses: epidemiology and risk assessment. *Applied Ergonomics*, **23**(4), 226–232

Poulsen, E. and Jorgensen, K. (1971) Back muscle strength, lifting and stooped working postures. *Applied Ergonomics*, **2**(3), 133–137

RCN (1992) *The Guide to the Handling of Patients*, 3rd edn, Royal College of Nursing/National Back Pain Association, London

Scholey, M. and Hair, M. (1989) Back pain in physiotherapists involved in back care education. *Ergonomics*, **32**, 179–190

Snook, S.H., Campinelli, R.H. and Hart, H.N. (1978) A study of three preventative approaches to low back pain. *Journal of Occupational Medicine*, **20**, 478–481

Stubbs, D.A., Buckle, P.W., Hudson, M.P., Rivers, P.M. and Worringham, C.J. (1983) Back pain in the nursing profession: 1, Epidemiology and pilot methodology. *Ergonomics*, **26**, 755–765

Videman, T., Rahaula, H., Asp, S. *et al*. (1989) Patient handling skill, back injuries and back pain: an intervention study in nursing. *Spine*, **14**(2), 148–156

18. *Clinical measurement*

INTRODUCTION

Clinical measurement is essential if the physiotherapist is to practise effectively. While there are many sophisticated and elaborate measurement tools available, the importance of simple measurements carried out carefully and correctly in the clinical setting should not be overlooked. All too frequently clinical practice tends to concentrate on assessing and treating the patient, but pays little attention to the accuracy of the measurements taken in the initial evaluation or in subsequent evaluations of therapeutic interventions.

This chapter will outline:

- reliability and validity of measurement
- the measurement of joint range
- the measurement of muscle strength
- other clinically useful measurements
- the measurement of Activities of Daily Living.

The importance of using tests that are reliable and valid is emphasized, as are the methods of recording examination results.

DEFINITION

Measurement is the quantification of an observation against a standard, whereas assessment also includes the process of interpreting the measurement (Wade, 1992). For a measurement to be of any value it must have been demonstrated to have validity and reliability.

Validity

A valid measure is one that has been proven to measure that which it is intended to measure. The term is also sometimes used to mean that the measurement obtained can legitimately be used to make inferences, i.e. validity may deal with how a measurement is used.

Reliability

Reliability refers to the consistency of a measurement of and to the extent to which two observations agree. The three main reasons why a test may be unreliable are:

- flaws in the instrumentation
- a lack of consistency of the variable in question in the patient
- errors made by the person taking the measurements.

Types of reliability

There are three main types of reliability:

1. *Inter-tester reliability*. The extent to which different people performing measurement of the same entity achieve agreement. In clinical practice this will be important when either two physiotherapists treat the same patient or when clinical data is shared. It should be the case that we can assume that if one physiotherapist performs a measurement, other physiotherapists using the same method on the same patient should get the same result. If this is the case the measurement can be said to have good inter-tester reliability. A good measurement will be one that has a clear procedure so that different practitioners will measure the same thing in the same way and produce the same results as each other.
2. *Intra-tester reliability*. This is how reliable a measure is when the same thing is measured by the same person on different occasions. Thus multiple measurements are taken by one person over a period of time and this will test reliability of the measure over time. Care must be taken over how much time elapses between measurements and whether the changes seen are related to inaccuracy in the measurements taken or to a changing clinical picture.
3. *Parallel reliability*. This reflects whether different forms of a test or instrument produce the same results. For example, is the range of motion obtained using one type of goniometer the same as when the same measurement is taken with a different type of goniometer?

Objective and subjective measurements

Physiotherapy measurement and assessment techniques are often criticized as being too subjective without due consideration being given to what is meant by the term subjective. *Subjective measurements* are those that are affected by the person taking the measurements, e.g. manual muscle testing, whereas *objective measurements* are those that cannot be affected by the person taking the measurements, e.g. muscle strength measured using a dynamometer.

However, even a seemingly subjective measurement such as observational gait analysis can be an objective measurement if enough different observers can be shown to agree, i.e. show inter-tester reliability. If different observers obtain the same measurements for the same subject, then they must be using a technique that effectively discounts any subjective aspects, such as observer bias, from the measurement. Thus any system of measurement can be objective, so long as it can be proven to be reliable.

It can be seen, therefore, that any measurement taken by the physiotherapist should be:

- reproducible
- valid
- reliable.

Outcome measures

There is a difference between clinical measurements and using such measurements as outcome measures. Outcome measures frequently use clinical measurement, but additionally utilize them to observe and document a relative change. Thus the outcome of a course of physiotherapy on a patient may be assessed as the difference between the end-point compared to the situation before treatment started. Outcome measures are always relative, whereas a clinical measurement can be absolute. For example, a knee may have a range of motion that is 10–75° which is a definite clinical measurement, but its range may have improved by 30° over a period of treatment, which is the outcome measure compared to the starting point. It is important that the measurements that we use are appropriate.

An outcome measurement that does not directly relate to the treatment given may lead to erroneous conclusions being drawn about the usefulness of that treatment procedure (Pynsent *et al.*, 1993).

MEASUREMENT OF JOINT RANGE OF MOTION

Measurement of range of motion (ROM) of joints is the commonest evaluative technique used by physiotherapists (Gajdosik and Bohannon, 1987). The objective measurement of joint motion is also known as goniometry.

Normal range of motion

Normal ROM values have been reported by many authors and act as a standard against which the physiotherapist can attempt to determine whether a patient has a limitation of motion. The most widely accepted guidelines of normal ROM are those published by the American Academy of Orthopaedic Surgeons (AAOS, 1965). This guide contains ROMs, standard zero positions and the method of measuring and recording range of motion at the different joints. Published normal ROM values vary according to the age, sex and population that they are based on. Thus a single guide to a set of normal values is unlikely to be applicable to both sexes over a lifetime (AAOS, 1965; Gerhardt and Russe, 1975; Boone and Azen, 1979).

The AAOS (1965) suggest that the best way to determine the normal ROM in a patient suffering from a unilateral condition is to compare range with that of the contralateral, unaffected side. Some authors question the validity of using the contralateral limb as a standard for comparison in anything other than very acute, unilateral cases. They argue that in more chronic conditions, altered patterns of use may develop and thus alter normal values. However, in a patient with a unilateral acute condition, comparison of the limbs provides a ready guide in the clinical setting to what is normal for that patient.

Goniometry

Moore (1949) reviews goniometry and suggests the use of the universal goniometer. Other types of goniometer have been developed, including joint-specific goniometers, fluid goniometers, pendulum gonio-

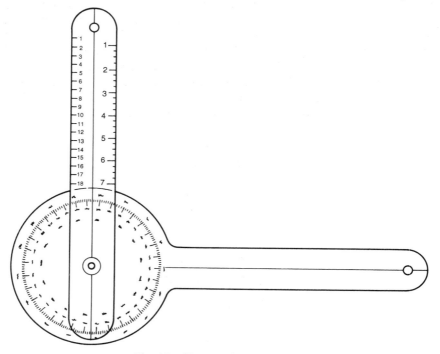

Fig. 18.1 Universal goniometer

meters and electrogoniometers. However, the universal goniometer is the one that is most likely to be encountered in the clinical setting.

Universal goniometer

This consists of a protractor with two arms attached to it, one fixed and one mobile (Figure 18.1). It is called the universal goniometer as it may be used to measure any joint. The ideal length of the arms is considered to be between 12 and 16 in (30 and 40 cm) to facilitate accurate placement of the goniometer over anatomical landmarks (Moore, 1949), although models with arms as short as 6 in (15 cm) exist. Hellebrandt *et al*. (1949) found that in the hands of a skilled operator the universal goniometer was the most accurate method of joint measurement.

Joint-specific goniometer

At some joints, especially the small joints of the hand, it is difficult to attain accurate measurements with a universal goniometer. To address this problem, joint-specific goniometers for some joints have been developed, but Hellebrandt *et al*. (1949) found that their accuracy was rarely better than that of the universal goniometer.

Pendulum goniometer

This was developed in the 1930s. It consists of a dial with a 360° scale to which is attached a weighted pointer. The goniometer is strapped to the limb segment. Both components of the goniometer are influenced by gravity, operating independently of each other. The dial is locked to the extreme of the range of motion, e.g. full extension, and the arc of movement described by the pointer. The device is most accurate when used with the subject in an upright position. Today it is most commonly used to measure the range of motion of the cervical spine via a dial strapped to the head and a moving pointer (Figure 18.2).

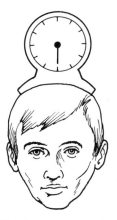

Fig. 18.2 Pendulum goniometer

Fluid goniometer

This was developed by Schenker in the 1950s. It consists of a 360° scale in a flat, fluid-filled tube containing an air bubble, similar to a spirit level. The goniometer is strapped to the limb and as the limb moves the scale rotates while the bubble remains stationary. The range of motion is read when the scale stops moving. The fluid goniometer has the advantage of being small and light and quick to apply and operating independently of the axis of joint rotation. However, as it is strapped to the limb its placement may be affected by soft-tissue variation and slippage may occur during movement.

Electrogoniometer

Although these are not commonly found in most physiotherapy departments their use is increasing, particularly in departments where research projects or gait analyses are carried out. An electrogoniometer converts angular motion into an electrical signal. At its simplest is the electronic potentiometer which is attached to rigid arms. The arms are strapped to the limb segments, movement of the arms causes the potentiometer to rotate and an electrical signal is generated and recorded. The main advantage of the electrogoniometer is that it can measure dynamic ranges of motion, as for example is the case in gait analysis. As the device is small and unobtrusive it can easily be worn under clothes. Nicol (1989) describes their use in a wide variety of clinical settings and predicts that their use will become more widespread.

Goodwin *et al*. (1992) compared the universal goniometer, a fluid goniometer and an electrogoniometer and found that the electrogoniometer produced the best results in terms of reducing inter-tester differences. However, they found considerable differences in the results obtained using the different types of goniometer and concluded that the interchangeable use of different types of goniometer in a clinical setting was inadvisable.

Placement of the goniometer

The most widely accepted technique for measuring joint range of motion places the centre of the goniometer over the axis of the joint, making the axis represent the centre of rotation of the joint and the goniometer arms, the distal and proximal limb segments. However, as the universal goniometer is a simple hinge joint whereas the joints of the body have axes of rotation that move as the limb goes through its range of motion, this is over-simplistic. With a hinge joint such as the elbow, bony landmarks can be located at the wrist, elbow and shoulder upon which to align the goniometer and thus to measure angular position. However, using the centre of the elbow joint (lateral epicondyle of the humerus) may not represent the precise line of the axis of the joint. If alignment of the goniometer axis does not coincide with anatomical bony landmarks, the joint's axis of rotation may lie above or below the location of the goniometer axis. The physiotherapist has then to decide which landmarks to follow and a subjective element is introduced to the measurement procedure. Variations in palpation of anatomical landmarks and goniometer placement will also introduce sources of error. Gerhardt and Russe (1975) recommend that the method of goniometer placement that provides the most reliable and reproducible results is to ignore the joint axis and place the goniometer arms relative to anatomical landmarks.

Technique of goniometer measurement

1. Unclothe the parts of the limb to be measured.
2. Choose a starting position that is as close to the joint's zero or resting position as pain or disability will allow.
3. Place the arms of the goniometer over anatomical landmarks proximal and distal to the joint.
4. Align the goniometer in the correct plane of motion.
5. If no suitable bony landmarks exist for the joint to be measured, the fixed arm of the goniometer should be placed parallel to the longitudinal axis of the fixed limb segment and the moving arm of the goniometer, parallel to the longitudinal axis of the moving limb segment.
6. If sequential readings are to be taken, e.g. when monitoring the effect of a therapeutic intervention, it may be useful to apply skin markings to assist with consistent placement of the goniometer.
7. Movement should be carried out actively, smoothly and slowly.
8. The movement should then be repeated passively and any differences between the active and passive ROM noted.
9. The joint range should clearly be recorded.

Repeated measures

Some authors, e.g. Galley and Forster (1987), advocate the taking of several readings and use of the average. However, Boone *et al.* (1978) studied the reliability of measuring ROM of six different joints in the upper and lower limb and found that a single set of measurements was as reliable as taking several sets of measurements and averaging the results.

Reliability of goniometer measurements

If one physiotherapist measures the same patient from time to time there will inevitably be some variation in the results recorded (intra-tester error). It is important that this variability is not mistakenly interpreted as the response to a therapeutic intervention. Smith and Walker (1983) found that, using the universal goniometer, intra-tester reliability was high compared with inter-tester reliability. Several authors have reported that reliability is increased by following a standardized protocol for the measurement technique (Ekstrand *et al.*, 1982; Stratford *et al.*, 1984). Nicol (1989) reports that the only method to test the accuracy of goniometric measurements is to refer to the skeletal positions using X-rays. Gogia *et al.* (1987) used X-rays to compare the measurements taken by two physiotherapists. They found good agreement between the measurements taken by the physiotherapists using the universal goniometer and those from estimation of the angles of the bone axes taken from the X-ray. Riddle *et al.* (1987) found that goniometer measurements of the shoulder were reliable and Watkins *et al.* (1991) reported good inter-tester and intra-tester reliability using the universal goniometer to measure the knee.

Linear measurements of joint range of motion

For joints where measurement with a universal goniometer is difficult, or where there is an absence of bony landmarks on which to place the goniometer arms, linear measurements may be the technique of choice. For example, lumbar flexion may be measured by measuring the distance between the fingertips and the floor, and cervical spine flexion the distance between the chin and the sternal notch. The measurement is recorded in centimetres. As long as the two points that are being used for the measurements are clearly recorded, this method has good reported accuracy (Gerhardt and Russe, 1975).

For measurements of the hand, alternative techniques such as tracing the contour of the hand with pliable wire and drawing around it, or directly making a tracing of the hand, will give an accurate record that can then be measured with a protractor if a numerical record is required.

Recording range of motion

The most widely accepted method of recording ROM is the 0–180 system recommended by the American Academy of Orthopaedic Surgeons. All motion is measured from a defined starting position of 0°. Movement proceeds towards 180°. Thus the extended anatomical position of an extremity will be 0°, e.g. the elbow in full extension (Table 18.1).

Table 18.1 Recording ROM by 0–180 system

Shoulder joint	Degrees (°)
Flexion	180
Extension	55
Abduction	180
Adduction	75
Internal rotation	90
External rotation	90

Other notation systems that exist are the 180–0 system, where movements towards extension approach 0° and those to flexion 180°; and the 360° system, where movements past the neutral approach 360°. Neither of these forms of notation are commonly used (Miller, 1985).

SFTR recording

This system of notation was developed by Gerhardt and Russe (1975) using the 0–180 system. It records according to the planes of movement, hence its name – Sagittal, Frontal (coronal), Transverse, Rotational recording.

The ROM for all movements in each plane is recorded. In each plane three numbers are used, one for each extreme of motion and one (the middle figure) to represent the starting position (this is usually 0).

Extension, movements away from the trunk, left lateral flexion and left rotation are recorded first and their opposites last. For example, recording shoulder ROM by the SFTR method (Table 18.2).

This system allows recording of positional deformities. For example, knee hyperextension may be recorded as S:10–0–130, a 20° valgus deformity as F:20–0. A further advantage of the SFTR system is that it clearly records starting positions. For example, a fixed flexion deformity at the knee would be recorded as S:0–30–130, the middle figure indicating that the starting position was not neutral. Although it is not widely used, the SFTR system allows standardized recording in a quick, clear and logical manner.

Table 18.2 Recording ROM by the SFTR system

Shoulder joint	Degrees (°)	
S	55–0–180	(extension/flexion)
F	180–0–0	(abduction in frontal plane)
T	45–0–135	(horizontal abduction/adduction)
R (in frontal plane 90°)	90–0–90	(rotation)

TESTS OF MUSCLE STRENGTH

Muscle testing may be either manual, using dynamometers, or using isokinetic devices such as the Cybex or Kin-Com.

Manual muscle testing

Manual muscle testing tests the strength of individual muscles at their maximum voluntary contraction. Over the years many different scales have evolved to grade the strength of muscles, most using the principles of gravity and applied external load to determine the ability of the patient to develop muscle tension (Lamb, 1985).

In the UK, the most commonly accepted method of grading muscle strength is the Oxford/Medical Research Council scale which grades on a 0–5 scale (MRC, 1976).

MRC scale
0   No contraction.
1   Flicker or trace of contraction.
2   Active movement through range with gravity counterbalanced.
3   Active movement through range against gravity.
4   Active movement through range against gravity and some resistance.
5   Normal power.

All movements are tested through their full range of motion. It may be seen that with the exception of grade 3, all readings require an element of subjective judgement by the physiotherapist performing the test. Some testers further subdivide grade 4 into 4−, 4, 4+, according to whether slight, moderate or strong resistance is applied. For a muscle to be assigned a grade 5 it must be equal to the contralateral muscle if the condition is unilateral, or judged the same as that of a person of the same age, sex and build as the patient. The muscle must also be able to perform in all of its roles, i.e. prime mover, synergist, antagonist, etc.

Kendall and McCreary (1983) advocate the following scale for manual muscle testing based on percentages:

0%   No palpable contraction.
5%   Palpable contraction but no movement.
20%   Move through small arc of movement, gravity eliminated.
50%   Hold against gravity.
80%   Hold against gravity and submaximal applied force.
100% Hold against maximum applied force.

This system is found fairly commonly in North American practice and literature, but is less commonly used in Britain. Both systems require that the tester has a thorough knowledge of the underlying anatomy and kinesiology.

Procedure for manual muscle testing

1. Test on a firm surface in a warm room, free from draughts.
2. The patient should be adequately unclothed.
3. The patient should understand the purpose of the testing.
4. All joints should be tested through their passive range of movement.
5. The patient should be in a position so that the body segment is stabilized before the muscle is tested.
6. The test should be carried out with resistance applied throughout the whole range of motion (in grades 3 and above), in a smooth manner and applied directly opposite to the line of pull of the muscle.
7. The patient should not be fatigued or in pain during the test.
8. Any restrictions due to lack of joint range, presence of pain or restriction of the starting position for the test should be recorded.
9. A chart of the results of the testing should be kept.

The test positions for individual muscles are detailed in MRC (1976) or in Kendall and McCreary (1983).

Reliability of manual muscle testing

Manual muscle testing is a widely used clinical procedure which relies on much subjective judgement by the tester. Frese *et al*. (1987) examined the reliability of manual muscle testing of gluteus medius and the middle fibres of trapezius in a clinical setting by eleven physiotherapists. These authors found that inter-tester reliability for right and left trapezius and gluteus medius was low, with the percentage of physiotherapists obtaining a rating of the same grade or within one third of a grade ranging from 50% to 60%. They concluded that manual muscle testing was of questionable value as an accurate clinical tool. However, Florence *et al*. (1992) found that manual muscle testing was reliable for the assessment of muscle strength in muscular dystrophy patients when consecutive evaluations were performed by the same physiotherapists.

Muscle testing by hand held dynamometry

Dynamometers or myometers rely on the patient imposing a force against the dynamometer, which is translated linearly via a gauge to give a reading. Generally, such equipment is easy to use and provides an immediate reading of the force generated. Many studies have been conducted to test its reliability, most reporting it to be a reliable measure of muscle strength with good intra- and inter-tester reliability (Bohannon and Endemann, 1989). Others disagree, however, finding large error rates in the inexperienced user (Lennon and Ashburn, 1993).

Other forms of dynamometer that may be utilized include fixed springs, cable tensionmeters or use of the 1 RM and 10 RM test (see Chapter 10).

Isokinetic testing

Isokinetic devices provide an objective measurement of dynamic and static muscle strength (Chapter 11).

OTHER USEFUL MEASUREMENTS

Measurement of leg length

Leg length is frequently measured, as inequality in the length of the legs may contribute to clinical problems such as back pain, scoliosis and running injuries. Opinion differs about how much of a discrepancy between the legs needs to exist before it becomes clinically significant. Subotnick (1981) suggests that a difference of as little as 3 mm requires correction, whereas Anderson (1972) states that a difference up to 19 mm is acceptable.

Tests may be performed to measure either real or apparent leg length. The patient usually lies supine in a straight line and linear measurements are taken with a tape measure and recorded in centimetres.

Beattie *et al.* (1990) investigated the validity of leg-length measure taken with a tape compared with those from X-rays. They found that the measurements taken with a tape were valid, especially when using an average of two readings, rather than a single measure.

Real leg length. Distance from the anterior superior iliac spine (ASIS) to the medial malleolus.
Apparent leg length. Distance from a point on the midline of the trunk, e.g. xiphisternum to medial malleolus.

Procedure for measuring leg length

Method 1
Locate the ASIS by approaching from below, i.e. sliding up the anterior surface of the thigh until the fingers contact the ASIS. The tape measure is placed at this point and extended to the distal tip of the medial malleolus.

Method 2

Stand the patient with their legs about a metre apart and with the observer having a clear view of the posterior aspect of their legs from the buttocks down. Blocks of wood are placed under the foot of the shorter leg until both the pelvis and the buttock creases are level. The amount of correction applied to achieve this is measured to show the amount of discrepancy that was present.

MEASUREMENT OF LUMBAR SPINE FLEXION

It is difficult to attain accurate measurements of lumbar spine flexion/extension due to the absence of bony landmarks and the presence of hip and pelvis movements in combination with the back moving. Schober's index measures lumbar movement by the skin distraction method. The index has been modified by Macrae and Wright (1969) to place a mark 5 cm below the lumbosacral junction and 10 cm above it while the patient is in neutral standing. The increase in distance between the two marks on full flexion gives a measure of spinal movement. Such measures have been reported to be a reliable measure of lumbar movement (Williams *et al.*, 1993).

Measurement of hip flexion deformity (Thomas's test)

With the patient lying supine, one hand is placed behind the lumbar spine to assess lumbar movement. The good hip is flexed fully ensuring that the lumbar lordosis is flattened. If the affected hip rises from the couch, this indicates the presence of a fixed flexion deformity and the angle may be measured. Thus the notation hip range 30–90° indicates a fixed flexion deformity of 30° and a hip that flexes to 90° (Figure 18.3).

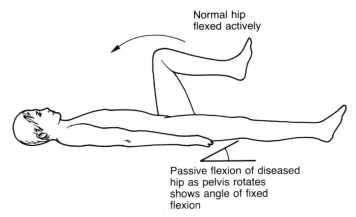

Normal hip
flexed actively

Passive flexion of diseased
hip as pelvis rotates
shows angle of fixed
flexion

Fig. 18.3 Thomas's test for measurement of hip deformity

Trendelenburg test

This will indicate weakness in the gluteal muscles or inhibition of their function, e.g. from pain. The patient stands on one leg while being viewed from behind. Normally, standing on one leg results in the gluteal muscles bringing the centre of gravity over the weight-bearing leg and tilting the pelvis to elevate the buttocks on the unsupported side.

The Trendelenburg test is positive when the buttock on the non-weight-bearing side fails to rise (Figure 18.4).

Leg circumference

Frequently the physiotherapist will wish to know the circumference of the limb in the calf to detect or monitor the presence of a thrombosis or to document muscle wastage, or in the thigh to record wasting or effusion. The method of recording thigh circumference uses three measures:

● Base of patella – measures bony circumference
● 10 cm above – measures joint effusion
● 15 cm above – measures muscle wasting.

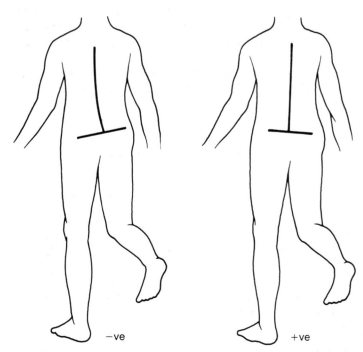

−ve　　　　　　　　　　+ve

Fig. 18.4 The Trendelenburg test to indicate weakness in the gluteal muscle

MEASUREMENT OF ACTIVITIES OF DAILY LIVING

Measurements of activities of daily living (ADL) are usually made by using standardized rating scales that measure physical impairment. Such scales are particularly suited to heavily disabled patients such as those suffering from neurological impairments. However, more modern scales have been designed to deal with less severe cases. Numerous different types of rating scales exist and it is beyond the scope of this chapter to review them all, instead the reader is referred to Wade (1992).

Most rating scales share common areas of study such as self-care, ambulation and hand function. An index that is widely used and which is applicable for patients with long term musculoskeletal and neuromuscular disorders is the Barthel index. Many other ADL scales exist and may be more appropriate for certain patients, but the Barthel index is quoted as an example of such a scale as it is widely used in UK rehabilitation centres.

Barthel index

This uses a rating by the physiotherapist, or person conducting the test, of ten aspects of daily living and produces an overall score of between 0 and 100. A score is assigned for each activity, according to whether the activity is carried out independently, with help or is not successfully completed. A score of 0 indicates complete dependence for all activities and 100 complete independence. The categories used are given in Table 18.3.

As this scale is aimed at people with relatively severe disability, it will not be sufficiently sensitive to assess the orthopaedic patient with problems confined to one limb. It is, however, useful in assessing more

Table 18.3 Barthel index ratings of 10 aspects of daily living

	Score	
ADL	Help	Independent
1. Feeding	5	10
2. Transfer from wheelchair to bed and return	5–10	15
3. Personal toilet	0	5
4. Transfer on and off the toilet	5	10
5. Bathing	0	5
6. Walking on a level surface	10	15
7. Walking up and down stairs	5	10
8. Dressing	5	10
9. Bladder control	5	10
10. Bowel control	5	10

widespread musculoskeletal problems such as rheumatoid arthritis. It is a scale with good reported validity and reliability that compares favourably with other rating scales (Granger *et al.*, 1979).

SUMMARY

- Clinical measurement is an important clinical tool that the physiotherapist needs to practise with skill and accuracy in both the examination and assessment of patients.
- It is important that the tests used when undertaking a measurement are reproducible, reliable and valid measures.
- Measurements may be used as assessment tools or as outcome measures to monitor the effect of an intervention.
- Joint range of motion is usually measured by goniometry. The different types of goniometer, the techniques for their use and their reliability are discussed.
- Joint range may also be measured by linear methods, particularly for the spine and hand.
- Joint range measurements may be recorded by the conventional 0–180 system or using the SFTR system.
- Leg-length discrepancies may be real or apparent and measured using a tape.
- Muscle strength may be tested by manual muscle tests using grading scales such as the MRC grades.
- Muscle strength may also be measured using specialized equipment such as hand held dynamometers or isokinetic dynamometers.
- Overall function may be measured using an ADL scale such as the Barthel index.

REFERENCES

AAOS (1965) *Joint Motion: Method of Measuring and Recording*, American Academy of Orthopaedic Surgeons, Chicago

Anderson, W.V. (1972) *Modern Trends in Orthopaedics*, Appleton-Century-Croft, New York

Beattie, P., Isaacson, K., Riddle, D.L. *et al*. (1990) Validity of derived measurements of leg length differences obtained by use of a tape measure. *Physical Therapy*, **70**(3), 150–157

Bohannon, R.W. and Endemann, N. (1989) Magnitude and reliability of hand held dynamometer measurements within and between days. *Physical Practice*, **5**(4), 177–181

Boone, D.C., Azen, S.P., Lin, C.P. *et al*. (1979) Reliability of goniometric measurements. *Physical Therapy*, **58**, 1355–1360

Ekstrand, J., Wilktorsson, M., Öberg, B. *et al*. (1982) Lower extremity goniometric measurements: a study to determine their reliability. *Archives of Physical Medicine and Rehabilitation*, **63**, 171–175

Florence, J.M., Pandya, S., King, W.M. *et al*. (1992) Intrarater reliability of manual muscle test (Medical Research Council scale) grades in Duchenne's muscular dystrophy. *Physical Therapy*, **72**, 115–126.

Frese, E., Brown, M. and Norton, B.J. (1987) Clinical reliability of manual muscle testing. *Physical Therapy*, **67**, 1072–1076

Gajdosik, R.L. and Bohannon, R.W. (1987) Clinical measurement of range of motion: review of goniometry emphasizing reliability and validity. *Physical Therapeutics*, **67**, 1867–1872

Galley, P.M. and Forster, A.L. (1987) *Human Movement*, Churchill Livingstone, London

Gerhardt, J.J. and Russe, O.A. (1975) *International SFTR Method of Measuring and Recording Joint Motion*, Huber, Bern

Gogia, P.P., Braatz, J.H., Rose, S.J. *et al.* (1987) Reliability and validity of goniometric measurements at the knee. *Physical Therapy*, **67**, 192–195

Goodwin, J., Clark, C., Deakes, J. *et al.* (1992) Clinical methods of goniometry: a comparative study. *Disability and Rehabilitation*, **14**(1), 10–15

Granger, C., Dewis, L. Peters, N. *et al.* (1979) Stroke rehabilitation: analysis of repeated Barthel Index Measures. *Archives of Physical Medicine and Rehabilitation*, **60**, 14

Hellebrandt, F.A., Duvall, E.N. and Moore, M.L. (1949) The measurement of joint motion. Part III reliability of goniometry. *Physical Therapy Review*, **29**, 302–307

Kendall, F.P. and McCreary, E.K. (1983) *Muscles – Testing and Function*, 3rd edn, Williams and Wilkins, Baltimore

Lamb, R.L. (1985) Manual muscle testing. In *Measurement in Physical Therapy* (ed. J.M. Rothstein), Churchill Livingstone, New York

Lennon, S.M. and Ashburn, A. (1993) Use of myometry in the assessment of neuropathic weakness: testing for reliability in clinical practice. *Clinical Rehabilitation*, **7**(2), 125–133

Macrae, I.F. and Wright, V. (1969) Measurement of back movement. *Annals of Rheumatic Diseases*, **28**, 584–589

MRC (1976) *Aids to the Examination of the Peripheral Nervous System*, Medical Research Council, Memorandum No. 45, HMSO, London

Miller, P.J. (1985) Assessment of joint motion. In *Measurement in Physical Therapy* (ed. J.M. Rothstein), Churchill Livingstone, New York

Moore, M.L. (1949) The measurement of joint motion. Part II The technic of goniometry. *Physical Therapy Review*, **29**, 256–264

Nicol, A.C. (1989) Measurement of joint motion. *Clinical Rehabilitation*, **3**, 1–9

Pynsent, P.B., Fairbank, J.C.T. and Carr, A. (1993) *Outcome Measures in Orthopaedics*, Butterworth–Heinemann, Oxford

Riddle, D.L., Rothstein, J.M. and Lamb, R.L. (1987) Goniometric reliability in a clinical setting: shoulder measurements. *Physical Therapy*, **68**, 668–673

Smith, J.R. and Walker, J.M. (1983) Knee and elbow range of motion in healthy older individuals. *Physical and Occupational Therapy in Geriatrics*, **2**, 31–38

Stratford, P., Agostino, V., Brazeau, C. *et al.* (1984) Reliability of joint angle measurements: a discussion of methodology issues. *Physiotherapy Canada*, **36**, 5–9

Subotnik, S.I. (1981) Leg length discrepancies of the lower extremity. *Journal of Orthopeadics, Sports and Physical Therapy*, **3**, 11–16

Wade, D.T. (1992) *Measurement in Neurological Rehabilitation*, OUP, Oxford

Watkins, M.A., Riddle, D.L. and Lamb, R.L. (1991) Reliability of goniometric measurements and visual estimates of knee range of motion obtained in a clinical setting. *Physical Therapy*, **71**, 90–97

Williams, R., Binkley, J., Bloch, R. *et al.* (1993) Reliability of the modified-modified Schober and double inclinometer methods for measuring lumbar flexion and extension. *Physical Therapy*, **73**, 26–37

INDEX

Abduction, 37
Acceleration,
    definition, 32
    Newton's law, 11
Actin, 53, 56
Activities of daily living,
    measurement of, 399
Adduction, 37
Adenosine triphosphate-creatine
    phosphate system, 103, 122, 123
Age,
    cardiovascular disease and, 109
    flexibility, 244
    gait changes and, 313
    muscle strength and, 198
    oxygen uptake and, 154
    postural sway and, 279
Akinesia, 80
Alcohol,
    consumption of, 107
Alexander technique, 293
Ankle,
    motion during walking, 302, 306
    muscles in walking, 304
    in standing, 166
Ankle-foot orthoses, 345
Antalgic gait, 315
Antenatal classes, 167, 168–171
Anterior horn cells, 69, 71, 72, 73
Anterior spinocerebellar tract, 84
Anterior spinothalamic tract, 84
Apparatus,
    testing, 161
Archimedes' principle, 31
Arm See Upper limb
Arthrokinematics, 44
Artificial limbs,
    pressure and, 27
Ascending tracts, 83–85
Asthma, 107
Astrand nomogram, 139, 140, 141
Astrand Submaximal cycle ergometer test,
133, 138, 139
Ataxia, 179, 184
    rehabilitation, 191, 207
Athetosis, 81

Atlanto-occipital joint,
    mechanics of, 14
Australian lift, 373

Babinski sign, 74
Back,
    rehabilitation programmes for, 163–165
Back pain, 163, 165
    from postural malalignment, 285
    functional restoration programmes, 165
    in health care context, 254
    in pregnancy, 170
    relaxation techniques for, 273
Back school, 164
Balance, 175–195
    activities for re-education, 182
    assessment, 178
    associated reactions and, 181
    centre of pressure, 178, 179
    classification, 175
    controlling, 278
    definition of, 175
    dynamic, 176
        rehabilitation, 183
    early stage re-education, 182
    factors affecting, 176, 177
    guidelines for assessment, 180
    performance of functional activity, 180,
        181
    platform stabilometry in, 178
    postural stabilization and, 178
    retraining, 181–183
        later stage, 183
        proprioceptive neuromuscular
            techniques in, 190
        rhythmic stabilizations, 183, 190
        slow reversals, 191
    Romberg test, 179
    sitting, 182
    standing, 176
    static, 175
        education of, 182
        rehabilitation, 183
    testing, 178
Balke protocol, 133

Ball and socket joints, 40
Ballismus, 81
Ballistic stretch, 249
Baroreceptors, 83
Barthel index, 399
Basal ganglia, 67, 76, 77, 78, 80, 81
  anatomy of, 78
  motor disorders affecting, 80
  muscle tone and, 73
Biaxial joints, 41
Biceps,
  as lever, 16
  paying out, 102
Bicondylar joint, 39
Biodex isokinetic dynamometer, 224
Blink reflex, 68
Blood doping, 132
Blood pressure, 107
  oxygen uptake and, 137
Body,
  axes, 188
  link segment model, 42
  planes, 34, 35, 188
Bone, 48
Borg scale, 133, 153, 154
Brachioradialis,
  mechanics of, 15
Bradykinesia, 318
Braiding, 187
Brain function in elderly, 109
Brain injury,
  rehabilitation, 89
Brave walk, 88
Breathing,
  deep, 166
Bridging, 167
Bruce protocol, 137
Bruce tests, 133
Buoyancy, 31
Buttock,
  isometric exercises, 166

Cable tensiometry, 225
Calcium in muscle activation, 55
Calf injury, 244
Canadian Home Fitness Test, 139, 143
Capillary density,
  strength training and, 202
Cardiovascular disease,
  age and, 109
Cardiovascular endurance, 219
Cartilaginous joints, 40
Casts, 249
  drop-out, 250
Cast braces, 346
Central nervous system,
  injury of, 67, 75
Centre of buoyancy, 31
Centre of gravity,
  definition, 32
  equilibrium and, 24

  stability, 25
Centre of mass, 32
Centre of pressure, 178, 179
Cerebellum, 81–82
  injury to, 76
Cerebral cortex, 76, 77
Cerebral shock, 76
Cerebrovascular accidents,
  rehabilitation, 247, 249
Cervical muscles,
  as levers, 14
Chemoreceptors, 83
Children,
  exercise prescription for, 110–111
  gait in, 314
Chondromalacia patellae,
  gait in, 317
Chorea, 80, 81
Circuit training, 173
  for cardiac patients, 112
Clasp-knife phenomenon, 73, 74
Clinical measurement, 385–401
  definition, 385
  inter-tester reliability, 386
  intra-tester reliability, 386
  objective and subjective, 386
  outcome measures and, 387
  parallel reliability, 386
  reliability, 386
  validity, 385
Clonus, 74
Closed kinetic chain exercise, 103, 104
Codman's exercises, 22
Cog-wheel rigidity, 75
Cold,
  use of, 246
Collagen time scale, 245
Colles fracture,
  group exercise classes, 161, 165
Concentric exercise, 102
Contraction,
  and muscle strength, 197
Contracture, 249
Cooper's 12 minute run test, 139, 142, 143
Coracobrachialis muscle,
  angle of pull, 16
Coronary heart disease,
  risk factors, 106
Corticospinal (pyramidal) tract, 73, 77, 79
Creep effect, 30
Crutches, 325, 330–332
  axillary, 330
  balance and, 182
  elbow, 331
  forearm/gutter, 332
  swing-through gait, 340, 341
  swing-to gait, 340, 341
  types, 330
  walking with, 336
Cryotherapy, 253
Cycle-ergometry, 220
  versus treadmills, 136
cystic fibrosis, 107

Daily living activities,
  measurement of, 399
Decerebate rigidity, 75
Degrees of freedom, 41
Delayed onset of muscle soreness, 200,
216, 234, 240
De Lorme boot, 208
De Lorme protocols, 209
Deltoid,
  fibres, 60
    force from, 7
Descending tracts, 77, 79
Diet and nutrition, 107
Distal phalanges,
  degrees of freedom, 41
Donjoy brace, 348
Dorsal root ganglion, 69
Dorsiflexion, 37
Dowager's hump, 283
Dynamic equilibrium, 24
Dynamics,
  definition, 32
Dynamometer,
  isokinetic, 102
Dystonia, 80

Eccentric exercise, 102
Eccentric training, 216–218
  loading, 217
Elbow,
  mechanics of, 18
  myositis ossificans, 100
Elbow crutches, 331
Elderly,
  cerebral function in, 109
  exercise prescription for, 108
  falls among, 314
  gait re-education in, 313
  health of, 109
  social contracts for, 109
Ellipsoid joint, 39
Endomysium, 51, 70
Endurance exercises, 219–221,
  examples of, 220
Endurance performance tests, 147–150
Endurance training, 202
Energy, 27, 122–124
  adenosine triphosphate-creatine
    phosphate system, 122, 123
  aerobic system, 103, 122, 123
  anaerobic, 103
  glycolysis-lactic acid system, 122, 123
  source of, 103
Epimysium, 51, 64
Equilibrium, 23–24
  definition, 33
  reactions, 176
Equipment,
  safe, 115
  testing, 161
European Directive on Manual Handling,
  361

Exercise,
  active assisted, 97
  active resisted, 98
  antenatal, 168
  auto-assisted, 98
  for back schools, 164
  benefits of, 113, 115
  for circuit training, 173
  classification of, 97–105
  concentric, 102
  dangers of, 115–116
  eccentric, 102
  for endurance, 219
  for flexibility, 247
  free active, 97
  group *See Group exercise*
  home programme, 171
  isokinetic *See Isokinetic exercise*
  isometric *See Isometric exercise*
  isotonic *See Isotonic exercise*
  kinetic chain, 103, 104
  motivational aspects, 116–117
  muscle contraction types and, 129
  muscle strengthening, 215
  postoperative, 167
  precautions of, 117
  prescription, 106–119
    for cardiac patients, 112
    for children, 110–111
    for elderly, 109
    general principles, 108
    for health related fitness, 150
    for respiratory patients, 111
    use of heart rate, 151
    for various patient groups, 108–113
  programme,
    critical analysis, 117
    holistic approach, 118
    planning, 107
    progression, 113, 114
  progressive resistance, 207
  rehabilitation, 107
  shoulder, 20
  soundness of, 115
  specialized, 166–171
  for strength programme, 205
  stretching, 255
  types of movement, 97
  for wrist fracture, 165
Exercise boards, 26
Exercise cycling,
  in relation to VO_2 max, 152
  oxygen consumption and, 151
Exercise groups,
  specialist, 166–171
Exercise testing,
  aerobic power, 130
  for ATP-CP system, 125
  long-term endurance, 130
  power test, 125
  reasons for, 121
  for short-term endurance, 128
  Wingate test, 128

Exertion,
　Borg scale, 153, 154
　perceived rate, 154
Extension, 37

Falls,
　among elderly, 314
Fasiculi,
　arrangement of, 54, 55
Felderkrais re-education of posture, 284
Fibrous tissue, 40
Fitness,
　activity specific, 121
　definition of, 120
　health-related, 121
　muscle strength and, 198
Fitness testing, 120–157
　aerobic endurance performance, 138
　cardiovascular, 121, 137
　　Harvard strip test for, 147
　endurance performance, 147–150
　jump tests, 125, 126
　maximum rate of oxygen uptake,
　　see Oxygen uptake, maximum rate
　musculoskeletal, 121
　neuromuscular, 121
　non-motorized treadmill test, 129
　one-repetition maximum, 127
　optimal conditions, 124
　preparation, 124
　questionnaire, 124
　respiratory, 148
　stair sprinting, 126
Flaccidity, 75–76
　associations of, 76
Flatback, 282
Flat-footed gait, 321
Flexibility, 243–257
　age and, 244
　body tissue and, 245
　definition, 243
　exercises for, 247
　factors affecting, 244
　increasing,
　　see also Stretching
　　methods of, 246
　　proprioceptive neuromuscular
　facilitation in, 252
　nervous system and, 245
　static and dynamic, 243
　training,
　　benefits of, 244
　types of, 243
Flexion, 37
Flight or fight response, 262
Fluid mechanics, 30–31
Foot,
　exercises, 166
　instability,
　　due to structural integrity loss, 344
　　due to structural malalignment, 344

　orthosis for, 342
　orthoses, 342–345
　weak muscles, 321
　weakness fo extensors/flexors, 343
　weak pronators, 343
　weak supinators, 342
Footware,
　friction and, 26
　high heeled, 27
Force, 3–10
　see also levers
　analysis of, 6
　as vector, 4
　composition of, 6
　　parallelogram method, 6
　concurrent systems, 5
　contact, 3, 4
　definition of, 32
　effect of, 4
　free-body diagrams, 10, 11
　gravitational, 23
　linear, 4
　non-contact, 3, 4
　parallel, 4
　　in equilibrium, 5
　parallelogram of, 7
　pendulum and, 22
　pressure and, 27
　resolution of, 8
　　by trigonometry, 9, 10
　resultant of, 8
Force couple, 4
Force platforms, 278, 308, 309
Free body diagrams, 10, 11
Friction, 26–27
　definition, 33
　limiting value, 26
　stability and, 25
Frontal (coronal) plane of body, 36
Frontal plane movement, 37
Frozen shoulder, 22
　home exercise programme, 171
Functional body movements, 189

Gait, 297–324
　see also Walking
　abnormalities,
　　correction of, 319
　age changes, 313
　analysis, 298, 304
　　cinematography, 310
　　electrogoniometry, 311
　　foot switches, 308
　　kinematic, 310
　　kinetic, 308
　　kinetic and kinematic, 312
　　observational techniques, 305
　　quantitative, 306
　　stroboscopy, 310
　　time-distance measurements 306
　　Vicon, Selspot and Coda techniques,
　　　311

Gait (*cont.*)
  analysis (*cont.*)
    video-taping, 311
  antalgic, 315
  assessment, 322
  benefits of clinical assessment, 322
  cadence, 298
    measurement, 307
  changes in disease, 314
  in children, 314
  cycle, 297, 298
  decreased step length, 319
  definitions, 297
  displacement of centre of mass, 302, 303
  distance and temporal factors, 299
  double support phase, 300
  flat-footed, 321
  following hip replacement, 316
  foot angle, 298
  four-point, 339
  ground contact, 299
  heel off, 300
  heel strike, 298
  hemiplegic, 89, 317
  in hip replacement, 316
  in hip rheumatoid disease, 316
  in knee replacement, 317
  mid-stance, 300
  mid-swing, 301
  orthotic patterns, 352
  in Parkinson's disease, 318
  patterns with walking aids, 337, 338, 339
  in postosteoarthritis of hip, 316
  push-off phase, 300
  reciprocal two-point, 340
  re-education, 162, 321–322
    in elderly, 313
  stance phase, 298, 299
  step length, 298
  step width, 298
  stride length, 298
    measurement, 307
  study methods, 304–313
  swing phase, 298, 301, 303
  swing-through, 340, 341
  swing-to, 340, 341
  three-point,
    with walking aids, 337, 338
  Trendelenburg, 304, 315, 321
  two-point,
    with stick, 337, 338
  with crutches, 335
Gastrocnemius muscle, 60
Glenohumeral joint,
  degrees of freedom, 99
Glenoid fossa, 47
Gluteus muscles, 50
  in walking, 303
  weakness of, 398
Glycolysis-lactic acid system, 103, 122, 123
Golgi tendon organ, 72
Gomposis, 40
Goniometry, 388–392

instruments (Goniometers), 388–350
  comparisons, 390
  electro, 390
  fluid, 390
  joint specific, 388
  pendulum, 389
  universal, 388
  placement of instrument, 390
  reliability, 392
  repeated measures, 391
  technique of, 391
Gracilis,
  angle of pull, 19
Grasp reflex, 89, 90, 91
Gravitational forces, 23
Gravity,
  centre of, 23
Ground reaction force, 12
Group exercise, 158–174
  advantages of, 158
  back class, 161
  circuit training, 173
  for Colles' fracture, 161
  disadvantages of, 159
  lower limb, 161
  planning classes, 160
  postoperative, 166
  role of physiotherapists, 159
  safety aspects, 160
  types of class, 161
Guillain-Barré syndrome, 191
Guthrie Smith frames, 211

Hamstrings, 16
Hand,
  resting splints, 252
Handling,
  causes of injury, 354
  ergometric approach, 354–384
  legislation, 361
    enforcement, 363
  manual,
    assessment of, 362
    avoidance, 362
    risk assessment, 358
  patient *See Patient handling and lifting*
  prevention of injury, 355, 361
  training for, 354
Harvard step test, 147
Head in standing, 278
Headaches,
  relaxation for, 273
  tension, 271
Head injury, 250
Healing, of soft tissue injury, 245
Health,
  promotion of, 120
Health and Safety at Work Act 1974, 361,
  362, 363
Health of the Nation, 106
Heart disease,
  rehabilitation programmes, 112

Heart rate,
    in exercise prescribing, 151
    oxygen uptake and, 137
    prediction of, 153
Heat,
    use of, 246, 253
Heel,
    pain in, 344
Hemiballismus, 80
Hemiplegic gait, 89, 317
Hinge joint, 39
Hip,
    abduction/adduction, 214
    abductors, 8
        mechanics of, 8
    action of muscles in walking, 303
    arthritis, 321
    flexed, 319
    flexion deformity, 397
    mechanics of, 8, 22
    myositis ossificans, 100
    rheumatoid disease, 316
    in standing, 277
    in walking, 301, 305, 319, 321
Hip bends, 167
Hip guidance orthosis, 348, 349, 350
Hip hitching, 167
Hip-knee-ankle-foot orthoses, 348
Hip replacement,
    gait following, 316
Hoffman's sign, 74
Hold-relax, 252, 253
Holistic approach to exercises, 118
Home exercise programme, 171–173
Hooke's law, 29
Hopping reaction, 90, 91
Humerus,
    fracture of, 205
    Head of, 47
Hydrostatic pressure, 31
Hydrotherapy, 98, 203
Hyperlordosis, 285
Hypermobility, 244
Hyperplasia, 202
Hyperreflexia, 74
Hypertrophy of muscle, 202
Hypotension,
    postural, 180
Hypotonicity, 75

Ice,
    use of, 246, 253
Iliopsoas muscle, 16
Incontinence, 167, 169
Inertia,
    Newton's law of, 11
Internal rotation, 37
Interoceptors, 83
Intervertebral disc,
    during lying, 281
    during sitting, 280
Inverted U hypothesis, 262

Isokinetics,
    development of, 224
Isokinetic contraction, 201
Isokinetic dynamometer, 102, 216, 224, 225
    in research, 239
Isokinetic exercise, 102, 224, 242
    advantages of, 235
    concept of, 226
    contraction model, 229
    damping, 227
        disadvantages of, 228
    definition, 224
    disadvantages of, 235
    in eccentric mode, 239
    evaluation, 230–231
        in diagnosis, 239
        parameters, 231
        power and work, 233
        torque, 231, 232, 233
    gravity correction, 230
    isokinetic mode, 229
    isometric mode, 229
    isotonic mode, 229
    maximum voluntary contraction, 234
    moment/force threshold values, 228
    muscle fibre type and, 227
    passive mode, 229
    preload setting, 227
    range of movement, 227
    ramping, 228
    regimens, 210
    in rehabilitation, 237
    reliability of devices, 235
    repetitions, 228
    treatment spectrum, 238
    use of diagonal functional movement
        patterns, 237
    validity of, 236
    velocity, 226
    visual feedback in, 236
Isokinetic system, 225
    passive and active, 226
Isometric contraction, 200
Isometric exercise, 101, 220
    advantages and disadvantages, 201
    protocol, 201
Isotonic contraction, 199
Isotonic exercise, 200

Jogging, 152
Joints, 37–47
    articulation of, 34
    ball and socket, 40
    biaxial, 41
    bicondylar, 39
    cartilaginous, 40
    closed packed condition, 45, 46, 47
    degrees of freedom, 41, 99
    formation, 34
    hinge, 39
    illipsoid, (condyloid), 39
    loose packed position, 45, 46, 47

Joints (*cont.*)
  movement of, 44
  non-synovial (synarthroses), 37, 40
  ovoid, 44
  pivot, 39
  plane, 39
  range of motion, 43
    goniometry, 388
    linear measurements, 392
    measurement of, 43, 387–393
    normal, 387
    recording, 392
    SFTR recording, 393
  restraining effect of muscle, 43
  roll motion, 44, 47
  saddle, 39
  sellar, 39, 45
  slide motion, 44, 47
  spheroidal, 40
  spin motion, 44, 47
  suture, 40
  synovial (disarthroses), 37, 38–40
Joint mobilization techniques, 101
Joint position sense, 185
Joint stability,
  flexibility and, 244

Kilo pounds, 128
Kin Com AP Isokinetic dynamometer, 224, 225, 233
Kinematics, 3
Kinetic chain exercise, 103
  open and closed, 103, 104
Kinetic energy, 27
Kinetics, 3
Knee,
  braces, 347
  drop out casts, 251
  hyperextension, 348
  isokinetic exercises, 226
  orthoses, 347–348
  osteoarthritis, 317
  rheumatoid disease of, 317
  in standing, 277
  suspension therapy, 215
  in walking, 302, 304, 305, 317, 319
Knee-ankle-foot arthroses, 346, 347
Knee bends, 167
Knee pain, 317
Knee replacement, 317
  progression regimen, 113
Knee rolling, 167
Kypholordosis, 282
Kyphosis, 282
  decrease in, 284
  increased, 283, 284

Landau reflex, 90, 91
Lateral flexion, 37
Lateral rotation, 37
Lateral spinothalamic tract, 84

Lead-pipe rigidity, 75
Leg,
  see also lower limb
  artificial, 27
  exercises classes for, 161
Leg circumference, 398
Leg injury,
  rehabilitation after, 161–163
Leg length, 396
Leg press training apparatus, 208
Lennox Hill brace, 348
Levers, 13–17
  angle of pull, 16
  definition, 33
  first-class, 14
  mechanical advantage of, 13
  second class, 14,15
  third class, 15
  torque and, 17
Lido isokinetic dynamometer, 224
Lifting,
  biomechanics of, 364
  bracing, 364
  causes of injury, 354
  ergometric approach to, 354–384
  foot position, 365
  grips, 365
  legislation, 361
  patients *See Patient handling and lifting*
  position of lifter, 364
  preparation for, 364
  preparation of area, 365
  prevention of injury, 355
  principles of, 364–368
  rules of, 366
  squat, 367
  stoop, 367
  training for, 355
Ligaments,
  creep effect, 30
Limb braces, 348
Limiting friction value, 26
Linear stress, 28
Link concept, 103
Link segment model, 42
Locomotion, 297
Locomotor development, 85, 86, 87, 88–89
Long leg orthroses, 346, 347
Lordosis, 279
  increased, 283, 284
Low back pain *See Under back pain*
Lower limb,
  eccentric exercises, 216, 217, 218
  eccentric force generation, 227
  isokinetic exercises, 226
    evaluation protocol, 231
  muscle strengthening exercises, 215
  patterns of movement, 190
  plyometrics, 218
  spasticity, 250
  wobble board re-education for, 186
Lumbar lordosis, 279
  increased, 283

Lumbar spine flexion,
   measurement, 397
Lying posture, 280

MacQueen protocols, 209
Manipulation, 100
   under anaesthesia, 101
Manual Handling Operations Regulations,
   1992, 361, 362, 363
Maximal exercise, 137
Maximum rate of oxygen uptake, 130–142
Maximum voluntary contraction, 234
Mechanical advantage, 33
Mechanics,
   centres of gravity, 23
   pulleys, 18–21
   torque, 17–18
   units of measurement, 12
Mechanics of movement, 3–33
   creep effect, 30
   definitions, 32–33
   energy, 27
   fluid, 30
   force, 3–11
   friction, 26
   gravity, 23
   levers, 13–17
   motion, 11–13
   pendulums, 21–22
   power, 28
   pressure, 27
   springs, 30
   stress, 28–30
   work, 28
Medial rotation, 37
Meditation, 269
Mental relaxation, 265
Metabolic cart, 135
Metatarsophalangeal joint,
   mechanics of, 15
Milwaukee brace, 6
Mitochondrial volume,
   strength training and, 202
Moro reflex, 90, 91
Motion, 11–13
   angular/rotary, 11
   curvilinear, 11
   energy of, 27
   linear, 11
   Newton's laws of, 10, 11
Motivation, 116–117, 159
   forms of, 117
Motor control, 82–83
Motor development, 85, 86, 87, 88–89
Motor neurons, 74
Motor units,
   strength and, 196
Movements,
   see also specific types etc
   axes of, 36
   control by vestibular system, 85
   descriptors, 36, 37

in exercises, 97
functional, 189
mechanical basis of, 3–33
   see also under mechanics
gravitational forces, 23
higher centre control, 76
musculoskeletal basis of movement, 50
see also Muscles etc
neurophysiological aspects, 66–93
oblique, 188
rotational, 188
skeletal basis See under Skeletal system
types of, 97–101
unwanted, 80
Mulley's guidelines, 333
Multiple sclerosis, 249
Muscles
   see also Skeletal muscles
   activation, 55
   active tension, 57
   classification, 60
   contraction types, 199
   generating force, 227
   hyperplasia, 202
   hypertrophy, 202
   isokinetic contraction, 201
   isometric contraction, 200
   isotonic contraction, 199
   length/tension, 51, 58
   motor units, 56
   negative work, 102
   passive tension, 57
   positive work, 102
   power and work, 198
   resistance to stretch, 72
   restraining effect of, 43
   role of 61, 63
   shape of, 61
   tension, 57
   training, 200
   types of work, 102–103
Muscle atrophy, 74
Muscle contraction, 101–102
   fibre type and, 227
   isokinetic, 102
   isometric, 101
   isotonic, 101, 199
Muscle fibres, 57
   arrangement of, 54, 55
   characteristic, 60
   endurance and, 220
   isokinetic exercises and, 227
   types of, 59, 220
Muscle power,
   loss of, 76
Muscle soreness,
   delayed onset of, 200, 216, 235, 240
Muscle spasm,
   ice for, 253
Muscle spindles, 69
   function of, 71
   motor innervation, 71
   sensory innervation, 71
   structure of, 70

Muscle strength, 196–223
  age factors, 198
  cross-section and, 196
  definition of, 196
  design of programme, 204
  in early stage rehabilitation, 202
  eccentric training, 216–218
  endurance exercises, 219–221
  exercise programme, 205
  factors in, 196
  fitness and, 198
  length and, 197
  motor unit activity and, 196
  power and work, 198
  psychological factors, 198
  re-education,
    isokinetic contraction, 201
    overload, 204
  role of nervous system, 197
  speed of contraction and, 197
  static or dynamic, 198
  testing, 394–396
    hand held dynamometry, 395
    isokinetic, 396
    manual, 395
  training, 199, 291
    auto-resistance, 207
    capillary density and, 202
    examples, 215
    free weights, 207
    IRM, 208
    manual resistance, 210
    mechanical resistance, 211
    methods, 205, 206, 207–221
    mitochondrial volume and, 202
    overload principle, 199
    physiological effect of, 202
    plyometrics, 218
    pulleys, 211
    repeated contractions, 210
    safety aspects, 221
    specificity, 203, 221
    suspension therapy, 211, 213
    weights, 216, 221
  types of, 198
Muscle tension in stress, 262
Muscle tone, 72–76
  disorder, 80
  high, 73
    see also Spasticity
  loss of, 76
  low, 75–76
  normal, 73
Muscular endurance, 219
Myofibrils, 51
  proteins, 53
  structure, 53
Myosin, 53, 56
Myositis ossificans, 100

Naughton protocol, 133
Nerve fibres,
  sensory, 69

Nerve palsy from crutches, 330
Nervous system,
  dysfunction of, 67
  flexibility and, 245
  muscle strength and, 197
  plastic adaptation, 89
  structure of, 66
Neurological injury, 184
Neurophysiological aspects of movement,
  66–93
Neutral equilibrium, 24
Newtons, 126
Newton's laws, 11–12
Non-motorized treadmill test, 129

Obesity, 107
Olivospinal tract, 77, 79
One repetition maximum, 127
Open kinetic chain exercise, 103, 104
Orthoses, 241–353
  ankle-foot, 345
  coronal plane deviations, 351
  definition, 341
  foot, 342–345
  gait patterns, 351
  hip guidance, 348, 349, 350
  hip-knee-ankle-foot, 348
  knee, 347–348
  knee-ankle-foot (long leg), 346, 347
  reciprocating, 348, 350
  sagital plane deviations from, 351
  transverse plane deviations from, 351
Osteoarthritis,
  age-related, 287
Osteokinematics, 48
Outcome measures, 387
Overload principle, 199, 204
Oxford protocol, 209
Oxford scale, 225
Oxygen consumption, 151
  during walking, 304
Oxygen uptake,
  anaerobic threshold, 134
  blood lactate, 135
  blood pressure and, 137
  heart rate and, 137
  maximum rate, 130–142, 202
    analysis of results, 133
    during walking, 304
    exercise cycling and, 152
    factors affecting, 130
    genetic factor, 130
    limiting, 131
    measurement requirements, 133
    metabolic factors, 132
    nervous control, 132
    predicting from age, 154
    submaximal tests, 138
    test contraindications, 137, 138
    tests predicting, 138
    training and, 131
    walking, jogging running and, 152

Oxygen uptake (*cont.*)
sex differences, 131
Oxygen utilization, 132

Pacing, 151
Pain,
in abnormal posture, 281, 290, 292
hyperlordosis and, 285
Pain management,
relaxation techniques, 273
Paraplegia,
rehabilitation, 191
Parawalker, 349
Parkinson's disease, 67
gait in, 318
muscle tone in, 73
rigidity in, 75
signs and symptoms, 80
Passive accessory movements, 101
Passive movements,
forced, 100
performing, 99
precautions, 99
Patient,
compliance, 173
Patient handling and lifting, 368–382
axillary hold, 369, 370, 378
banned procedures, 380
in bed, 371–377
sitting up, 371
bed to chair, 376
combined lift, 374
comparison of procedures, 381
counterbalance, 371
cradle (orthodox), 380, 381
double wrist hold, 368
drag lift, 380, 381
elbow hold, 369, 370, 378
finger hold, 369
holds, 368–370
neck hold, 380, 381
rocking pivot transfers, 379
shoulder/Australian lift, 373, 376
single handed, 376
sitting to standing, 377
slings, 368
through-arm lift, 375, 376
two-sling lift, 372
use of rocking/body weight, 370
waist belt hold, 370, 371
Patient handling sling, 368
'Paying out', 102
Pelvic floor,
exercise, 167, 169
re-education, 169, 170
Pelvic girdle, 42
Pelvic tilt, 167, 279
walking with, 320
Pelvis,
in walking, 305
Pendular movement, 22
Pendulums, 21–22

Peripheral nerve injury, 76
Peripheral nervous system,
injuries to, 67
Physical Activity Readiness Questionnaire,
124
Placing reactions, 90, 91
Plane joint, 39
Plantar reflex, 37, 74
Plastic adaptation of nervous system, 89
Platform stabilometry, 178
Plyometrics, 218–219
Poron insoles, 344
Posterior column, 84
disorder of, 179
Posterior spinocerebellar tract, 84
Postoperative breathing,
relaxation and, 272
Postoperative exercise class, 166
Postural hypotension, 180
Postural pain syndromes, 281
Postural reflex, 68
Postural stabilization, 178
Postural sway, 278
measurement, 177
Postural tone in stress, 263
Posture, 275–296
see also Standing, Sitting etc
abnormalities, 282–287
causes, 285
environmental factors, 286
genetic factors, 285
occupational, 287–289
pain in, 292
physiological factors, 286
recurrence of, 293
re-education, 290
alignment, 276
awareness of, 291
changes in, 279
definition, 275
during pregnancy, 170
dynamic, 275, 276
dysfunction, 281
energy expenditure and, 287
examination of, 289
flatback, 282, 283
flat neck, 284
forward head, 284
good, 276
hyperlordotic, 282, 283, 285
idiopathic abnormalities, 286
malalignment, 285
measurement of, 289
movement and, 276
occupational disorders, 287–289
from VDU tasks, 288
upper limb, 287
poor, 281
re-education, 289
abnormal tensions, 290
Alexander technique, 293
alternative therapies, 293
Feldenkris method, 294

Posture (*cont.*)
  re-education (*cont.*)
    functional integration in, 295
    increasing range of motion, 290
    neuromuscular control, 291
  relaxed, 282, 283
  retraining neuromuscular control, 291
  static, 275, 276
  swayback, 282, 283
  working, 287
Potential energy, 27
Power, 28
  definition, 33
Pregnancy,
  see also Antenatal classes
  posture and back care advice, 170
Pressure, 27
  hydrostatic, 31
Primitive reflexes, 88, 90, 91
Progressive resistance exercise, 207
Pronation, 37
Proprioception, 83, 175
  deficit, 184
    in neurologically injured patient, 184
    testing for, 184
  re-educating, 185
    methods, 187
    wobble board, 185
Proprioceptive neuromuscular facilitation,
    188–193
  definition, 188
  for balance re-education, 183
  in flexibility, 253
  technique, 98
Proprioceptive stimuli, 189
Proteins in myofibrils, 53
Protein synthesis, 202
Pulleys, 18–21
  combination, 20, 21
  functions of, 20
  in muscle strength training, 211
  single fixed, 19
  single movable, 20
Pulmonary function,
  VO_2 max and, 132

Quadriceps,
  myositis ossificans, 100
Quadriceps femoris, 208
Quadriceps muscle,
  nervous control of, 69
  strengthening, 113

Radial deviation, 37
Receptors, 82–83
Reciprocating gait orthosis, 348, 350
Rectus abdominus,
  fibres in, 55
  strengthening, 168
Reflexes, 68
  abnormal activity, 88–89

  control of, 68
  primitive, 88, 90, 91
Rehabilitation,
  abnormal reflex activity in, 88–89
  after brain injury, 89
  back, 163–165
  balance, 181
  of cardiac patient, 112
  competition in, 158, 159
  endurance exercise in, 219
  flexibility and, 244
  for children, 110
  from stroke, 89
  isokinetic exercises in, 237
  lower limb injury, 161
    weight bearing, 161, 162
  muscle strength re-education, 202
  patient self-responsibility, 158
  programmes, 107
  proprioception in, 184
  pulmonary, 111, 112
  stroke, 99, 182
Relaxation, 261–274
  accelerated progressive, 267
  after childbirth, 170
  autogenic training, 269
  behavioural manifestations, 265
  biofeedback, 270
  clinical use of, 272
  cognitive manifestations, 264
  definition, 261
  meditative, 269
  mental, 265
  physiological manifestations, 264
  progressive, 266
  reciprocal physiological, 267
  summary of techniques, 272
  training, 265
Relaxation response, 264
Relaxation themes, 270
Relaxed passive movements, 98
Repetitive strain injury, 287
Respiratory patient,
  exercise prescription for, 111
  oxygen uptake and, 132
Resting splints, 251
Reticulospinal tract, 77, 79
Rheumatoid arthritis,
  gait in, 316, 317
  relaxation and, 272
  resting splints for, 251
  serial casts for, 250
Rhythmic stabilizations, 183, 190, 191
Rib cage, 42
Righting reflexes, 90, 91
Rigidity, 74–75, 80, 81
  associations of, 75
  causes, 75
  cog-wheel, 75
  decerebrate, 75
  lead-pipe, 75
Rockport one mile walk test, 139, 142, 144
    145, 146

Rollators, 328
Romberg test, 179, 278
  interpretation of, 180
Rooting reflex, 68, 88
Rose-Parker insole, 345
Rotator cuff injury, 191
Rubrospinal tract, 77, 79
Running,
  oxygen uptake and, 152

Saddle joints, 39, 45
Sagittal (median) plane of body, 36
  movement, 37
Sarcomere, 52,
Sarcoplasm, 51
Sargent jump-and-reach (verticle jump),
  125
Saving reactions, 176
Scoliosis, 279, 282, 283
Self-esteem, 159
Self-paced treadmill walking test, 149
Sensory ataxia, 184
Sensory fibres, 69
Serial casting (splinting), 249–252
  applications, 250
  clinical applications, 252
  drop-out, 250
  resting splints, 251
Shear modulus, 29
Shear stress, 28
Sherrington's law of reciprocal inhibition,
  268
Shoulder,
  auto-assisted exercises, 20
  discomfort in, 291
  frozen, 22
  pendular movements, 22
  stability exercise, 205
Shoulder girdle,
  exercises, 171
Shoulder joint, 17
  rehabilitation, 191
Shuttle running test, 149
Shuttle walking test, 149, 150
Sit and reach test, 253–254
Sitting, 279–280
  balance, 182
  posture problems, 288
Skeletal muscles, 50
  A-band, 51
  activation of, 55
  active tension, 57
  as agonist, 63
  as antagonist, 63
  as synergist, 63
  attachments, 64
  biomechanical action, 61
  classification, 60
  components, 51
  concentric contraction, 62
  contractility, 50
  contraction,

mechanism of, 55
  sliding filament theory, 55
  types of, 61
distensibility, 51
eccentric contraction, 62
elasticity, 51
fasiculi, 51, 54
fibres, 51, 57
  types of, 59, 60
fixators and stabilizers, 63
gross structure, 54
H-zone, 53
I-band, 53
insufficiency, 59
irritability, 50
isokinetic contraction, 62
isometric contraction, 62
isotonic contraction, 62
length/tension, 57, 58
location of, 61
M-band, 53
microscopic structure, 53
motor unit, 56
passive tension, 57
properties of, 50
range of motion, 64
role of, 62, 63
sarcomere, 51
spurt or shunt, 61
structure of, 51–55
ultrastructure, 51
Skeletal system, 34–49
  planes and axes, 34, 35, 36
Slow reversals, 191
Smoking, 107
Social contacts for elderly, 109
Soft tissue injuries, 239
  eccentric contractions following, 217
  healing, 245
Soleus muscle fibres, 60
Sorbothane insoles, 344
Spasticity, 73–74
Specific gravity, 31
Spinal cord,
  ascending tracts, 68, 83, 84
  descending tracts, 77
  injuries, 67
    balance re-education in, 182
    muscle tone in, 73
    rehabilitation, 117
Spinal discs, 164
Spinal reflex, 76
Spinal shock, 76
Spine,
  flexion exercises, 164
  passive stretching for, 249
  in standing, 279
Spino-olivary tract, 84
Spinotectal, 84
Spinoreticular tract, 84
Splints,
  resting, 251
Springs, 20, 211

Stability, 24
  factors affecting, 24
Stable equilibrium, 23
Stair climbing, 204
Stair sprinting, 126, 127
Standing, 279
  balance, 176, 177
  postural alignment in, 277
Standing broad jump (Horizontal jump),
  126
Stepping exercises, 220
Stepping reflex, 68, 88
straight partial sit-up, 168
Strain,
  definition, 33
Strength, 196–223
  see also Muscle strength
Stress, 28–30
  see also Relaxation
  definition, 33
  inverted U hypothesis, 262
  position of, 263
  response to, 261
Stress/strain diagram, 28, 29
Stretch,
  ballistic, 249
Stretching,
  exercises, 255
  methods of, 247
  passive, 248–252
    clinical application, 249
    method and precautions, 248
  patient advice, 247
  rationale for, 243
Stretch reflex, 68, 90, 91
  heightened, 74
  loss of, 76
  in rehabilitation, 203
Strokes, 67
  muscle tone in, 73
  reduction of, 106
  rehabilitation, 89, 99, 182
Supination, 37
Suspension therapy, 211, 213
  axial fixation, 213
  re-education with, 213
  safety precautions, 214
  vertical fixation, 213
Suture joint, 40
Swayback, 282, 283
Swedish knee cage, 348
Swimming as endurance exercise, 220
Symphyses, 40
Synchrondroses, 40
Syndesmosis, 41

Tabes dorsalis, 179
Tectospinal tract, 77, 79
Tension headache, 271
Thigh,
  isometric exercises, 166
Thomas's test, 397

Three-point pressure system, 6
Toes,
  in walking, 306
  weak extensors/flexors, 343
Tonic neck reflexes, 90, 91
Torque, 17–18, 231
  definition, 33
Torque curves, 18
Transverse (horizontal) plane of body, 36
Transverse plane movement, 37
Treadmills, 151
  as endurance exercise, 220
  protocols, 133
  versus cycle-ergometer, 136
Tremor, 81
Trendelenburg gait, 304, 315, 321, 398
Triceps,
  as levers, 14
Trigonometry, 9
  troponin, 55
Trunk in standing, 278
Trunk braces, 348
Twelve minute walking test, 148

Ulnar deviation, 37
Unstable equilibrium, 24
Upper limb,
  forces through, from walking aids, 336
  movement patterns, 190
  muscle strengthening exercises, 215
Upper motor neuron lesions, 253
Upright stance, 177

Vastus lateralis and medialis, 7
Vaulting in walking, 320
Vector, 32
Velocity, 32
Vertebral column,
  in standing, 279
Vestibular system, 87
  controlling movement, 85
Vestibulospinal tracts, 77, 79
Vision,
  postural sway and, 278
Visual display units, 288

Walking, 121
  see also gait etc
  ankle motion in, 302
  'brave', 88
  circumduction, 320
  definitions, 297
  energy expenditure in, 304
  hip joint in, 301
  knee in, 302
  learning, 298
  muscle activity during, 303
  oxygen uptake in, 153
  ranges of joint motion, 301
  role of friction, 26

Walking (*cont.*)
  speed of, 306
    measurement, 307
    with walking aid, 339
  truncal rotation, 301
  vaulting in, 320
  with flexed hips, 319
  with flexed knees, 319
  with frame, 329
  with pelvic tilt, 320
Walking aids, 88, 180, 325–341
  see also walking sticks, walking frames,
    crutches etc
  effects of using, 326
  forces through upper limb, 336
  gait patterns with, 337
  height of, 333
    importance of, 334
  metabolic cost of, 335
  pre-walking exercises, 336
  stability and, 25
  supply of, 325
  three-point gait, 337, 338
  types of, 326
Walking frames, 327–330
  folding, 328
  forearm supporting (Gutter, Pulpit), 329

pattern of walking with, 329
  reciprocal, 328
  rigid, 327
  selecting, 330
  wheeled, 329
Walking reflex, 90, 91
Walking sticks, 325, 326, 332
  contralateral use, 334,
    with one stick, 337
  ipsilateral use, 334
    two-point gait, 337
  tripods and tetrapods, 332
  two-point gait, 340
Watkins protocol, 209
Westminster pulley, 211, 212, 216
Wheelchair athletes, 117
Wingate test, 128
Wobble board, 185
Work, 28, 33
Wrist,
  fracture, 161, 165
  resting splint, 252

Yates splint, 345, 346
Young's modulus, 29

THE NEW LIFE LIBRARY

PALM READING

PALM READING

A PRACTICAL GUIDE TO CHARACTER ANALYSIS
AND DIVINATION

STACI MENDOZA AND DAVID BOURNE

HERMES HOUSE

This Paperback edition published by Hermes House

Hermes House is an imprint of
Anness Publishing Limited
Hermes House
88–89 Blackfriars Road
London SE1 8HA

Publisher: Joanna Lorenz
Project Editor: Debra Mayhew
Designer: Nigel Partridge
Illustrator: Anna Koska
Photography: John Freeman

Printed and bound in China

© Anness Publishing Limited 2000
Updated © 2002

1 3 5 7 9 10 8 6 4 2

Publisher's note:
The reader should not regard the recommendations, ideas and techniques
expressed and described in this book as substitutes for the advice of a qualified medical
practitioner or other qualified professional. Any use to which the recommendations, ideas and
techniques are put is at the reader's sole discretion and risk.

Contents

Introduction	6
A History of Palmistry	8
General Aspects of the Hand	10
Elemental Hands	16
The Major Lines of the Hand	18
Maps of Time	32
Lines and Signs of Special Interest	34
The Percussive or Palm Edge	37
The Mounts of the Hand	38
The Fingers and Thumb	54
From Theory to Practice	60
Further Information	63
Index	64

INTRODUCTION

The "hands-on" approach used in palmistry makes it one of the kindest and friendliest methods of divination – used in revealing information about a person's future and characteristics. All the time you are looking into the palm, to interpret the lines and markings found there, you are touching and holding someone's hand. This intimate and caring gesture can have a profound effect in lifting any impersonal barriers to communication: it allows the person to feel comfortable and cared-for as an individual enabling them to open up and bring out whatever is really on their mind at the time.

◄ The palmistry hand is an ancient teaching tool.

Before you begin to read someone's palm, it is important to look at both the hands to judge the various changes between childhood and adult life. Take note of whether the person is right- or left-handed, in order to establish the "major" and "minor" hands. The palm-reader reads the major hand (the one used to write with) to find what an individual has made of their life up to

▼ *The hands are used as a gesture of elation.*

▼ *Hands are used to express protection and tenderness.*

6

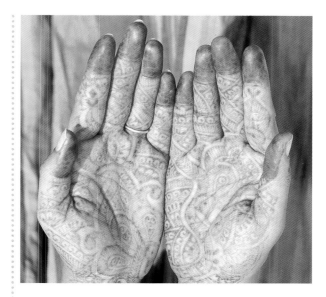

▲ *The hennaed Hand of Fatima is an historical Middle Eastern symbol of creation on the hands of women.*

the present time, and what is in store for them in the future. The minor hand reveals the past history and family background and shows what talents or assets are inborn.

Before you begin your reading, choose a peaceful setting that is comfortable both for you and the other person. Try to clear your mind, and then proceed to look clearly at the hands, always listening to your sixth sense or intuition when relaying the information you can uncover. You will also find it helpful to have a notebook (to jot down your observations), a magnifying

glass (to see the palm's markings more clearly) a ruler, and a pair of compasses. Put them within easy reach so that you do not have to interrupt your reading once you have begun.

This book is an introduction to the concept of palmistry, illustrating in detail what to look for in the hand and providing guidance in interpreting what you find. The final section takes you through a complete reading with practical guidelines to help you order your thoughts and communicate them effectively.

Once you have discovered just how much information is on the palm, you are bound to look at hands in a different light. Palmistry is a wonderful way to discover more about yourself as well as your friends and family.

▼ *A handshake or simple touch can mean many things.*

A HISTORY OF PALMISTRY

Palmistry is one of the oldest and most universally practised of all the forms of divination. Its ancient roots lie in the East, where records exist in China from as early as 3200 BC showing that palmistry was regularly practised as a means of divination. In India, Hindus also practised palmistry from a very early date, and developed it as a system called *hastarika* (the "study of the hand with its forms and lines"). Like its counterpart from China, it is still in prominent use today.

From these two ancient sources, palmistry migrated via the trade routes through Persia and the Middle East into the West. The ancient Greeks have left us records of their practice of

▼ The Chiromancer *by Piero della Vecchia is a 17th-century study of the courtiers' fascination for palm reading.*

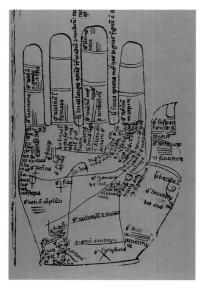

▲ *Early maps of the palm show us historical approaches to palm reading.*

chiromancy (a word which comes from *kheir*, "hand" and *manteia*, "divination"). Aristotle, Pythagoras and Anaxagoras all expounded the benefits of the study of this practice.

Perhaps the most famous exponents of palmistry are the nomadic tribes who roamed across Europe for centuries practising their popular trade.

During the 15th and 16th centuries printed books began to be widely circulated. *Die Kunst Ciromantia* by Johann Hartlieb and *Chyromantiae* by Barthelemy Cocles laid the foundations, and palmistry became a documented practice all over Europe.

After this golden age, interest in palmistry waned until the middle of the 19th century: a period that saw a great revival of interest in all things esoteric. It was restored to public attention by notable figures such as d'Arpentigny, Desbarroles and "Cheiro" (Count Louis von Hamon), who set the salons of London alight with his charismatic consultations given in his Indian room.

It was Cheiro who brought together the various philosophies into a coherent whole. Adding his own ideas, he enabled the practice of palmistry to grow. Since then, it has developed to fit the constraints of the modern world where palmists still practise what is perhaps the most truly human-orientated form of divination.

▶ The Fortune Teller *by Jean Antoine Watteau portrays the spread of palmistry among European gentry.*

GENERAL ASPECTS OF THE HAND

Before studying the lines and markings of the palm itself, some general indications of personality can be drawn from some aspects of the whole hand. This information should be collected as a whole to give you a general picture of the person whose palm you are interpreting. More advanced palmists will also take into consideration the undertones of the skin and the texture of the hand itself.

THE SHAPE OF THE HAND

Ascertain the overall shape of the whole hand by simply holding the hand up with the palm facing you and using an imaginary outline to gauge its shape. The more this process is practised the easier it is for the eye to recognize the shapes. There are four main shapes:

◀ *Conical hand.*

◀ *Pointed hand.*

THE POINTED OR "PSYCHIC" HAND

Narrow hand; middle finger peaks higher than others

Individuals with this shape hand tend to have keen intuitive faculties and a sixth sense, hence its name. Usually very good-looking, they strive for perfection around them and within themselves.

THE CONICAL OR "ARTISTIC" HAND

Gently rounded shape

This shape is so called because people with this type of hand are extremely visual, and have a tendency to artistic and visually-based pursuits; they are sensual by nature. They want to see all the beauty in life, and they see life as something to be enjoyed.

▶ *Square hand.*

THE SQUARE OR "USEFUL" HAND

Square shape

This shape belongs to individuals who need to be needed. They have a logical pattern to their thinking, and usually have a good mechanical sense. They are often very busy physically.

THE SPATULATE OR "NECESSARY" HAND

The palm widens out from a narrow base

These people get things done. They will do whatever is necessary to succeed, and are persistent and bright enough to carry it off. These individuals hate to waste time.

◀ *Spatulate hand.*

THE PROPORTIONS OF THE HAND

Differences in length between the palm and the fingers can usually be seen with the naked eye but if necessary simply measure the difference using a ruler. By holding the ruler lengthwise alongside the whole hand you can assess at a glance the proportions of fingers to palm.

PALM IS LONGER THAN FINGERS

This indicates people who have difficulty in saying no to their whims; people of ideas and dreams who can conjure up great schemes but need to watch out for the "Oh, I'll put it off until tomorrow" syndrome. They are creative, and may be artists or musicians.

◀ *Palm is longer than fingers.*

PALM IS SAME LENGTH AS FINGERS

Very balanced individuals have this balanced hand. They find it relatively easy to cope with the highs and lows in life and are usually stable in character, both mentally and physically. They are determined individuals who have the ability to see things out to the end and have a logical approach to life. They are very fortunate in that they suffer few health problems, mental or physical.

◀ *Palm is same length as fingers.*

▼ *A creative aptitude and dextrous ability is needed to play musical instruments. A person with these abilities is likely to have a palm longer than their fingers.*

▲ *Some talented people use their hands to make beautiful works of art.*

PALM IS SHORTER THAN FINGERS

These individuals will always use their gut intuition to guide them through life. They are very imaginative, spiritual and sensitive personalities. They have a delicate constitution and may suffer health problems.

▶ *Palm is shorter than fingers.*

11

FINGER SHAPES

When assessing the shape of the fingers, examine them with the palm facing you looking only at the overall shape of the fingertips while ignoring the shape of the fingernails. Many people will have a mixture of finger shapes so look for the shape which occurs most frequently. Also take into account the settings, spacings and patterns upon the fingers and thumb, all of which are dealt with in a later chapter.

POINTED FINGERTIPS

Finicky, precise personalities, these people have a good eye for colours, shapes and designs. They are refined, with a highly developed aesthetic sense and good taste, evident in their dress and homes.

CONICAL FINGERTIPS

These individuals carry certain instinctive beliefs about themselves and possess a great inner knowledge of other people's circumstances and concerns. They are generally wise souls with a gentle nature, always willing to lend a hand and help out. They are usually very attractive.

SQUARE FINGERTIPS

These people prefer to lead a simple life, with simple pleasures to keep them happy. They are excellent workers, who are always able to make money easily, and so do well in the field of business. They are always fair in their approach and in their dealings with others.

▲ *The homes of people with pointed fingertips are often stylish and tasteful, reflecting their good taste.*

SPATULATE FINGERTIPS

Highly intelligent and witty, these people have a dry sense of humour and are mentally versatile. They enjoy travelling. They are very active and will generally go for a career in which they will work non-stop around the clock. They are adaptable, capable of handling most situations and other people.

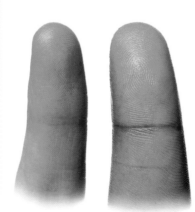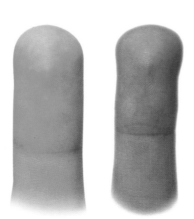

◀ *Most people will have more than one shape of fingertip on their hand. Base your reading on the shape which occurs most frequently. Finger shapes (left to right): pointed, conical, square and spatulate.*

HAND THICKNESS

Tilt the hand sideways to gain some idea of its depth and suppleness. The suppleness is difficult to ascertain at first, but with practice you will get used to the feeling of different types of hands and will be able to gauge this with more confidence.

VERY THICK AND VERY HARD

These individuals tend to behave in a very rough and tough manner, following their own basic needs. Their thought processes tend to be crude.

THICK AND HARD

These people have very basic needs: food, shelter and love. Free from ambition, they have no desire to keep up with the rat race.

▲ *Very thick hand.*

▼ *Thick hand.*

THICK AND MEDIUM HARD

These people are good workers, always reliable and trustworthy. They will tend to work hard throughout their lives. Life does not come too easily for them, but they usually enjoy it.

THICK AND MEDIUM SOFT

These people work hard and play hard; they really want to enjoy themselves with other people. They need to be needed and like to be useful and relied on.

THICK AND SOFT

These people are artists, poets or musicians, but are not usually very good workers. They tend to dream and ponder on life, rather than getting their hands dirty.

THIN AND VERY HARD

These individuals know exactly what they want. They possess strategic skills and can be quite calculating. They may be accused of being cold, but in fact it just takes time to get to know them.

THIN AND HARD

People with this kind of hand tend to be selfish by

▲ *Thin hand.*

▼ *Very thin hand.*

nature and self-opinionated. They are possessive and stubborn and do not make friends easily, although the alliances they do make are usually for life.

THIN AND SOFT

These people love to have a good time, and are always the last to leave a party. They do not have much willpower and find it difficult to say "no" to people, so they are susceptible to physical temptations such as affairs or one night stands.

THIN AND VERY SOFT

These individuals have a keen intuition, but also have a tendency to focus on the negative, which may lead them to react harshly or snap at other people. They can even be prone to problems such as depression, morbidity and paranoia.

THE SIZE OF THE HAND

Small, average or large hands may appear a strange distinction at first. When ascertaining the size of the hand, however, you should consider it in relation to the person's size and build. Ask yourself if it is in proportion. For example, a small person who has small hands would be considered to have hands of average size. The hands of a tall and broadly built man might be larger, but if they look dainty compared to the rest of his body you should consider them as small.

▼ *People with large hands are often surprisingly dextrous.*

VERY SMALL HANDS

Individuals with proportionally very small hands tend to be free thinkers. They often have a strong sense of moral politics and will stick firmly to their beliefs. Consequently, they like to fight for the underdog against dishonesty and injustice. This passion, however, may lead to a tendency not to listen to the other side of a story, so their support can be misguided. If a man has exceptionally small hands in relation to his size, it can indicate a cruel side to his nature.

▼ *Very small hands are not only petite but often delicate to the touch.*

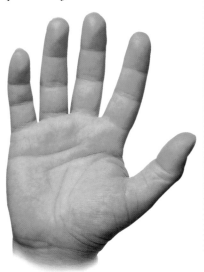

SMALL HANDS

Individuals who have proportionally small hands are ideas people. They often come up with bright and broad-reaching ideas, but need others to help carry them through. They make very good committee members and fundraisers because they possess an ability to gather the support and enthusiasm of others. They are usually very dear and sweet in nature, and would not hurt a fly.

▼ *Small hands are petite but have larger fingers than the very small hand.*

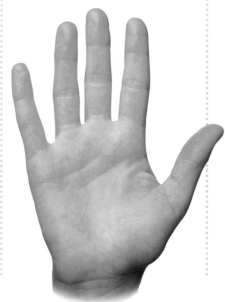

AVERAGE HANDS

People with average size hands are down-to-earth individuals who usually possess good common sense and moderate views on life. Such people have balanced, healthy attitudes and are good-natured when dealing with other people. Any mental or physical problems are usually easily overcome.

▼ *The average hand may not catch your eye but it is a good, solid shape with pleasing proportions.*

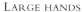

LARGE HANDS

People with large hands have a, perhaps surprising, aptitude for doing fiddly things with great patience, and may use this talent to earn their living. They have excellent analytical talents and are mentally strong and good-natured. They can usually be found figuring out detailed projects or pursuing hobbies.

▼ *Large hands are wider than average hands.*

VERY LARGE HANDS

These individuals are very bright mentally. They love trivia and mental exercises which sharpen their minds. They can be the mavericks and trendsetters and are unlikely to accept the status quo. They often possess a great strategic ability. They like to be constantly in charge of life, and object to being told what to do.

▼ *Very large hands are unusual.*

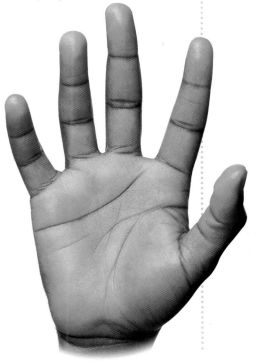

ELEMENTAL HANDS

There are four basic types of the elemental hand: water, air, fire and earth. Assessing this aspect gives an overview of a person's character based on astrological principles. This then complements the detailed interpretation of the palm and fingers. The elemental hand often corresponds with the individual's astrological sign.

▼ *The Water Hand indicates an artistic, sensitive character.*

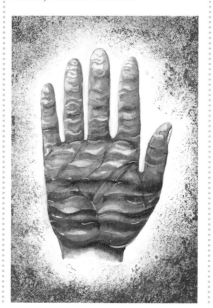

THE WATER HAND

A delicate structure with long fingers and a long palm. Features a fine mesh of linear markings

Often, though not always, the water hand belongs to those people born under one of the three water signs: Cancer, Pisces and Scorpio. They possess emotional natures, with very sensitive and sensual personalities. They are often artistic, enigmatic, esoteric and intuitive. The water hand usually accompanies an attractive face with large, intense eyes and soft lips. These people have a love of music, art and culture, and seek relationships: they need to belong to someone or something, in order to feel content with their lives.

THE AIR HAND

A robust hand structure with long fingers and a fleshy, though square, palm. Well-defined linear markings

The air hand frequently belongs to those born under one of the three air signs: Libra, Gemini and Aquarius. They are intellectual in their pursuits and possess well-

▲ *The Air Hand indicates a strong-minded and well-balanced character.*

balanced minds. They are literate and not overly sensitive to visual stimuli. They need facts and figures to help guide their decision making, and they look for mental challenges. They are strong-minded individuals with a powerful sense of self. For their relationships to work and not become bogged down or boring, these people need to retain their independence and personal freedom. They need a strong sense of self to be content in life.

THE FIRE HAND
A lively hand structure with short fingers and a long palm. The hand features lively linear markings.
Often, though not always, the fire hand belongs to those people

▼ *Individuals with a Fire Hand tend to be fiery characters who like responsibility as long as it involves action and excitement.*

born under one of the three fire signs: Aries, Sagittarius and Leo. These individuals have strong instincts, and act on gut feelings. Often they are assertive and quick to rise, but also quick to cool down.

They are usually very active individuals who prefer action to excuses; they need to be kept busy, otherwise they are easily bored. These people can handle the stresses of a battle but cannot always cope with the more mundane responsibilities of life. They like to be life's leaders and do not enjoy following on behind others.

THE EARTH HAND
A heavy, thick hand structure with short fingers and a coarse, square palm. The lines on the palm are few but are deeply incised
The earth hand frequently belongs to people born under one of the three earth signs: Capricorn, Taurus and Virgo. They are well-balanced, practical and logical by nature and so make good problem solvers. They have an inherent wisdom. They are sure-footed in their dealings with people and have a useful

▲ *Individuals with an Earth Hand are down-to-earth, reliable characters with a useful ability to detect fakes and swindlers.*

ability to detect fakes or liars. Once committed to a personal relationship, they are normally very devoted.

Earth types need a purpose in life and usually feel a need to be relied upon in order to be content. They are not afraid of, nor do they mind, good, honest hard work in order to reach their goals.

THE MAJOR LINES OF THE HAND

There are six major lines of the hand. At least three will be found on every hand, and some people will have all six. Remember that, whether you find only three or all six, the hand must be interpreted as a whole. So, while you are considering these major lines, you will still be considering the shape and size of the hand, the mounts, other lines, and the fingers. The major lines are normally assessed by the palm-reader in their order of importance, so you should follow the order given below.

You may find it difficult to decide whether a particular line is actually missing or is merely very faint. Try to ask a professional reader for advice if this is likely to cause anxiety.

LIFE LINE OR "VITAL LINE"
As the most vital line in the hand, the life line is usually read first by the palm-reader. It deals with the length of life, the strength and generalities of life and family ties. It is never absent.

HEAD LINE OR "CEREBRAL LINE"
This line deals with the mind, indicating weak or strong mentalities, possible career directions, and intuitive and creative faculties. Serious mental illness may be indicated by its absence, although this condition is very rare.

HEART LINE OR "MESAL LINE"
This line deals with love and the emotions. It indicates degrees of contentment and happiness in life, and shows the kinds of relationships people have with others. The longer the line and the more it reaches towards the Jupiter finger, the longer a

◀ The major lines of the hand are the basis for the palm-reader's first impression. The palmist will then assess the finer lines and markings to fine tune their reading.

18

relationship is likely to last. Its absence is rare, but it can be a grave omen. If you detect this line be tactful with your subject and ask a professional reader for advice.

FATE LINE OR "LINE OF LUCK"

This line deals with career, work and ambition. It is concerned with the directional force of life,

▶ *These three palms illustrate the differences in the length and position*

social standing and the public aspects of people's lives. It frequently takes the place of the line of the Sun.

LINE OF THE SUN OR APOLLO, ALSO KNOWN AS THE "LINE OF FORTUNE AND BRILLIANCY"

This line deals with luck, talent, and money. It augurs success and, possibly, fame. Any visible Sun

of the major lines on the hand. Look very carefully: are you able to

line is good luck. The longer it is, the greater the luck. It often takes the place of the fate line.

LINE OF MERCURY. ALSO CALLED THE "HEALTH LINE" OR LIVER LINE

This line deals with health issues which may be hereditary. It is often absent, but this is auspicious as it indicates good health.

recognize the lines on each of the three hands illustrated?

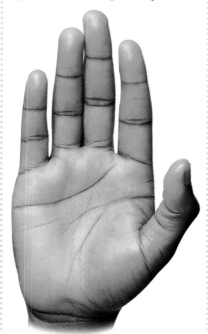
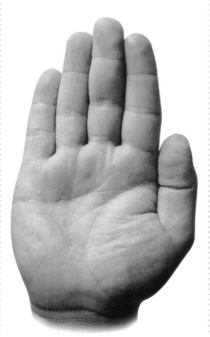
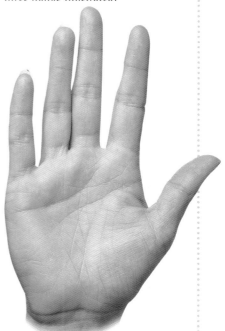

THE LIFE LINE

The life line is the measure of vitality and life force. It deals with the length and strength of life, family ties and the

▼ *The main life line may be supplemented by the Mars line, worry lines and loyalty lines.*

generalities of life. It starts above the thumb and is then read downwards towards the wrist, where it ends.

You may notice one or more thin horizontal lines cutting directly across the life line. These indicate slight obstacles at a given time period.

1 LIFE LINE CLOSE TO THE THUMB

This indicates someone who has a close relationship with their family and is content with family life. They are happy spending time at home, with no great urge to travel around the world, and are not ambitious. It can also mean a heavy family commitment. These people are sensitive and cautious.

2 LIFE LINE LIES TOWARDS THE MIDDLE OF THE HAND

If the life line skirts the thumb in a wide arc and rests more in the middle of the hand, it indicates a person who wants to achieve great things and break new ground. They will have a keen sense of adventure and want to travel.

3 LINE ENDS TOWARDS THUMB SIDE OF WRIST

When the life line veers around the thumb to end at the side of the wrist, this indicates an individual who yearns for home, and wants to end their years on home ground.

4 LINE ENDS TOWARDS OPPOSITE SIDE OF WRIST

When the end of the life line veers away from the thumb towards the opposite side of the hand, this indicates a person who will emigrate, or move away from their family, culture or country.

5 LINE STARTS AT BASE OF JUPITER

When the life line starts at the base of the mount of Jupiter, below the index finger, the individual seeks a change of lifestyle for the better. They are strong-willed and very ambitious, willing to conquer obstacles to achieve their goals.

6 LINE STARTS AT SIDE OF HAND NEAR JUPITER

This is similar to the previous position, and indicates people who are ambitious and will

▲ *The position of the life line indicates those with an adventurous spirit and those happy to remain close to home.*

single-mindedly achieve success. They are proud characters who make good leaders.

7 LINE CUTS CLOSE TO THUMB

When the line cuts very close to the thumb, this indicates someone who is living a restricted life: they may be under a strong religious or cultural influence at home. It may also mean that they have been imprisoned at some point, or have lived in the same town all their life.

8 BREAKS IN THE LIFE LINE

Breaks signify starts and stops in life resulting from big changes such as marriage, divorce, or the death of a close relative.

9 DOUBLE LIFE LINE

A double life line signifies a dual existence, and can mean any of three things: the person may be one of a pair of twins; they may have a guardian angel watching over them; or they may lead a double life, such as a mother who cares for four children during the day and works in a club at night.

10 EFFORT LINE

When the life line veers upwards towards the mount of Saturn, a person is putting great effort into their life, working hard and not taking no for an answer. But this does not necessarily mean success.

11 SUCCESS LINE

When the life line veers upward towards the mount of the Sun, below the third finger, this indicates great success and good financial fortune. It may also be a sign of fame.

12 MARS LINE

This line, running inside the life line, indicates that the person has a guardian angel or protective spirit looking after them here on earth; this is usually a very close relative or friend who has died.

13 WORRY LINES

Horizontal lines creased in the pad of the thumb indicate stresses and worries. The deeper the lines the more serious the worries. If the lines are many and faint, the individual is prone to anxiety and is nervous by nature.

14 LOYALTY LINES

Vertical lines creased in the pad of the thumb indicate loyalty to family and friends.

▼ *Family-oriented individuals are likely to have a life line situated close to their thumb.*

THE HEAD LINE

This line deals with the mind, indicating weak or strong mentalities, possible career directions, and intuitive and creative faculties. When reading

▼ *The head line deals with an individual's strength of character.*

the head line, you should look first to see whether the line is thin (faint) or thick (wide and deep). A thin line indicates a person who is highly strung emotionally, and can at times be volatile or unstable. A thick line indicates someone who is methodical. They seem to lack enthusiasm and drive at times; they are solid and sound but stubborn in nature and cannot be swayed by the opinions of others.

1 LINE STARTS HIGH NEAR JUPITER

A head line that starts on the mount of Jupiter, under the index finger, indicates an ambitious individual who is a self-starter and very focused. They can be very competitive and determined in their efforts to achieve their goals in life.

2 HEAD AND LIFE LINES TIED

This usually indicates an individual beset by doubt and confusion which leads to a lack of independence. They are very involved with the family. They have little self-confidence and

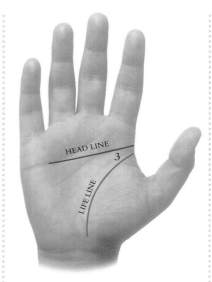

▲ *The head line may be positioned separate from the life line.*

doubt their own abilities. This is often as a result of their upbringing or early environment.

3 HEAD LINE SEPARATE FROM LIFE LINE

When the head line is clearly segregated from the life line, it indicates an individual who possesses great independence and self-will. This is a free thinker who does not follow others but will make their own road in life.

▲ *An eminent scientist like Albert Einstein is likely to have a double head line. Not surprisingly, it is a rare mark.*

4 LINE CUTS STRAIGHT ACROSS PALM

When the head line touches both sides of the palm, this indicates a self-centred individual who has the ability to stay very focused on their needs and goals. They must have secure material foundations, such as house ownership, insurance policies, and a healthy bank balance. The good news is that they are good citizens and will often give to charity.

5 LINE DIPS DOWNWARD IN PALM, MAKING AN ARCH

This indicates a very sensitive sensitive individual with a highly developed intuition. They are imaginative and perhaps rather eccentric. This may lead them to artistic pursuits or into a profession that involves caring for others.

6 THE "WRITER'S FORK"

A head line that splits in three directions at the end (like a chicken's foot) indicates an intuitive nature, similar to the previous example.

People with this characteristic are inclined to communicate cerebral ideas, thoughts and information to others by using their hands. They are likely to be writers, journalists or computer programmers.

7 THE "COMMUNICATOR'S DIVIDE"

When the head line divides into a two-pronged ending, it indicates a person who has great communication skills, and is an able public speaker. They may work in radio or television, as a professional after-dinner speaker or be an actor or singer.

8 THE DOUBLE-HEADED LINE

Two head lines lying close together indicate an individual with outstanding mental abilities. The double-headed line is so rare that when it does exist the person is likely to be in a very studious and analytically-driven career, such as an eminent mathematician, scientist or other scholar.

▼ *If the head line divides at the end into two prongs it indicates an individual with a good public presence.*

THE HEART LINE

This line deals with all matters relating to the heart: love, romantic interests, and relationships with family and friends. It also indicates the

▼ *The heart line is an indicator of the nature of your emotional life.*

nature of those relationships: whether they are likely to be steady or stormy, short term or long term. In general, the heart line deals with a person's emotional and romantic life.

1 LINE LOW IN THE HAND
The heart line starts below the underside of the knuckles in the palm. These individuals are lovers; feminine and romantic, they believe in love and are looking for "fairy-tale" perfection. In matters of the heart they can become perfectionists and expect too much from their partners. This can lead to them feeling let down.

2 LINE HIGH IN THE HAND
The heart line starts on or above the underside of the knuckles in the palm, and appears to rest near the base of the fingers. These individuals are extremely sensitive to what others think of them and can be quite destructive with their self-criticism. They tend to be very reserved with their emotions.

▲ *A fairy-tale view of romance is the individual's driving force in life when their heart line drops low in the hand. They may be disappointed.*

3 END OF LINE VEERS TOWARDS MIDDLE OR INDEX FINGERS
This indicates a dominant and demanding person, who expresses their emotions bluntly. Someone whose line veers towards Saturn will have a very contented family life. They keep family members close to them both physically and in their heart.

Veering towards Jupiter, the heart line indicates a person who is very successful in love. They will end their years in love, and being loved.

4 SHORT, HIGH LINE
This person's loyalties and morals are expendable: they believe that sex is the answer. If they are feeling unloved or neglected, or if someone pays a lot of attention to them, they can give themselves over too easily.

5 SHORT, LOW LINE
This individual cannot be faithful; they think that sex is a game or a sport. They

▼ *The heart line can indicate who is likely to love and be loved into old age.*

are very self-indulgent and assume that if they do not talk about their activities then no harm will be done. So they are quite self-deceptive too.

6 HEART LINE DROPS DOWN TO TOUCH HEAD AND LIFE LINES
This person wants the best of both worlds: to be happy both at home and at work. They need love but also have a strong sense of independence. At times they can feel divided between family and career. More often they will find a way to juggle both and find a happy medium in their own way.

▲ *If the heart line touches both the head and life lines, the individual will be good at juggling the often conflicting demands of their family and career.*

7 LINE RUNS STRAIGHT ACROSS THE PALM
A heart line which runs in an almost straight line right across the palm of the hand indicates a humanitarian with a great sense of purpose in life. This type of person will put a great deal of time and effort into working hard for the good of the community. He or she will experience great luck in life due to their selfless nature.

THE FATE LINE

This line deals with career, work, ambition, and the direction of a person's life, beginning from their childhood. It also indicates the

▼ *Career, work and degree of ambition can all be deduced by the position of the fate line.*

person's faith in their own abilities. It is read from its starting point at the bottom of the palm, near the wrist, and leads up the palm towards the base of the fingers. It ends in varying positions between the index and middle fingers.

1 FATE LINE RUNS VERTICALLY UP THE MIDDLE OF THE PALM

This indicates an individual who has a keen sense of direction and purpose in their life and career. They have known what they want in life from an early age and are very likely to achieve their aims. This line could also indicate that they will pass down their trade or business to their family members.

2 BREAKS IN THE FATE LINE

When the fate line is full of breaks, starts and stops, it indicates an individual who has had many changes in their life and career. Since they have never really been able to become fully involved in one project or in a single career, they may well only be able to achieve mediocre success.

▲ *Some individuals have a keen sense of purpose in their life. This is apparent when the fate line is deep, unbroken up the middle of the palm.*

3 FATE LINE BEGINS BY VEERING AWAY FROM, OR TOUCHING, THE LIFE LINE

Family commitments were important early in the life of this individual. Before they reach middle-age they will have great family responsibilities or will be heavily involved in a family business. They tend not to want to travel and will tend to stay close to home in later life.

4 LINE STARTS ON LUNAR MOUNT, ON EDGE OF PALM
When the fate line starts from the side of the palm opposite the thumb, it normally indicates an independet individual who will break away from the family

▼ *If a strong fate line veers upwards towards the Jupiter finger the individual's leadership qualities and drive for success will be to the fore.*

traditions in their life and career. It may also indicate someone who is likely to move overseas to work.

5 LINE OF MILIEU
If a separate line runs vertically alongside the fate line for a short distance, it indicates outside pressures or responsibilities for the person concerned. These may be slowing down the individual's way forward, so that they are thwarted in their aims and ambitions. Fortunately this normally lasts only for that period of time.

6 INFLUENCE LINES
If short lines shoot off in a feathering motion from the fate line and veer upwards, they indicate positive influences from other people, whether business associates or family and friends. When the feathery lines veer downwards, they indicate negative influences from other people.

7 LINE ENDS VEERING UPWARDS INTO JUPITER FINGER
This indicates a great ambition and the individual will be successful. They possess great

▲ *Too much pressure could be adversely affecting an individual's work, if a line of milieu runs along their fate line.*

determination and keen leadership qualities. The will to win is second nature.

DOUBLE LINE
A double fate line means one individual with two careers. For example, this could be a person may be running two companies or have one job in the daytime and a different one in the evening. Either way, this dynamic person will have great reserves of energy.

THE SUN LINE

Also known as the line of Apollo, this line deals with luck, success in life, talents, and money. The Sun line is seldom seen below the

▼ *The Sun line is a fortunate line to have as it indicates almost certain good fortune and a charmed life.*

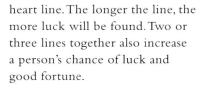

heart line. The longer the line, the more luck will be found. Two or three lines together also increase a person's chance of luck and good fortune.

1 SHORT LINE

When the Sun line is short (anything less than 1cm/½in is considered short), it indicates an individual who has so far been unable to achieve their goals and dreams in life, never quite getting that big break. As any Sun line is a good omen, this individual stands a good chance of finding fortune later in life.

2 CRESCENT-SHAPED LINE VEERS TOWARDS THUMB

This individual works very hard to achieve their goals. No one else does the work for them, and, as a result of this self-reliance, they are quite capable of holding on to their achievements.

3 CRESCENT-SHAPED LINE VEERS AWAY FROM THUMB

This line is a good omen if the individual works in the public eye

▲ *Good fortune will come naturally to the lucky individuals who have a Sun line on their palm.*

as it indicates public prestige, and possibly eventual fame, in recognition of their talents.

4 LONG LINE RUNS VERTICALLY

A long Sun line (anything over 2.5cm/1in is considered long in the hand) indicates a gilded life, with good luck falling into one's lap. Life will come very easily and successfully for this individual, and happiness will follow them all their days.

THE MERCURY LINE

This line is also known as the health or liver line as it deals with health issues, including those that

▼ The Mercury line is also referred to as the health or liver line. It can indicate good or poor health or an intuitive nature.

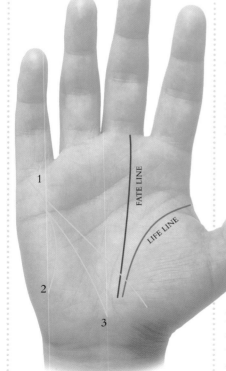

are hereditary. It can be a useful prompt to the individual to look after his or herself. If no Mercury line is present do not worry, it is a good sign as it indicates a disposition to very good health. The Mercury line also indicates those who have a keen intuition.

1 MERCURY LINE CUTS ACROSS LIFE LINE

This indicates a weakened constitution. It is a clear sign of hereditary illness in the family, such as diabetes, heart disease or arthritis. This individual needs to pay special attention to their health in order to combat the likelihood of this kind of illness. (It must be stressed that this line does not show the possibility of a fatal illness.)

2 INTUITION LINE

A Mercury line which runs in a reversed crescent shape indicates a strong sixth sense. It is always present in the hand of the highly intuitive, such as clairvoyant individuals who use their natural insight to guide them through life rather like a compass. These are people who

▲ A long and happy life is predicted when the Mercury line does not touch the life line.

have a deep interest in all things esoteric. They are peace-loving and do not like loud noises, big changes or chaos.

3 MERCURY LINE DOES NOT TOUCH LIFE LINE

This is an extremely fortunate line to have in the palm. It indicates very good health and longevity, together with success in business ventures, and good financial fortune.

MARKS ON THE THREE PRINCIPAL LINES

As you read along the major lines of the palm – the life, head and heart lines – you are likely to come across various markings

▼ *Life, head and heart lines*

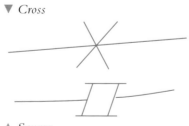

created by the many small lines that cross or abut them. These may include distinctive shapes, such as stars and squares, which will help you in your reading. It must be said that none of these markings imply anything grave or fatal, they simply indicate stresses or irritations, indeed some are signs of protection or good fortune. Similar markings appear on the mounts of the palm, and their meanings are explained in the later section on the mounts.

BARS AND DOTS

These signify interruptions or hindrances that are preventing the individual from moving forward in a given area. There will be hard work involved in recovering momentum, and willpower must be kept up throughout this period of interruptions.

CROSS

This is a sign of a more significant or longer lasting problem such as divorce or the loss of a job or home. On the life line, one cross may indicate a

▲ *Bars (above) and dots (below).*

non-fatal accident in early life or childhood. Two crosses indicate an individual who is sensuous in nature and willing to learn from others. Lots of crosses at the end of the life line can indicate poverty or ill health in old age.

SQUARE

This is a wonderful marking to have; it represents protection and good health, and indicates "getting away with it", or being saved at the last minute. A square containing a cross is a sign of preservation: there will be danger but it will not be harmful. Any square on the life line is good as it indicates safety from danger.

▼ *Cross*

▲ *Square*

▲ *Chain*

CHAIN
This marking indicates confusion, and come on to the lines when someone is trying to do too much at once and spreading their energy too thinly.

ISLAND
An island is a sign that the person's energy temporarily diverges in two directions. The mark shows that the individual has bitten off more than they can chew, but also shows they have the ability to pull it all back together. On the life line, islands indicate serious but treatable illnesses.

▼ *The islands on the palm offer a temporary respite from a situation.*

▲ *A tassel on one of the lines indicates that our reserves of strength have been dissipated.*

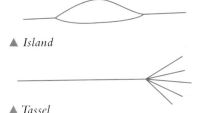

▲ *Island*

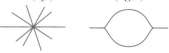

▲ *Tassel*

TASSEL
This mark appears at the end of a line and indicates a scattering of the power of the line. On the life line it shows weakened health or life force at the end of life. On the heart line it indicates weakened relationships or an absence of relationships. On the head line it shows that the person is weakened mentally, and possibly even mentally confused.

▲ *Fork*

FORK
A fork in a major line shows increased possibilities of success in life, love or career.

STAR
On the life line, this can indicate the gain (birth) or loss (death) of a relative. For each gain there is a loss and vice versa.

CIRCLES
On the life line, a circle could indicate problems with the eyes.

▼ *Star (left) and Circle (right)*

▼ *The celestial stars are mirrored in the palm.*

MAPS OF TIME

Palms can be divided into time maps that are used to give the palm-reader a clearer idea about when events and situations will take place. They are drawn by dividing the life, fate or head lines on the palm into short sections that roughly correspond to periods of years to give a time frame for events in a person's life. The periods are often gauged by marking the increments of time directly on to the hand with a pen guided only by the naked eye. If you do not have much experience of reading these maps, it is probably safer to mark the increments directly on the hand using a pair of compasses.

OLDER MAP OF TIME

The older map is used to interpret the ages in the life and head lines. The line is broken down

◄ The older map of time gives the most detailed readings. It can show when important life-changing events are likely to take place.

◄ The newer map of time is a less accurate, but quick and easy, guide to the timing of life events.

into sections which represent increments of ten years. Starting from a middle point at the heart of the thumb pad, divide the life line as shown in the diagram. Begin the reading at the start of the life line, just above the thumb, and move downwards.

▲ *The hand can give a time frame for events in a person's life.*

NEWER MAP OF TIME

This is used to interpret the ages on the life line only. It is a more general, and therefore less precise, way of measuring time on the palm and is often used as a quick reference guide to the timing of events in a life. The life line is broken down roughly into increments of six years. Although more generalized than the older map, this works on the same basis. By reading from the start of the life line, the palm-reader gets a quick idea of the timing of events in the individual's life. If necessary, they can then use the older map of time to look at certain points in depth, giving a more accurate picture of when events are likely to take place.

AGEING MAP OF TIME

This guide to ageing is extremely useful in judging the timing of events and occurrences indicated on the fate line, such as changes of job or career, or a spell of good fortune. The Sun line is also divided into a time frame. It works on a similar basis to the

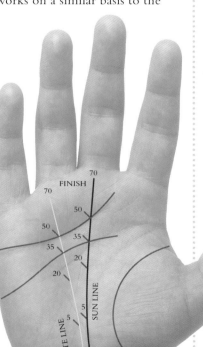

▲ *Important life-changing events, such as the birth of a baby, can be predicted using the Older and Newer Maps of Time.*

other two maps but, here, you take the point at the beginning of the fate line, just above the wrist, and work towards the base of the fingers. The increments on this map are uneven. The ages are: 5, 20, 35, 50 and 70. The Sun line can be read in a similar fashion.

◀ *The ageing map of time can pinpoint important events in an individual's life such as a change of career or spells of good luck.*

LINES AND SIGNS OF SPECIAL INTEREST

Special-interest lines complement the major lines of the palm. They are each unique in their meaning and everyone has at least one, although some people have all of them. They give valuable additional information to

▼ *Individuals with a line of Mars present will often work in the field of safety and protection of others.*

◄ The line of Mars is a sign of courage.

the reader about an individual's character, personality and situation, whether it is an indication of courage, psychic ability or a long-lasting romantic commitment.

THE LINE OF MARS
Situated on the side of the palm, between the heart and head line, this line indicates great courage. People who have this line make excellent protectors – it is often to be found on courageous military leaders.

THE RASCETTES, OR "BRACELETS OF LIFE"
These lines are found running across the underside of the wrist just below the palm of the hand, and this is one of the most important areas to check for longevity. The lines can be very

◄ The rascettes indicate longevity.

faint or very deep, or somewhere in between. Their depth does not matter; it is the number of lines that counts:

- 1 rascette is equal to 15–35 years of life
- 2 rascettes are equal to 35–55 years of life
- 3 rascettes are equal to 55–85 years of life
- 4 rascettes are equal to 85–105 years of life.

THE GIRDLE OF VENUS
This can either be one continuous line forming the shape of an upturned crescent moon, or made up of two lines. This marking indicates an individual who can

► The girdle of Venus is a sign of a soft nature.

they are watching a sad film on the television.

SYMPATHY LINES

These lines are always straight and are angled upwards. They indicate a caring nature. They can be found on the hands of nurses, doctors and people who feel a strong need to alleviate pain and suffering in others.

MEDICAL STIGMATA

This mark, which is found on the hand of a healer, is made up of no fewer than three lines, with a slash cutting through the middle of them. People with these lines have a healing touch or healing hands. They may be doctors, nurses or other professional carers.

RING OF SOLOMON

A ring around the mount of Jupiter, which starts at the side of the index finger

◀ *The ring of Solomon can be a sign of psychic ability.*

and sweeps around to end between the first and second fingers, indicates wisdom and a deep interest in the occult, the supernatural and other psychic phenomena. On a less psychic level, the ring of Solomon indicates someone who has good leadership skills, is very good at managing people, and will usually achieve success in life.

RING OF SATURN

This semicircular mark beneath the middle finger is rarely found. Whether it is continuous or made up of two or more lines, it seems to isolate and overemphasize the negative Saturnian qualities. Someone with this line will tend to be too serious about life and its problems, and this may lead to depression at times.

▶ *The ring of Saturn is a very rare marking.*

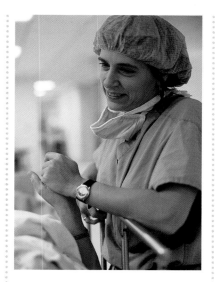

▲ *Those who care for others will have sympathy lines on their palms.*

empathize with the sorrows of others. Unfortunately, this can lead to them becoming too involved and this can sometimes lead to depression. On a lighter note, these individuals often need to reach for the tissue box when

◀ *Sympathy lines are straight, never curved.*

◀ *Medical stigmata indicate professional carers.*

◀ Look for the mystic cross on the hands of tarot or palm-readers.

THE MYSTIC CROSS

THE MYSTIC CROSS

This marking indicates someone who is naturally talented, possessing a sixth sense and well developed intuition. This is an individual who is keenly interested in the occult sciences, such as the tarot, palmistry, runes and magic.

PLAIN OF MARS

This is the area in the centre of the palm of the hand. The plain of Mars shows how sensitive an individual is. In most people, this area will appear concave. If it is slightly raised, or even flat, it is therefore called "high". Only if it is very indented is it defined as "low".

AVERAGE PLAIN OF MARS

When the area is slightly concave, it indicates that the person is balanced emotionally, with good sensibilities and a practical approach to life.

SHALLOW PLAIN OF MARS

Individuals with this kind of palm can be stubborn, proud and overbearing. They are single-minded and can do one thing at a time very well. Their lack of sensitivity, however, can make them unaware of the problems of other people around them.

DEEP PLAIN OF MARS

These individuals will always try to help others, and are highly sensitive to other people's opinions and feelings. They feel other people's pain very personally, and make strong efforts not to upset or offend anyone. If the plain of Mars is too low, it can indicate a tendency towards depression.

▼ Deep plain of Mars (below) and shallow plain of Mars (bottom).

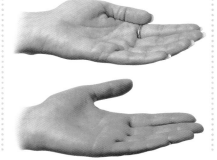

▲ 99.9% of people have marriage lines present which represent serious relationships as well as marriage.

LINES OF MARRIAGE OR UNION

There may be one line, indicating one serious involvement and commitment, or two or more lines indicating additional emotional involvements. The longer the line horizontally, the longer the relationship.

LINES OF MARRIAGE

▶ Lines of marriage or union.

THE PERCUSSIVE OR PALM EDGE

The edge of the palm is a unique area of the hand. It is not a part of the palm itself but rests at the outside edge of the hand. When interpreting this area, take into account that this is the aspect of the individual's personality that is projected to the outside world and shows how others may view them. Have the palm facing you when considering this area.

◀ Independent percussive edge.

▶ Creative percussive edge.

INDEPENDENT PERCUSSIVE

This person is independent and follows their own instincts. They will usually be leaders. They have good intuitive faculties.

ACTIVE PERCUSSIVE

This person is usually always busy, with a very active social life. Ever the perfectionist, they search for the best that they can achieve. However, they can have a highly strung nature and be prone to nervousness. These active people are often very physically attractive.

◀ Active percussive edge.

CREATIVE PERCUSSIVE

This individual tends to be colourful, with a great imagination and creative tendencies; they set their own trends. They have a knack for creating attractive domestic surroundings.

PHYSICAL PERCUSSIVE

This person has excellent physical health with a physique designed for sport and endurance. They are usually involved in physical activities, such as gardening, walking and sport. They need to feel useful and productive.

INTELLECTUAL PERCUSSIVE

This person prefers mental activity to physical. They are problem solvers with an analytical nature. They tend to be weak physically and need to take rests inbetween bouts of exertion.

▼ Physical percussive edge (left) and intellectual percussive edge (right).

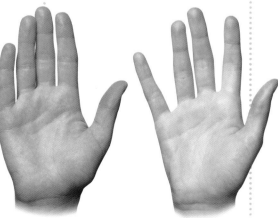

THE MOUNTS OF THE HAND

The fleshy mounds present in different segments of the palm are called the mounts on the hand. Personality types and even physical traits are indicated by the dominant mount. The predominance of one particular mount is found by looking at the palm from various angles and taking note of which mount is raised higher than the others. As the mounts of Venus and the Moon are the widest, it is important that you compare the height rather than the width to establish dominance. A later section, The Mounts Combined, looks at the effect of having two mounts of equal dominance and explores the personality types indicated by each combination.

When looking at the mounts on the palm, the reader usually works clockwise beginning with the mount of Venus.

▼ *The mounts of the hand indicate personality types and can even predict physical characteristics.*

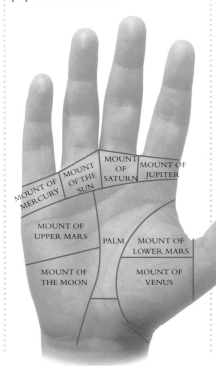

MOUNT OF MERCURY
MOUNT OF THE SUN
MOUNT OF SATURN
MOUNT OF JUPITER
MOUNT OF UPPER MARS
PALM
MOUNT OF LOWER MARS
MOUNT OF THE MOON
MOUNT OF VENUS

PREDOMINANT MOUNT OF VENUS

Physical traits The Venusian is likely to be of above average height with a round face, large, clear eyes, small mouth, thick lips, white teeth and small ears. They will have a high instep, small ankles and long thighs. Venusian men will usually keep their hair until late in life.

▲ *Alma-Tadema's painting,* The Years at the Spring, *has a sensual feel and shows a woman with large, clear eyes and a small, full mouth. She is a typical Venusian, who tend to be life's romantics.*

Health Strong and healthy, with a cheerful disposition.
Mental and moral character Venusians are happy and sensually inclined people who love life and social interaction. They are the souls of kindness, and hate quarrelling, strife and warfare.

PREDOMINANT MOUNT OF THE MOON

Physical traits Individuals with a predominant mount of the Moon tend to be tall, with a round head and broad forehead. They have very fine hair and hardly any body hair. Large, yellowish teeth, a small mouth with thick, full lips and large, round bulging eyes are all typical characteristics of these individuals.

▼ *Lunar subjects can be ethereal, both emotionally and physically. The moon maiden in this painting by Botticelli,* La Primavera, *has a delicate air.*

Health Lunar subjects are constantly anxious about their health. They suffer from poor circulation, have bursts of energy followed by a need for rest, and may experience problems with their kidneys or bladder.

Mental and moral character Moon subjects are charming and live their life to the full. They are fond of travel and new and exciting experiences, but also enjoy relaxation. They have a fickle nature, and will tend to start a new job before completing the previous one.

PREDOMINANT MOUNT OF MARS

Physical traits Martians tend to be of above average height with a strong bone structure; most noticeably they have prominent cheekbones. They have a large mouth and eyes, thin lips, small, yellowish teeth and small ears. Their head may be proportionally small and the nose may be beak-like. Their voice is powerful and attracts attention.

Health Their fiery temperament may lead to a fever. With this heated character, Martians are at a great risk of accidental injury while arguing.

▲ *The character in Lord Leighton's painting,* La Nanna, *looks strong and striking. She has a typical Martian's strong bone structure, large eyes and prominent nose.*

Mental and moral character Martians are amorous individuals by nature with a generous personality and enjoy social occasions. They can be domineering, and will not listen to reason. They are courageous but not vicious.

PREDOMINANT MOUNT OF
MERCURY
Physical traits The Mercurian is
small in stature with good bone
structure. They stay young-
looking longer than others. Their
hair is curly and the skin is soft.
They have deep-set, penetrating
eyes, a long pointed chin and
large hands with long thumbs.

▼ *Alma-Tadema's* Portrait of Alice
Lewis *captures a Mercurian's
intelligent air. Their quick and
enquiring mind and even temper
makes them excellent judges of other
people's character.*

Health The Mercurian is
susceptible to a weak liver and
digestive organs. They often have
a nervous temperament.
Mental and moral character
Quick in thought and action,
Mercurians are skilful at all
games, good students of
mathematics and medicine and
excellent in business. They are
great judges of human character.
Usually of an even-tempered
nature, they love the closeness of
family life. Their acuity and
enjoyment of others makes them
natural observers and born actors.

PREDOMINANT MOUNT OF THE
SUN/APOLLO
Physical traits Solar subjects are
usually above average height
and shapely. They tend to be
muscular and fit, and seldom
stocky. Their hair is soft and
wavy, their mouth is normal-
sized, and they have beautiful,
large, almond-shaped eyes.
Health Apollonians, or solar
subjects, have good general
health. Their eyes are their weak
point. Their below-average
eyesight may make them prone to
silly accidents like tripping over
the carpet.

▲ *Sun types can often be found
watching people with great interest.
Auguste Renoir's painting,* Femme
à la Rose, *also shows the solar
subject's soft hair and large, almond-
shaped eyes.*

Mental and moral character
Solar subjects have versatile
minds, with clear, logical thought
processes and understanding.
They love everything that is
beautiful in art and nature but
are also, in contrast, very
competitive and assertive, always
wanting to be ahead of the pack.
They make an ardent and
trustworthy friend but, beware,
they can be bitter enemies.

PREDOMINANT MOUNT OF SATURN

Physical traits Saturnians are tall and thin. They have a long face with a pale complexion. Their eyes are deep set and slope downwards so that they appear sad. They have a wide mouth with thin lips, prominent lower jaw, and fine teeth.

▼ *Sensitive Saturnians are often deeply burdened by the sadder side of life. Their physical characteristics emphasise this. Modigliani's painting,* Frans Haellens, *shows the sloping eyes, long face and pale complexion.*

Health These people are susceptible to problems with their legs and feet. They are not keen on drinking plain water, so dehydration may be a problem.

Mental and moral character Saturnians have a certain sadness to their lives. Conservative and suspicious by nature, they dislike taking orders. They are very prudent, born doubters, good problem solvers, and are interested in the occult sciences. They enjoy country life and love solitude. They spend little and save more, but are passionate gamblers. They like dark colours.

PREDOMINANT MOUNT OF JUPITER

Physical traits Jupiterians have a strong bone structure. They are of average height, usually with attractive curves, and they have a stately walk. They tend to have large, deep-set eyes and thick, curly hair. They have a straight nose, full mouth, long teeth, a dimple at the base of the chin and ears close to the head. Jupiterian men may lose their hair at an early age.

Health Jupiterians have a tendency to suffer with digestive problems and will often be overweight.

▲ *Jupiterians have thick curly hair, although the men tend to lose it at an early age. Lord Leighton's* Music *shows a typical Jupiterian straight nose, full mouth and dimpled chin.*

Mental and moral character Destined for public life, Jupiterians have confidence in themselves and can be selfish. They like eating out, most social functions and spend money too freely. They love peace, believe in law and order and are, to a degree, conservative.

THE MOUNTS COMBINED

In some palms two mounts are equally raised. This combination gives you an additional insight into the character of the person whose hand you are reading.

Jupiter and Saturn Excellent luck ahead.

Jupiter and Sun Fame and fortune.

Jupiter and Mercury Love and success in business and science.

Jupiter and upper Mars Bravery and success as a commander.

Jupiter and Moon Imagination.

Jupiter and Venus Pure and respected love towards others.

▲ *Venus (above) and Saturn (below)*

▲ *Jupiter (above) and Mars (left)*

Jupiter and lower Mars Cautiousness.

Saturn and Sun Deep artistic tendencies.

Saturn and Mercury Love of science and nature.

Saturn and upper Mars Argumentative temper.

Saturn and Moon A gift for the occult sciences.

Saturn and Venus Vanity and pride.

Saturn and lower Mars A self-critical and reserved nature.

Sun and Mercury Brilliant talker.

Sun and upper Mars Leadership instincts.

▶ *The Moon*

Sun and Moon Imaginative.

Sun and Venus Love of cultural interests.

Sun and lower Mars Cheerful.

Mercury and upper Mars Logical and strategic.

Mercury and Moon Inventive mind.

Mercury and Venus Prudent and sensible in love.

Mercury and lower Mars Perseverance.

Moon and Venus Looking for the ideal in love.

Upper Mars and Venus Mentality typical of a soldier.

▲ *The Sun (above) and the planet Mercury (below)*

LINES AND SIGNS ON THE MOUNTS

Each mount usually features lines and markings such as crosses, squares, or very strong horizontal lines. These signs give the reader a deeper insight into the person's character than can be found by assessing the dominant mount in isolation. When examining the mounts for these lines and signs, use a magnifying glass to give better definition.

LINES AND SIGNS ON THE MOUNT OF VENUS

Flat, hard mount This marking indicates an individual who has grown cold to love, due to difficulties in past relationships.

Two or three lines This indicates an individual who suffers with ingratitude in love. They believe that they can always do better, hence they can be inconstant in relationships.

MOUNT OF VENUS

▼ *2 or 3 lines*   ▼ *Island*

▲ *Strong, horizontal lines*

▲ *Mixed lines*

▲ *St Andrew's Cross*

▲ *Star*

Strong horizontal lines This indicates someone who has an overpowering influence on members of the opposite sex.

Mixed lines This person's disposition will be of a powerfully passionate nature.

Islands Islands in the lines are a sign of someone who has a tendency to feel guilty in love.

▼ *Venus equals love in many languages, even in the language of palmistry.*

St Andrew's Cross A large cross of this type is a sign that there will only ever be one true love in this person's lifetime.

Small cross This indicates a very happy and joyous love affair.

Star by the thumb This indicates a wonderful marriage that will last a lifetime.

Star at base of mount This indicates misfortune for the individual due to the opposite sex, such as divorce or a partner's extreme overspending.

Square at base of mount This person will live a sheltered and protected life.

Triangle This is the mark of someone who is calculating in love: they may marry for money to get ahead.

Grille This is a sign of someone with a dreamy and gentle nature.

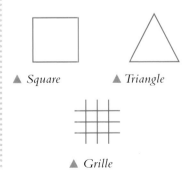

▲ *Square*   ▲ *Triangle*

▲ *Grille*

LINES AND SIGNS ON THE MOUNT OF THE MOON

Long line with a line crossing it Indicates a tendency towards aching bones which could lead to rheumatism.

Cross Indicates a tendency towards heart trouble.

Many lines Indicate a tendency towards insomnia.

Horizontal line The person will be likely to travel.

A voyage line (an angled, horizontal line which reaches up towards the heart line) This individual might suddenly abandon everything to go on a long voyage, or might go to live in another country for reasons of love.

▲ *Horizontal line*

▲ *Ill formed cross* ▲ *Cross*

▲ *Many lines* ▲ *Voyage line*

Mixed lines This, together with a chained heart line, indicates inconsistency in love – the person cannot make up their mind, in matters of love.

Cross This indicates an individual with a superstitious nature.

▼ *When we look at the moon, we are captivated by the mysteries of life.*

Large cross This can indicate an individual who has a tendency to brag a lot.

Cross on upper part of mount This indicates the possibility of trouble with the intestines.

Cross in middle of mount This indicates a tendency towards suffering from rheumatism.

Cross on lower part of mount This cross indicates a tendency towards trouble with the kidneys or possibly with the bladder.

Square This mark signifies protection from bad events throughout a person's life. The greater the number of squares, the greater the luck the individual is likely to have.

Triangle The triangle indicates an individual who has great inner wisdom and creativity.

Grille The grille indicates a tendency towards nerve trouble.

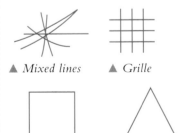

▲ *Mixed lines* ▲ *Grille*

▲ *Square* ▲ *Triangle*

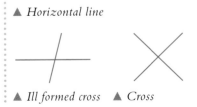

LINES AND SIGNS ON THE MOUNT OF UPPER MARS

One line Indicates an individual with great courage.

Several lines This is someone who may have quite a volatile temper; they can get confused by love, so that they are unable to have a contented relationship.

Horizontal line or lines This indicates a susceptibility to bronchial troubles.

Spot A spot indicates that the individual has been wounded in a fight at some point.

Circle This indicates that the person has been wounded in, or around, the eye.

Square This marking indicates an individual who experiences

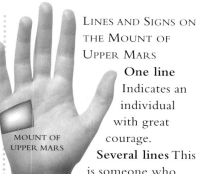

MOUNT OF UPPER MARS

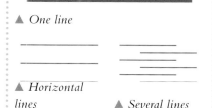

▲ *One line*

▲ *Horizontal lines*　　　▲ *Several lines*

▲ *The god Mars signifies courage and loyalty in abundance; characteristics that are evident in individuals with a line on the mount of upper Mars.*

uncannily good protection from bodily harm.

▲ *Spot*　　　▲ *Circle*

▲ *Square*　　　▲ *Triangle*

Triangle This indicates an individual who is strategically minded, and is especially adept at military operations.

LINES AND SIGNS ON THE MOUNT OF LOWER MARS

Ill-formed cross This marking may indicate that the individual seriously considered suicide in their youth.

Star on horizontal line This indicates an individual who has experienced a great misfortune, or the death of a close relation or friend, in their youth. A marking on the mount of lower Mars is not as auspicious.

MOUNT OF LOWER MARS

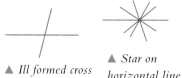

▲ *Ill formed cross*　　　▲ *Star on horizontal line*

LINES AND SIGNS ON THE MOUNT OF MERCURY

MOUNT OF MERCURY

One line This is a good marking to have. Unexpected financial good fortune will come to this lucky person in the form of a windfall, a lottery win, or an inheritance.

One deep line This marking shows great scientific aptitude: this person is set to carry out valuable research or make an important scientific discovery.

Three or more lines together These multiple lines indicate an individual who has a great interest in medicine and its various, related schools of study.

Mixed lines This individual is, financially, very shrewd and good at saving money to the extent that they may have great difficulty in spending it.

Mixed lines below the heart line The opposite of the example, above, this individual is so generous that they have a tendency to spend too much on others. They should try to curb their desire to spend.

Cross This person has a tendency to deceive, though not always in a

▲ *One line*

▲ *One deep line*

▲ *Three lines*

▲ *Mixed lines*

▼ *Good fortune in the form of a lottery win or an unexpected inheritance may come to a person who has a line on the mount of Mercury.*

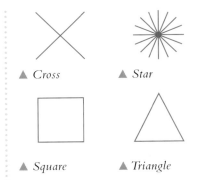

▲ *Cross*

▲ *Star*

▲ *Square*

▲ *Triangle*

negative manner. Sometimes you will see this mark on the palms of actors or sales executives, people who sometimes need to present an image which is not their own. However, people who are prone to lying a great deal can also have this marking.

Star This individual definitely has difficulty telling the truth. More often than not they will be dishonest in their dealings.

Square The individual with this marking is blessed. They will be saved or preserved from heavy financial losses. This is a wonderful marking to have.

Triangle This individual is shrewd in politics and in their dealings with others. They tend to listen first and then respond, and they will usually do so with tact and diplomacy.

MOUNT OF
THE SUN

LINES AND SIGNS
ON THE MOUNT OF
THE SUN

One line This marking is a fortunate one and indicates the likelihood of gaining great wealth.

Two lines These lines indicate real talent but, unfortunately, without achieving much success.

Many horizontal lines This person has artistic tendencies and could be successful in the creative worlds of painting, for example.

Cross This marking indicates the likelihood of success.

Star The star indicates that fame may be nigh but this is only after the individual has taken many risks to achieve this goal.

▼ *Markings on the mount of the Sun may also indicate a vain individual.*

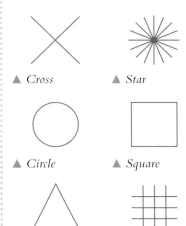

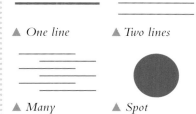

▲ *One line*   ▲ *Two lines*

▲ *Many*   ▲ *Spot*
horizontal lines

Spot A spot indicates that a person is in danger of losing their reputation. They must be watchful and careful if they are to avoid this.

Circle This is a very rare mark and indicates great fame.

Square The square indicates an individual who has a great commercial mind.

▲ *Cross*   ▲ *Star*

▲ *Circle*   ▲ *Square*

▲ *Triangle*   ▲ *Grille*

▲ *The brilliant yellow sunflower blooms rapidly and echoes the fast-growing success and likelihood of fame often indicated by the signs on the mount of the Sun.*

Triangle This marking indicates a selfless individual who wants to assist in the success of others.

Grille The grille can indicate that an individual is inclined to vanity because of their good fortune and fame.

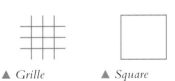

MOUNT OF
SATURN

LINES AND SIGNS ON
THE MOUNT OF
SATURN

One line
A single line
signifies that
an individual
will benefit from
very good luck.

**One long, deep
line** A long, deep line
indicates a peaceful
ending in old age,
perhaps passing away quietly
while sleeping.

Three or more lines This
marking indicates bad luck. The
more lines on the individual's
hand, then the more bad luck
they are likely to face.

Circle A circle is a good marking
to have. It indicates good luck,
and protection from most
troubles in life.

Square A square signifies good
protection from accidents. For

▲ *One line*

▲ *One long, deep
line*

▲ *3 or more lines* ▲ *Circle*

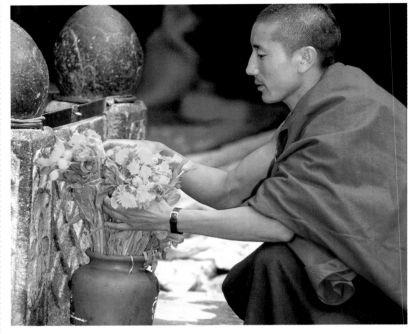

▲ *Most lines on the mount of Saturn
are good omens indicating fortune or
inner wisdom and strength. A
triangle, in particular, is likely to be
found on the hand of someone
spiritual and calm who radiates an
inner peace.*

example, this individual could
emerge from an accident without
a scratch.

Triangle A triangle indicates an
individual who possesses great
inner wisdom and strength.

Grille A grille is a negative
marking here. It indicates
someone who is likely to lose
their luck, especially in old age.

▲ *Grille* ▲ *Square*

▲ *Triangle*

MOUNT OF JUPITER

LINES AND SIGNS ON THE MOUNT OF JUPITER

Two lines This marking indicates an individual whose ambitions are divided; they are likely to be confused over which path to follow.

Line crossing heart line This indicates that the individual is likely to suffer misfortunes in love.

Cross The cross is a desirable marking. It indicates a very happy relationship where commitment is usually involved.

Cross and star This is the "soulmate" marking: it shows that the individual has found or will find their partner for life.

Star The star marking on the mount indicates a satisfying and sudden rise to fame in life.

Square This indicates an individual who has a natural

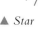

▲ *Line crossing heart line*

▲ *Star*

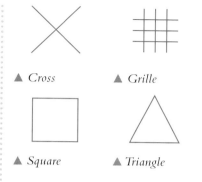

▲ *Cross*

▲ *Grille*

▲ *Square*

▲ *Triangle*

capacity to lead or command. They may follow a military path or be a teacher.

Triangle This marking indicates an individual who is extremely clever and diplomatic. It might be found on the palm of a successful business executive politician or world leader.

Grille This marking indicates an individual who tries too hard to please everyone.

▼ *A cross and a star together on the mount of Jupiter indicate someone who will meet a life-long partner, perhaps even their "soulmate".*

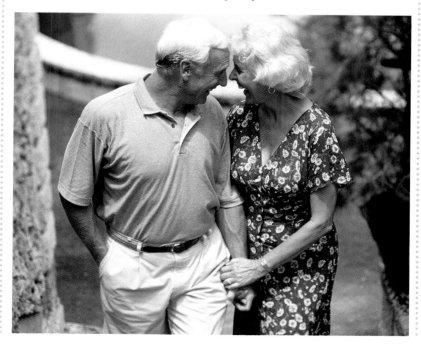

THE MOUNTS AND FINGER TYPES COMBINED

The dominance of a particular mount, in combination with the shape of the fingers, shows up additional aspects of an individual's character to the palm reader. These additional aspects are explored in the tables below and on the following pages.

THE MOUNT OF VENUS

Fingers:	pointed	conical	square	spatulate
normal mount	Believes in love and romance.	Materially minded. Desires security.	Loves family and loves life.	Good comrade and a loyal friend and family member.
raised mount	Imaginative with a creative mind.	Inconsistent in decision making.	Sensual taste in food, music and love.	A love in every port. Not good at stable relationships.
concave mount	Above love. Aloof. Prefers mental activity to physical.	Artistic and creative in nature.	Indifferent to the opposite sex.	Finds opposite sex an encumbrance and is easily annoyed by them.

THE MOUNT OF THE MOON

Fingers:	pointed	conical	square	spatulate
normal mount	Good imagination.	Artistic and visually creative.	Loves poetry and romantic literature.	Loves nature. Hates confined spaces.
raised mount	Emotionally fragile and easily offended.	Extravagant and a spendthrift.	Lacks commonsense. A daydreamer.	Often violent. Prone to overreacting emotionally.
concave mount	(This is never seen.)	An actor or good public speaker.	Has a humdrum existence. Accepts their lot in life.	Never asks "What if?"

Look carefully to see if the mount is average, raised or concave and compare with the finger shape that appears most frequently on the individual's hand. Finger shapes generally fall into one of four main categories – pointed, conical, square and spatulate. You can find illustrations of the four main finger shapes in an earlier section: General Aspects of the Hand.

THE MOUNT OF UPPER MARS

Fingers:	pointed	conical	square	spatulate
normal mount	Courage of the martyr.	Courage of the patriot.	Courage of the soldier.	Courage of the explorer.
raised mount	Religious persecutor. Judgemental of other people.	Vain. Needs lots of attention.	Scheming in business and in love. A manipulator.	A ruffian. A real rogue or heartbreaker.
concave mount	Cowardly by nature and afraid of verbal conflicts.	Cowardly by nature and afraid of verbal conflicts.	Cowardly by nature and afraid of verbal conflicts.	Cowardly in battle and afraid of physical pain.

THE MOUNT OF MERCURY

Fingers:	pointed	conical	square	spatulate
normal mount	Intuitive, even quite psychic.	Eloquent in speech. Stylish by nature.	Great inventor.	Great discoverer.
raised mount	Dreamer of new religions and philosophies.	An inventor of practical things for personal use.	Dangerous schemer, both in business and in love.	Adventurer who stops at nothing.
concave mount	Humane and caring. Loves all life forms.	Has physical or mental difficulties.	No business ability. Not inclined to be self-employed.	Active physically. Always busy.

THE MOUNT OF THE SUN/APOLLO

Fingers:	pointed	conical	square	spatulate
normal mount	Dreamer in life and love.	Idealistic artist or writer. A puritan.	Artist of high standard. A perfectionist.	Drawn to excitement and intrigue.
raised mount	Genius. Eccentric by nature.	Talented in all areas of interest. Competitive.	Stifles real talent. Afraid of own success.	Untalented braggart. Always busy.
concave mount	Art has no place in their life. Practically minded and down to earth.	Clever rather than gifted. Knows how to get ahead by useful associations.	Doesn't care for intellectual pursuits.	Dislikes cultural pursuits. Prefers physical activity.

THE MOUNT OF SATURN

Fingers:	pointed	conical	square	spatulate
normal mount	Poetic in speech and manner.	Morbid. Tends to stress the negative and ignore the positive.	Loves solitude and tranquillity. Doesn't like crowds.	Loves nature and loves life. Has a positive attitude and is fun to be around.
raised mount	Morbid. Tends to stress the negative.	Loves fine art and history.	Dislikes humankind. Prefers nature.	Aggressive and touchy but very loyal.
concave mount	Cynical. Doesn't want to trust others.	Realist in art. Looks for the practical solution.	Indifferent to most things in life.	Cares little for society. Prefers a select group of friends.

THE MOUNT OF JUPITER

Fingers:	pointed	conical	square	spatulate
normal mount	High religious ideals. Looks for the best in others.	Proud with a very loyal nature.	Proud with a very practical nature.	Enterprising in business. Calculating in love.
raised mount	Superstitious. Curious about the esoteric arts.	A perfectionist in artistic and cultural pursuits.	Vain. Sensitive to others' opinions.	Boastful. Not sensitive to others' feelings.
concave mount	Lacking in respect. Too self-interested.	No respect for others. Has difficulties dealing with people.	No self-respect. Allow others to run their lives.	Vulgar. Their actions can offend.

THE MOUNT OF LOWER MARS

Fingers:	pointed	conical	square	spatulate
normal mount	Cautious. Likes to wait and see.	Stoical. Has a very serious nature.	Patient. Enjoys helping others.	Ignores pain and fear. Has a strong personality.
raised mount	Unhealthy. Prone to colds and flu.	Hard-hearted. Afraid to love.	Passively cruel. Secretly likes to see others struggle.	Cruel and cold. Unchangeable.
concave mount	Sensitive soul. Very sweet minded.	Easily offended. Cares too much for the opinion of others.	Afraid of moral and physical pain.	Cowardly by nature. Dislikes conflict.

THE FINGERS AND THUMB

The settings, spacings and patterns on the fingers and thumb show the palm-reader the public persona of the individual. This is the personality that they choose to show to the outside world.

FINGER SETTINGS

The setting of the fingers on the palm varies. They form a distinct shape at the point where they meet the palm.

▼ Fingers set straight across. *This person is self-confident, with plenty of drive. They can be pushy and feel that whatever they do must be automatically right.*

▲ Fingers forming a pitch roof. *This person has a grudge against other people and an inferiority complex. They lack trust in others and self-confidence in themselves.*

◄ Fingers set straight across; little finger dropped down. *When only the little finger is low-set, it is an indication of someone who lacks self-confidence.*

◄ Fingers arched. *These are the fingers of a well-balanced individual, with moderate and tolerant views.*

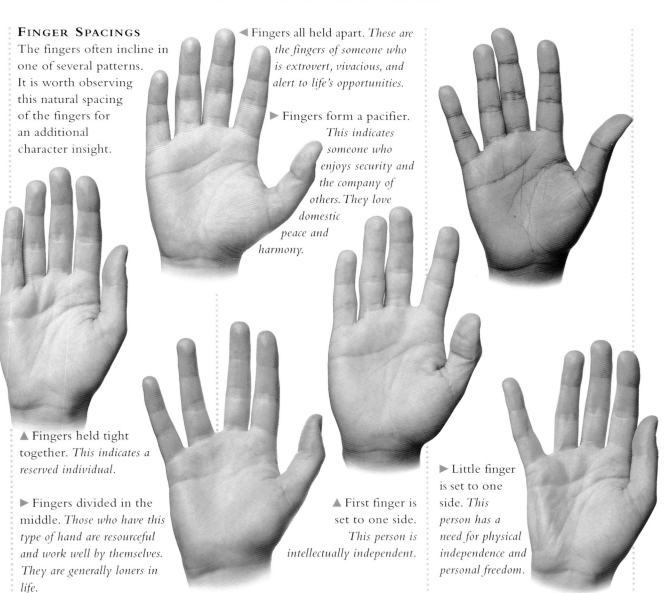

FINGER SPACINGS

The fingers often incline in one of several patterns. It is worth observing this natural spacing of the fingers for an additional character insight.

◀ Fingers all held apart. *These are the fingers of someone who is extrovert, vivacious, and alert to life's opportunities.*

▶ Fingers form a pacifier. *This indicates someone who enjoys security and the company of others. They love domestic peace and harmony.*

▲ Fingers held tight together. *This indicates a reserved individual.*

▶ Fingers divided in the middle. *Those who have this type of hand are resourceful and work well by themselves. They are generally loners in life.*

▲ First finger is set to one side. *This person is intellectually independent.*

▶ Little finger is set to one side. *This person has a need for physical independence and personal freedom.*

THE SETTINGS OF THE THUMB

The thumb is, if you like, the leader of the hand; it covers the fingers tightly when we clench our fists.

The characteristics of the thumb indicate the strength of a person's conviction and their powers of logic. Measure and compare the phalanges (joint settings) of the thumb. The top section relates to willpower and the second section to reasoning

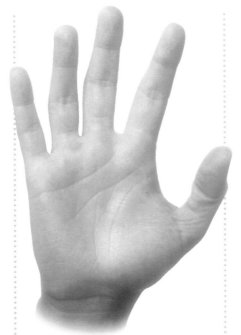

▲ The wider the opening of the thumb (its placement) the more open and trusting the individual.

◄ It is desirable to have a fairly even balance of willpower with reasoning capability. Where one of these areas is lacking, it suggests that the individual will usually be weak in that area: a thumb with a comparatively short first section, for example, would indicate poor willpower. The third indicates the level of an individual's desire and the tendency to act on it.

WILL

REASON

DESIRE

▲ A thumb that is placed close in to the hand indicates an introverted and mistrusting character.

and logic. The mount of Venus will usually be equal to the first and second sections measured together.

A thumb that is set low in the hand (close to the base of the wrist) indicates a practical and cautious person. A thumb that is set high in the hand indicates an individual with a passionate approach to life.

THE PHALANGES OF THE FINGERS

It is important to look for overall balance between the phalanges (joint settings). The first phalanx deals with the mind; the second deals with personal ambitions; and the third deals with desires. If one phalanx in any finger appears excessively long, use the table, opposite, to interpret this characteristic.

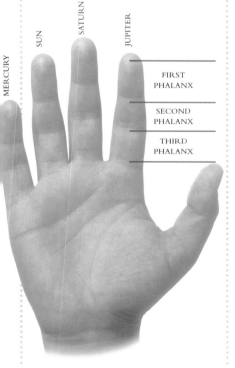

MERCURY SUN SATURN JUPITER

FIRST PHALANX

SECOND PHALANX

THIRD PHALANX

	Jupiter	Saturn	Sun/Apollo	Mercury
long first phalanx	Superstitious. Is interested in the esoteric.	Emotionally fragile. Needs peace of mind.	Artistic tendencies. Has a strong visual sense.	Lying disposition. Finds honesty difficult.
long second phalanx	Vain. Needs the approval of others.	Cautious in business. Respects money.	Inspiration held in check. Sober ego.	Manipulative with a strategic nature.
long third phalanx	Love of power taken too far.	Greedy. Never satisfied with their lot.	Foolish. Needy and clinging.	Always wants what they don't have.

◀ *The phalanges of the Jupiter, Saturn, Apollo and Mercury fingers show up an individual's intellect, personal ambition and desires. Check carefully to see if one phalanx is excessively long in relation to the others.*

▶ *An elongated first phalanx on the Saturn finger can indicate a sensitive individual who feels life's knocks deeply and longs for peace of mind.*

FINGERTIP PATTERNS

When looking at the fingertip and thumb-tip patterns make sure you have good lighting. Use a magnifying glass if it helps. You should interpret the pattern that occurs most frequently.

THE "PEACOCK'S EYE"

This is a rare pattern to find. If it is on the Sun/Apollo finger it guarantees protection from accidental death. On other fingers, it shows a high degree of perception or intuition.

▼ *The peacock's distinctive tail echoes the swirling fingertip pattern called, of course, the peacock's eye.*

▲ *The lines of the tented arch erupt upwards like smoke from the centre of a volcano.*

THE "TENTED ARCH"

This is the least common pattern, and is usually only found on the index finger, if at all. Four examples on ten fingers would be a high count. It suggests emotional sensitivity which verges on instability. This person is very sensitive to stimuli and needs peaceful surroundings. Artistic and idealistic, they have impulsive tendencies. They are highly strung and predisposed to nervous disorders.

THE WHORL

This pattern can be found on any finger or the thumb, but is most often seen on the Sun/Apollo finger. On the thumb, it indicates stubbornness and dogmatism; someone who will not back down even when they have been proved wrong.

On the Sun/Apollo finger, it shows a fine sense of discrimination, with fixed likes and dislikes in such things as clothes and food. It indicates a nonconformist who individualizes everything. They are prone to nervous digestive troubles, heart disease and other nervous disorders.

▼ *The circular swirls of the whorl look like the centre of a whirlpool.*

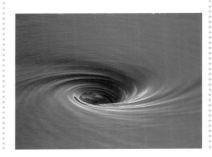

THE ARCH

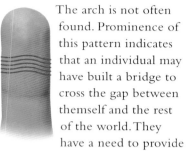

The arch is not often found. Prominence of this pattern indicates that an individual may have built a bridge to cross the gap between themselves and the rest of the world. They have a need to provide security for the family and the community. Dedication and loyalty are their watchwords. Their chosen path is that of the saint. They have a predisposition to digestive weaknesses, ulcers and blood disorders.

▼ *The old stone bridge has a gentle curve like the arch.*

▲ *The swirling "S" shape of the composite loop resembles a snake poised to attack.*

THE COMPOSITE LOOP

These two loops reflect two paths to choose from, and indicate an indecisive individual who will weigh up a problem for hours. The pattern is most often found on the thumb or index finger: indecisiveness will be greater if it is on the thumb. The person has a practical and material mind, but can be inflexible, repressive, critical and resentful. There is a predisposition to malignant conditions and mental troubles.

THE LOOP

This is the most commonly found of all patterns and is also known as the "Lunar loop" because it points in the direction of the Moon side of the hand. This marking indicates adaptability and versatility in the face of changing circumstances.

People with this pattern predominating are emotionally responsive and not confined by a narrow viewpoint. They have broad horizons and liberal ideas.

▼ *The sharp curve of a rip curl echoes the shape of the loop.*

FROM THEORY TO PRACTICE

The palm-reader analyses the fingers, the lines and markings on the palm and the actual size and shape of the hands. Anyone can learn to interpret these features, but because there is a lot of information to remember it will take a lot of time and practice. A good palm-reader is one who also brings intuition and common sense to their reading. You are dealing with a whole person, and that includes their feelings. This section is designed to help you understand what will be going through your mind as you read a palm, and how to communicate it with sensitivity.

You may want to gather together a magnifying glass, a pen to mark up lines on the hands and a book to jot down your observations as you go along. A ruler can be useful to work out the proportions of the hand accurately and a pair of compasses will help you mark on the maps of time, if desired.

◀ *In the private time before the palm reading begins, clear your mind and gather your thoughts, so that you can focus your whole attention and intuition on your client.*

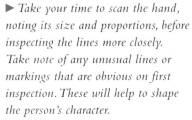

▶ *Take your time to scan the hand, noting its size and proportions, before inspecting the lines more closely. Take note of any unusual lines or markings that are obvious on first inspection. These will help to shape the person's character.*

60

- To begin, take a moment before the other person enters the room. Clear your mind and take several deep breaths.

- Invite the other person in. Focus on your unity with them. Maintain a calm silence until you feel ready to begin. Then explain briefly that the lines and mounts can change, so that everyone has control over their own life. Ask your subject if they are right- or left-handed and how old they are.

- While you are looking at the hands, keep half your attention focused in on your intuition and half of it focused out towards the hands. Maintaining a calm silence, scan the hand noting the relative strength of the mounts and the length of the fingers, including the relative lengths of the three phalanges.

- Also look for the main lines, noticing from where they originate and to where they carry the energy.

- Look for special marks: healing marks, the girdle of Venus, the ring of Solomon, marriage or union lines.

- Ask yourself what the main themes of this person's life are.

- After looking at, and listening to, everything, take a deep breath and let the information come together in your mind.

- You don't have to force it. It will come to you naturally.

◀ It may help to use a magnifying glass initially, to ensure that you can see all the lines and markings clearly. It will be especially useful in detecting the really fine lines of the typical water hand.

▶ Don't hesitate to take notes during the reading. There is a lot to take in, and your notes will be a useful prompt when you draw all the information, and your observations, together later on.

• When you have a general idea of what you're going to say, assess the person and decide on the best way to express yourself to them; remember, BE KIND!

• Continue to keep half your attention focused in on your intuition and half focused towards the other person.

• Tell them what the palm told you, speaking slowly and clearly. Let your intuition guide you as you decide where to begin. Ask if they understand you. Every hand is unique, every reading takes place at a unique meeting point of time, space and mind. There is no set pattern to follow.

• As long as you touch on the meanings of all the major lines and mounts in the context of the hand, you will do fine.

• Ask yourself: does this person understand what I am saying? Is what I am saying appropriate to their life now? Listen for the answers to these questions in your heart, not in your head.

• Be sure that what you say is exactly what you mean. Open up your intuition and listen to what the palm is telling you. Think carefully and relay this information clearly and with sensitivity.

• Ask the person if they have any further questions. See if the questions can be answered directly from the palm.

• As long as you make it clear that the hands represent probabilities, not certainties, and that the person's lines can often alter with time, you can answer the questions.

• If a person asks about death, or seems to be asking you to take responsibility for major life decisions, do not answer the question directly. Bear in mind that you are not a therapist.

• Above all, take your time and keep yourself open and receptive to your intuition.

• Once you have gone over all that you have seen in their hand, bring the reading slowly to a close.

• Take time once your client has left to go over your reading in your mind. Was there any thing else you could have said?

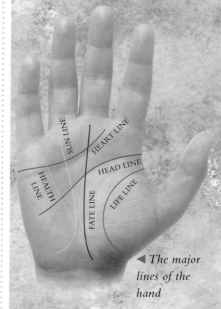

▶ *An old woodcut of a palm shows the intricate nature of this ancient practice, and also how little it has changed over the years.*

◀ *The major lines of the hand*

SUN LINE
HEART LINE
HEALTH LINE
HEAD LINE
FATE LINE
LIFE LINE

FURTHER INFORMATION

FROM THE AUTHOR
Palmistry has been studied and practised by all the female members of my family for many generations. As a result of this long matrilineal line I have arrived at this point in my own study of palmistry. Together with my work with David, my co-author, and my work at my shop *Way Out There and Back*, this lineage has enabled me to research and develop this form of divination on a practical level for use in many arenas in conjunction with our modern, everyday lives. It is my aim to offer people the springboard to gain insights and learn from the wisdom of the East and the New Age as well as my own experience of working in this field, to help them to further their own studies of palmistry.

I would like to thank the students who have studied with me, both past and present. And finally, I would like to thank all the readers of this book. Good luck!

Staci Mendoza

FIND OUT MORE
Personalised palm charts are available from:

Way Out There and Back
20 Evans Gardens
Arcade Road
Littlehampton
West Sussex
BN17 5AP
tel. 01903 722666
email: wotab@mistral.co.uk
URL: www.wotab.co.uk

PICTURE CREDITS
The majority of photographs in this book were taken by John Freeman. The publishers would also like to thank the following picture libraries for supplying images:
Images Colour Library pages 2, 8 (bl, tr), 21 (tl, br) 25 (tr), 42 (tm), 44 (bm)

INDEX

air hand, 16

bars and dots, 30

chains, 31
circles, 31
conical hand, 10
crosses, 30

earth hand, 17
elemental hands, 16–17

fate line, 19, 26–7
fingers: phalanges, 57
    settings, 54
    shapes, 12
    spacings, 55
fingertip patterns, 58–9
fire hand, 17
forks, 31

head line, 18, 22–3
heart line, 18–19, 24–5
history, 8–9

islands, 31

Jupiter, mount of, 40, 49, 53

life line, 18, 20–1
lines: major lines, 18–29
    on mounts, 43–9

maps of time, 32–3
marks on lines, 30–1
marriage lines, 36
Mars: Mars line, 34
    mount of, 39, 45, 51, 53
    plain of, 36
medical stigmata, 35
Mercury: Mercury line, 19, 29
    mount of, 40, 46, 51
Moon, mount of, 39, 44, 50
mounts, 38–53
mystic cross, 36

palm edge, 37
percussive edge, 37
phalanges, 57
plain of Mars, 36
pointed hand, 10
proportions of hand, 11

rascettes, 34
readings, 60–2

Saturn: mount of, 40, 48, 52
    ring of, 35
shape of fingers, 12
shape of hand, 10
signs, on mounts, 43–9
size of hand, 14–15
Solomon, ring of, 35
spatulate hand, 10
special-interest lines, 34–6
square hand, 10
squares, 30
stars, 31
Sun: Sun line, 19, 28
    mount of, 40, 47, 52
sympathy lines, 35

tassels, 31
thickness of hand, 13
thumb, 56
time maps, 32–3

Venus: girdle of, 34–5
    mount of, 38, 43, 50

water hand, 16